Trainee Guide

Painting

Level 1

National Center for Construction Education and Research

Wheels of Learning
Standardized Craft Training

Prentice Hall
Upper Saddle River, New Jersey   Columbus, Ohio

Prentice-Hall, Inc.
Upper Saddle River, New Jersey 07458

This information is general in nature and intended for training purposes only. Actual performance of activities described in this manual requires compliance with all applicable operations procedures under the direction of qualified personnel. References in this manual to patented or proprietary devices do not constitute a recommendation for their use.

Printed in the United States of America

10 9 8 7 6 5 4

ISBN: 0-13-771239-1

Prentice-Hall International (UK) Limited, *London*
Prentice-Hall of Australia Pty. Limited, *Sydney*
Prentice-Hall of Canada, Inc., *Toronto*
Prentice-Hall Hispanoamericana, S. A., *Mexico*
Prentice-Hall of India Private Limited, *New Delhi*
Prentice-Hall of Japan, Inc., *Tokyo*
Pearson Education Asia, *Singapore*
Editora Prentice-Hall do Brasil, Ltda., *Rio de Janeiro*

Preface

This volume is one of many in the *Wheels of Learning* craft training program. This program, covering more than 20 standardized craft areas, including all major construction skills, was developed over a period of years by industry and education specialists. Sixteen of the largest construction and maintenance firms in the U.S. committed financial and human resources to the teams that wrote the curricula and planned the national accredited training process. These materials are industry-proven and consist of competency-based textbooks and instructor guides.

The *Wheels of Learning* was developed by the National Center for Construction Education and Research in response to the training needs of the construction and maintenance industries. The NCCER is a nonprofit educational entity affiliated with the University of Florida and supported by the following industry and craft associations:

Partnering Associations

- ABC Texas Gulf Coast Chapter
- American Fire Sprinkler Association
- American Society for Training and Development
- American Vocational Association
- American Welding Society
- Associated Builders and Contractors, Inc.
- Associated General Contractors of America
- Carolinas AGC, Inc.
- Carolinas Electrical Contractors Association
- Construction Industry Institute
- Design-Build Institute of America
- Merit Contractors Association of Canada
- Metal Building Manufacturers Association
- National Association of Minority Contractors
- National Association of Women in Construction
- National Insulation Association
- National Ready Mixed Concrete Association
- National Utility Contractors Association
- National Vocational Technical Honor Society
- Painting and Decorating Contractors of America
- Portland Cement Association
- Steel Erectors Association of America
- University of Florida
- Women Construction Owners and Executives, USA

Some of the features of the *Wheels of Learning* program include:

- A proven record of success over many years of use by industry companies.
- National standardization providing "portability" of learned job skills and educational credits that will be of tremendous value to trainees.
- Recognition: upon successful completion of training with an accredited sponsor, trainees receive an industry-recognized certificate and transcript from NCCER.
- Each level meets or exceeds Bureau of Apprenticeship and Training (BAT) requirements for related classroom training (CFR 29:29).
- Well illustrated, up-to-date, and practical information. All standardized manuals are reviewed annually in a continuous improvement process.

Acknowledgments

This manual would not exist were it not for the dedication and unselfish energy of those volunteers who served on the Technical Review Committee. A sincere thanks is extended to:

Cindy Athey
Gary Barrett
James M. Chase
Dean von Holdt
Morris Hursh
Doug Kearney
Michael Leierer
Glenn M. Renner
Carl Scruggs
Jim Woodward

Contents

Careers in the Painting Trade

Module 07101

Painting Trainee Task Module 07101

CAREERS IN THE PAINTING TRADE

PAINTING AND DECORATING CONTRACTORS OF AMERICA

NATIONAL CENTER FOR CONSTRUCTION EDUCATION AND RESEARCH

Objectives

Upon completion of this module, the trainee will be able to:

1. Briefly describe the history of the painting trade, including how the Clean Air and Lead Abatement laws and regulations have changed the paint industry.
2. Identify career opportunities available for people in the painting trade.
3. Explain the purpose and objectives of an apprentice training program.
4. Describe the desired characteristics of professionals in the painting trade.

Prerequisites

Successful completion of the following Task Modules is required before beginning study of this Task Module: Core Curricula.

Required Trainee Material

Trainee Module

Course Map

This course map shows all of the *Wheels of Learning* task modules in the first level of the Painting curricula. The suggested training order begins at the bottom and proceeds up. Skill levels increase as a trainee advances on the course map. The training order may be adjusted by the local Training Program Sponsor.

Course Map: Painting, Level 1

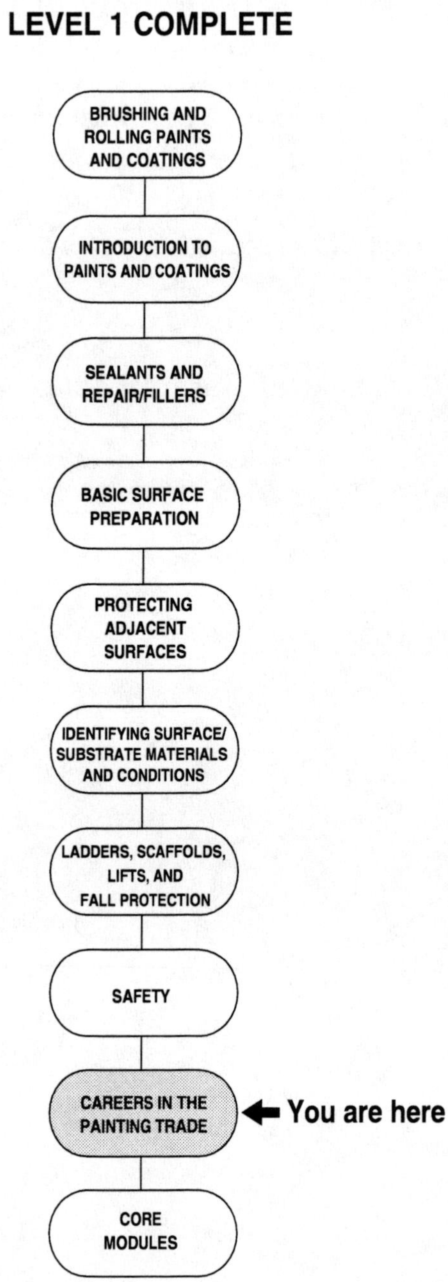

CMAP101.EPS

TABLE OF CONTENTS

Trade Terms Introduced In This Module

Additive: A substance added to a coating material to adjust, enhance, or improve the emulsion, suspension, drying, application, weathering, or other properties of the material.

Binder: The nonvolatile portion of a paint which serves to bind or cement the pigment particles together. Acrylics, alkyds, and epoxy esters are examples of binders.

Pigment: Material in the form of fine powders which, when dispersed in a liquid vehicle to make paint, impart color and other essential properties to a paint such as opacity, hardness, durability, corrosion resistance, and increased coverage.

Polymer: A substance, the molecules of which consist of one or more structural units (monomers), repeated any number of times; vinyl resins are examples of true polymers.

Pot life: The usable life of two-part paint once it is activated.

Resin: General term applied to a wide variety of transparent and fusible products. They may vary widely in color. Higher molecular weight synthetic resins are generally referred to as *polymers*. In a broad sense, the term is used to designate any polymer that is a basic material for coatings and plastics.

Solvent: A liquid, usually volatile, which is used in the manufacture of paint to dissolve other materials and which evaporates during drying and does not become part of the dried film. Solvents are also used to dissolve and remove oil, dirt, grease, soil, and waxes from metal surfaces. Mineral spirits, VM&P naptha, zylol, and isopropyl alcohol are examples of solvents.

Substrate: Surface to be painted. Any surface to which paint, coating, or wallcovering is applied.

Vehicle: The liquid portion of paint in which the pigment is dispersed. It is composed of binder and solvent.

Volatile organic compounds (VOCs): Solvents, propellants, or other substances (except water) that evaporate from coatings while drying.

White lead: White pigment obtained from lead sulphide ore by a process of fuming or burning. Lead is no longer legal to use in most coating materials.

1.0.0 INTRODUCTION

Recent studies show that in the coming years, the painting industry needs to grow by 12,500 to 15,000 workers per year. A career in the painting trade can be financially rewarding and provide satisfaction for the skilled craftsperson who takes pride in a job well done. Work on the job, combined with the appropriate formal training, provides numerous career paths for professional growth in this thriving, technical industry. This module introduces the painting trade. A brief history of the painting trade is covered first, followed by a description of the jobs and career path opportunities within the trade. The work ethics and attitudes of the successful painting/decorating professional are also covered.

2.0.0 TRADE HISTORY

2.1.0 EARLY HISTORY

Craftspersons in the painting trade have a long history. Paint has been used as a form of decoration since prehistoric times. Early cave dwellers used plants, clay, and water to make paint. They used this paint to decorate their bodies and the walls of caves (*Figure 1*). Most of what these early people painted were lines that formed pictures. The people of ancient Egypt used paint in tombs, palaces, and on temple walls using colors which they prepared from the soil. By 1500 B.C., they imported dyes such as indigo and madder. From these they made blue and red **pigments**. Pigments impart color and other properties to paint. The Egyptians were the first to use protective coatings. They applied forms of pitch and balsam to seal their ships. They also developed and used water-based paints produced from freshly-burned lime (*whitewash*) with milk curds as a **binder**. A binder serves to bind or cement the pigment particles together upon drying of a paint.

101F01.TIF

Figure 1. The Painting Craft Dates Back To Prehistoric Times

The Greeks, Romans, and others copied Egyptian painting practices. The early Greeks developed painting into an art form. They not only decorated flat surfaces, but painted human beings, wood panels, and vases. Paint was widely used by the Romans, who are credited with the introduction of white lead as a pigment in 430 B.C. The ancient Romans used stencils to paint borders on wall surfaces. They also painted stone and plaster to look like marble and granite. The Roman Empire collapsed in the 400s A.D. After that, the art of paint-making became lost to the western world until the middle ages (500 to 1450 A.D.) when the English and other Europeans began making and using paints to paint churches, public buildings, and the homes of the wealthy. By the time America was discovered at the end of the 15th century, most of the Indian tribes used dyes and pigments to color baskets and rugs. Early painting methods and materials first used by the settlers of America were influenced greatly by the methods and materials used in the European countries from which they came.

In Asia, the first pigment was developed before 6000 B.C. Coloring components included natural ores and organic pigments. Binders included color crayons sometimes made from boiled rice. **Vehicles**, the liquid part of paint in which pigments are dissolved or dispersed, were made from gum arabic, egg white (albumin), gelatin, and beeswax. The Chinese are credited with the use of lacquer as early as the Chou Dynasty (1122 to 221 B.C.). They used lacquer to decorate carriages, harnesses, and weapons. By the second century B.C., the Chinese were decorating their buildings inside and outside with lacquer. Among other Asians, the Japanese and Koreans also used lacquer extensively. During the Ching Dynasty, the Chinese developed iron oxide to produce the color red.

Paint manufacturing began in Europe around the 1700s. Then, manufacturers ground pigments and oils on a stone table with a round stone. This method was also used by early American colonists. These early paint manufacturers tended to be independent and worked secretively. Their paints were made for limited private use. The Industrial Revolution in England changed this. In the late 17th century and early 18th century, power-driven machinery was brought into the paint-making process. At this time, **white lead**, a white pigment obtained from lead sulphate, became more widely available. These new machines and white lead allowed for the development and production of protective, anti-corrosion paints used for protecting metal structures such as bridges. The first varnish factories were started in Europe beginning with England in 1790.

In the 18th century, there was a general increase in the availability of vehicles and pigments. White lead was put to more uses. Also, there was an extensive extraction of linseed oil from the flax plant. Paint grade zinc oxide was also developed. During this period, paint ingredients were mixed and used on the job. By the 19th century, paint manufacturing changed. For the first time, the two basic ingredients, vehicle and pigment, were mixed together before distribution. For the remainder of the 19th century, no significant changes occurred in paint formulas.

2.2.0 RECENT HISTORY

The evolution of materials technology throughout the 20th century has produced a proliferation of paint and related products (*Figure 2*). Materials technology began to advance during the first half of the century primarily as a result of World Wars I and II. World War I caused major changes in the manufacture of paint because linseed oil was restricted to miliary use. This drove the paint industry to develop new coatings that were not based on linseed oil. Today, linseed oil is found in only a few paints and in some putty and caulking compounds. World War II also resulted in numerous new materials being developed for use in military equipment. These required that chemists find new finishes for use in painting and protecting these materials. During this same period, the dye and plastics industries also began extensive research and development of new painting-related products. The dye industry created new pigments; while the plastics industry developed new **polymers**. Polymers are substances in which the molecules, consisting of one or more structural units, are repeated any number of times. Synthetic **resins** were produced as new binders for paint that provided significant improvement in weathering, water resistance, toughness, elasticity, and resistance to chemical exposure. These binders included alkyds, phenolics, chlorinated rubbers, vinyls, latexes, and acrylics. Resins are natural or synthetic substances which, when heated, are soluble in drying oils and solvents.

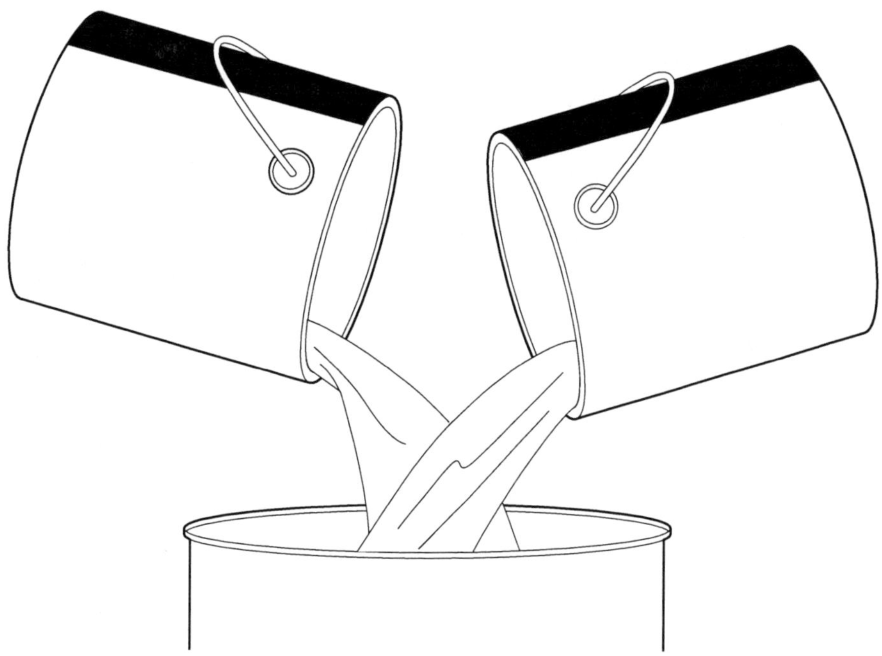

101F02.TIF

Figure 2. More Advances In Painting Materials Have Occurred During The Latter Part Of The 20th Century Than During All Of Prior History Combined

Shortly after World War II, new paints containing epoxy and urethane resins and zinc-rich coatings began to be produced in the United States. Over time, these coatings were enhanced by the development of high-quality **solvents** and **additives** that improved their application and performance properties. Solvents are liquids used in paints to dissolve pigments and other materials. Solvents evaporate during drying. These high-performance materials became the main coatings for protecting steel in industrial environments. Waterborne (water-based) coatings using acrylic latex binders were first introduced in the early 1950s. Their use gained in popularity during the 1950s and 1960s because they were easy to use and nontoxic.

The introduction of numerous new paint materials during the 1930s, 1940s, and 1950s meant that the painter had more choices and, therefore, more difficulty in selecting the right paint for the job. This greatly increased the need for professionalism in the trade and placed an increased emphasis on the painter's need to acquire knowledge and skills about the characteristics and use of these new products. The use of high-performance paints also meant a high degree of difficulty in the application of materials. The painter had to learn how to mix and catalyze chemically curing paints; how to work within the limits of **pot life** (useful life once activated); how to apply paints at extremely tight tolerances of thickness, measured in the thousands of an inch; and how to perform numerous other tasks more technical in nature than ever before.

During the latter half of the 20th century and up to the present, advances in painting materials and methods of application have been driven by increased concerns about health, safety, and the environment. In the United States, the passage of the Clean Air Act in 1970, and its amendments in 1990, mandated the containment of all sources of air pollution, including carbon-based solvents generally classified as **volatile organic compounds (VOCs)**. Oil-based paints use VOC solvents such as mineral spirits to keep them liquid until they are applied. Upon evaporation into the atmosphere, VOCs react with nitrous oxides (combustion compounds from automotive emissions and the burning of fuels) and sunlight to form ozone and air pollutants. Ozone is an unstable form of oxygen that is highly reactive. Ozone in the Earth's upper atmosphere (stratosphere) protects the earth and its inhabitants from harmful ultraviolet radiation. However, ozone trapped within the Earth's troposphere (*Figure 3*) contributes to smog that can be hazardous to humans, animals, and plant life. Emissions of VOCs from painting and various other sources can also contribute to poor indoor air quality.

Today, because the Environmental Protection Agency (EPA) is requiring the elimination of air pollution, the environmental laws and regulations that limit the amount of VOCs in paint are increasingly restrictive. This means that less VOC-type solvents can be used, causing the focus of most new materials technology to be toward the development of better waterborne paints. Today's waterborne paints and coatings are made in a variety of resin types, such as epoxy, polyurethane, alkyd, vinyl, latex, acrylic, zinc-rich silicate, and others. They are as effective on outdoor surfaces as they are on indoor surfaces. Special coatings are made

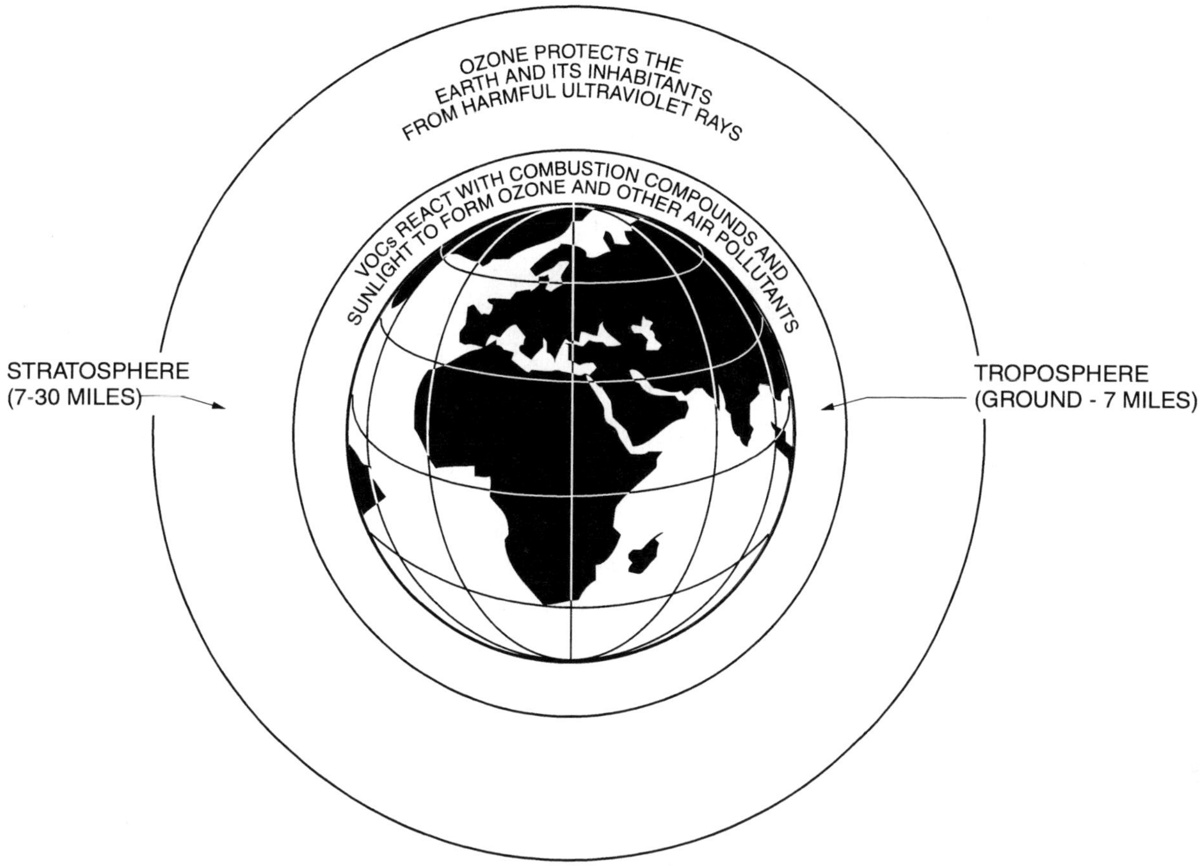

Figure 3. Ozone Trapped In The Earth's Troposphere Is Hazardous To Humans, Animals, And Plant Life

to protect metal, concrete, and other industrial **substrates** (surfaces being painted) in mildly harsh environments. Other types developed to comply with the air pollution laws include high solids and 100% solid materials. These are often difficult to apply, resulting in a number of paint manufacturers training and licensing painters to be exclusive applicators of their product(s).

Health concerns about the harmful effects of lead on our environment and bodies have also caused many changes in today's paint. Several decades ago, it was found that lead and chromate pigments contained in paint products are hazardous. Brain and organ damage can occur in children who play with or eat chips of lead-based paint. Because of this, lead abatement laws introduced during the 1970s restricted the use of lead in many materials, including paint. In 1978, the Consumers Product Safety Commission banned the use of lead in all household paints. Paint manufacturers responded to this ban by developing numerous waterborne and other types of paints as replacements. Currently, lead-based paint is only being used on bridges and other special industrial applications where its ability to inhibit rust and corrosion is still needed. Most industrial coating manufacturers are working to reduce or eliminate the use of lead in their paint.

2.3.0 WALLCOVERING HISTORY

The history of wallcovering is much less detailed than that for painting. The use of woven grasses or animal hides hung over openings is an ageless practice. Early forms of wallcovering made of woven grass and fiber materials were used to both decorate and protect walls. The Chinese made use of painted rice paper. In some areas, they covered windows and doorways, like draperies, to keep out the cold. Decorating using hangings dates back to about 200 B.C. The wallcovering industry progressed slowly. Weavers produced handmade tapestries. Artists painted pictures. In the early 1400s, people in several countries began making "dominos." These were rectangular pieces of paper or fabric printed with colored paint, and glued to the walls. The designs were prepared on wooden blocks. Today, we call this *block printing*. Dominos and stencils, along with hand painting, became a very popular paper-decorating practice. Flocking dates back to France in the 1600s, when imitation velvet cutouts were bonded to walls. The next advancement bonded the flocking material to the paper first, then the paper was bonded to the walls. This was an early form of textured wallcovering.

Up until the 1500s, all wallcoverings were hand painted. The textile and paper industries became involved with wallcoverings as people began bonding decorative paper, cloth, and other materials to walls. Inventions in the printing industry aided the wallcovering industry. In America, the first wallpaper printing machine was used in 1739. In 1839, a four-color press was introduced that needed only one machine setup, rather than the four previously required. Development of continuous roll printing machines, refined inks, and improved papers also aided the advancement of the wallcovering industry. Americans made wallpaper very popular in the 1920s and 1930s. Following World War II, numerous other materials were adapted to wallcoverings, including plastic, vinyl, and mylar sheeting.

3.0.0 CAREERS IN THE PAINTING TRADE

Recent technological advances in both materials and equipment have dramatically changed the craft of painting and decorating. Because of these changes, many of the methods and equipment used just a few years ago no longer apply. Today, the painting trade is more focused on the application of machine technology. Current advances, and those that will occur in the future, require that the painting craftsperson be knowledgeable about and posses the skills to use an ever increasing range of technologically advanced materials and equipment (*Figure 4*).

Careers in the painting trade are many and varied. There is a large existing base of structures that need painting or decorating. In addition, every time a structure is constructed, it needs painting or decorating. To get an idea of how vast the painting trade is, picture the city or town where you live. Then think about the fact that every building is usually painted on the inside with many painted on the outside. Expand that view

Figure 4. Painters Must Have The Skills To Use Technologically-Advanced Tools And Equipment

to include the entire country. New buildings and structures that need painting and decorating are being constructed every day; the existing ones need periodic repair and repainting and/or decorating. From this perspective, the opportunities in the trade appear limitless.

3.1.0 WHO EMPLOYS PAINTERS/DECORATORS?

In general, a painter (painter/paperhanger), or one with knowledge of painting/wallcovering materials and skills, will find employment in one of four different areas. These areas are shown in *Figure 5*.

During a painter's career, he or she may change jobs several times and may move from one area into another. For example, a master painter/paperhanger who has been working for a painting contractor may take a job as a paint and/or wallcovering manufacturing representative or applications engineer. In this new position, representing the manufacturer, the craftsperson provides assistance to painting contractors by recommending products for specific applications or by solving problems encountered with existing or new painting systems. The craftsperson may also move on to become a contractor or start a business that sells painting/wallcovering supplies. Regardless of the area of employment, the successful craftsperson will have considerable knowledge of painting/wallcovering practices, tools, materials, and supplies.

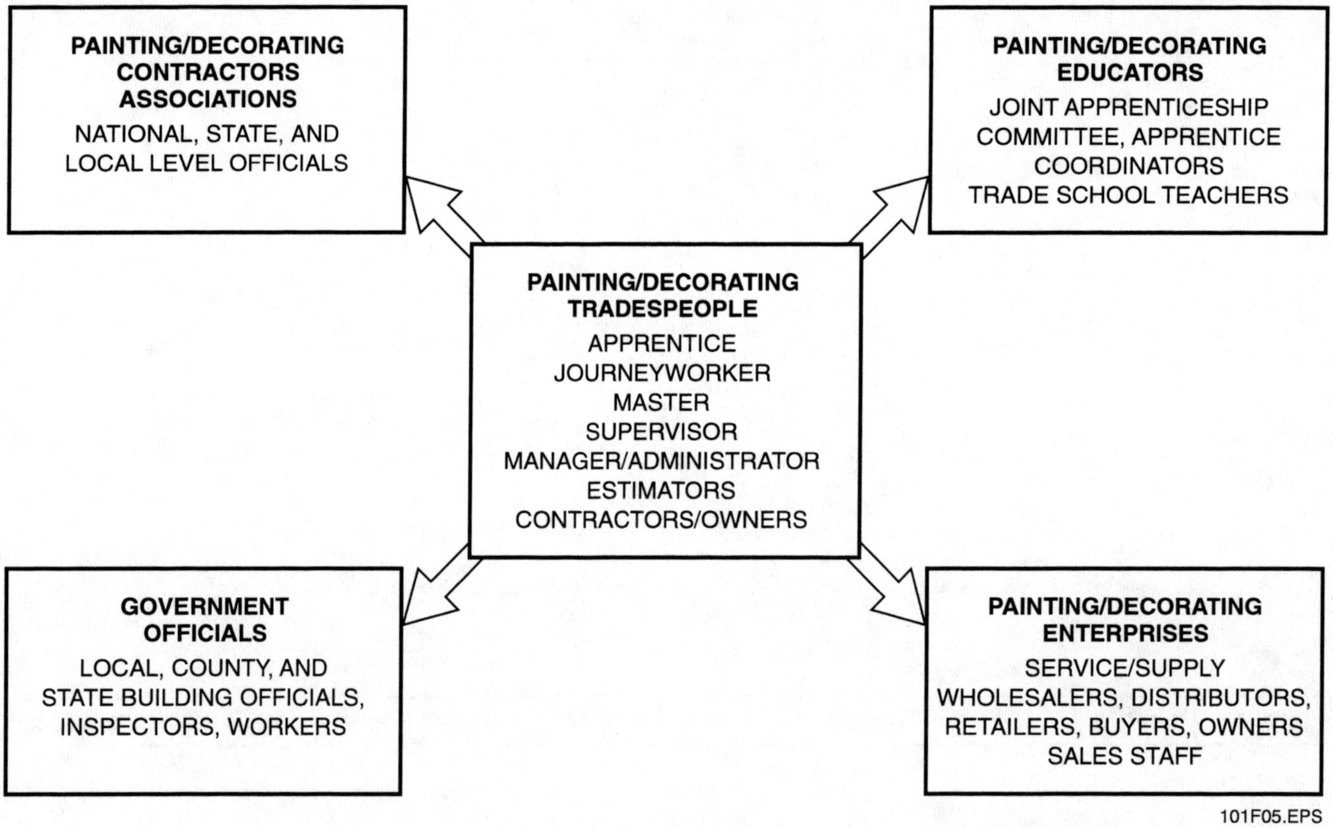

Figure 5. Career Paths In The Painting Trade And Related Fields

3.2.0 ADVANCING IN THE PAINTING TRADE

Some people have learned painting simply by working for several years as a helper to an experienced painter. However, those responsible for training painters recommend a formal apprenticeship program. The apprentice trainee can move up through several levels in the painting trade:

- Apprentice
- Journeyworker
- Master
- Supervisor
- Manager/Administrator
- Estimator
- Contractor/Owner

3.2.1 Apprentice And Apprenticeship Program

This training program is part of a nationally-recognized apprenticeship system approved by the U.S. Labor Department. In combination with your on-the-job training, it will help you learn the skills you need to qualify as a journeyworker in the painting trade.

Apprentice training goes back thousands of years; its basic principles have not changed in that time. First, it is a means for individuals entering the craft to learn from those who have mastered the craft. Second, it focuses on learning by doing; real skills versus theory. Although some theory is presented in the classroom, it is always presented in a way that helps the trainee understand the purpose behind the skill that is to be acquired. That is why this program, and others in the *Wheels of Learning* series, are said to be "competency-based."

Apprentice programs as they exist today have some specific characteristics:

- Apprentices must be no less than 16 years old.
- Opportunities must be available to all.
- The program must incorporate both training and on-the-job experience.
- A minimum of 144 hours per year of related training is given.
- There is a progressively increasing schedule of wages.
- There is proper supervision with adequate facilities.
- Job performance and related instruction are periodically evaluated.
- Successful completions are formally recognized.

A Youth Apprentice Program is also available that allows students to begin their apprentice training while still in high school. A student entering the program in eleventh grade will have completed one year of the *Wheels of Learning* three-year painting trade program by graduation. In addition, the program, in cooperation with local craft employers, allows students to work in the trade and earn money while still in school. Upon graduation, the student can enter the industry at a higher level and with more pay than someone just starting the apprentice program.

This training program is the same one used by National Center for Construction Education and Research learning centers, contractors, and colleges across the country. Students are recognized through an official transcript and can enter the second year of the program wherever it is offered. They may also have the option of applying the credits at a two-year or four-year college that offers a painting/decorating-related degree or certificate program.

3.2.2 Journeyworker Painter/Paperhanger

After successfully completing an apprenticeship, a trainee becomes a journeyworker. The term *journeyworker* originally meant to "journey" away from the master and work alone. A person can remain a journeyworker or advance in the trade. Journeyworkers may have additional duties such as supervisor or estimator. With larger companies and on larger jobs, journeyworkers often become specialists. Some may become proficient in applying high-performance coatings. Others may install special wallcoverings or fabrics, or paint graphics or murals. Still others may specialize as abrasive blasters.

3.2.3 Master Painter/Paperhanger

A master craftsperson is one who has achieved, and continuously demonstrates, the highest skill levels in the trade. The master is a mentor and teacher of those to follow. Master painters often start their own business and become contractors/owners.

3.2.4 Supervisors And Foremen

Large painting projects require supervisors or foremen who oversee the work of crews made up of apprentices and journeyworkers. They are responsible for assigning and directing the work of the painting crew members.

3.2.5 Manager/Administrator

Management and administration deal with controlling the scope and direction of the business. Larger contracting firms may have one, or several, manager/administrators. A production manager, sometimes called an *expeditor*, controls and guides the production of the painters. This person is responsible for worker output and must determine the best methods to use and the way to apply workers to accomplish the job. The production manager is often a troubleshooter and problem solver. An office manager is one who is responsible for a contractor's support operations, such as accounting, finance, and secretarial work. If a contractor has a retail sales branch, a painter might become a sales manager. A sales manager is responsible for managing all the sales personnel and decides where and how to best sell paint, wallcoverings, and related materials.

The administrator's function is to decide the direction and method of doing business. An administrator might determine the amount of work to be done by a business and the overall methods of achieving it.

3.2.6 Estimator

Estimators work for painting contractors. They make careful estimates of the materials and labor required to do the job. Based on these estimates, the contractor submits bids for jobs. Estimating requires a complete understanding of both paint/wallcovering application and the materials and supplies required. Only experienced painters who possess good math skills and the patience to prepare detailed, accurate estimates are employed to do this work. This is a highly responsible position since errors in estimates can result in financial losses to the contractor. Depending on the size of the business, the job of estimating is often done by the owner, manager, and administrator.

3.2.7 Contractor/Owner

Painting contractors/owners are master painters or others who have established a contracting business. Generally, they hire apprentices, journeyworkers, and master painters to work for them. Depending upon the size of the painting business, contractors may work with the crew or they may manage the business full-time. Very small painting contractors may have only one or two people to do everything, including managing the business, preparing estimates, obtaining supplies, and applying the paint or wallcovering systems on the job. If an individual owns the contracting business, it is called a *sole proprietorship*. If it has two or more owners, the company would be a *partnership*. Large contractor organizations are often organized as *officially state-chartered corporations*.

4.0.0 PROFESSIONALISM

There are many crafts in the construction industry. All are important, but almost none require more professionalism than the painting trade. Successful painters are those who believe in themselves and what they do for a living.

4.1.0 PRODUCTIVITY AND TRADE ETHICS

A professional painter performs on the job at a high level of productivity. In order to be successful, the professional must have the skills to use current trade materials, tools, and equipment to produce a finished product of high quality in a minimum period of time. A painter must be adept at adjusting methods to meet each situation. The successful painter must continuously train to remain knowledgeable about the technical advancements in trade materials and equipment and to gain the skills to use them. A professional painter never takes chances with regard to personal safety or the safety of others.

Honesty and personal integrity are important traits of the successful professional. A professional prides himself or herself in performing an honest day's work for a day's pay, and by striving to always be punctual and dependable. Each job is completed in the professional way, never by cutting corners or reducing materials. A valued professional maintains work attitudes and ethics that protect property such as tools and materials belonging to employers, customers, or other trades from damage or theft at the shop or jobsite. A professional never uses alcohol, drugs, and/or profanity on the job. Appendix A in the back of this module provides a list of the characteristics and values most people associate with ethical behavior.

4.2.0 CUSTOMER RELATIONS

4.2.1 Appearances Count

When people meet you for the first time, they form their critical first impression of you based, to a large extent, on how you look. Industry studies have shown that personal appearance is a factor that most employers and customers consider important.

Ask yourself these questions:

- Do you get enough sleep and look alert?
- Do you practice good personal hygiene?
- Do you wear a neat, clean uniform including a hat, proper safety shoes, shirt, and pants/ coveralls?
- Do you smile, display confidence and polite respect for your employer, the customer, and fellow employees?

If your answer to each question is "Yes" – Congratulations! You are well on your way to consistently making a good impression of both you and your company. If you answered "No" to any of the questions, you have identified an area that can change how people feel about you and your company. Get to work on making an improvement. You will be glad you did.

In addition, your vehicle is a traveling billboard for your company (*Figure 6*). How it looks and how you drive it can greatly influence the public's impression of your company. Make a positive impression on the road.

- Is your truck clean and in good repair?
- Are your driving habits courteous?

Figure 6. Your Vehicle Is A Traveling Billboard For Your Company

4.2.2 Treat Your Customers With Respect

Treating customers with respect is always important, but in residential work it is even more critical. Admitting you into their home or business is a gesture of faith by customers (*Figure 7*); faith that you will do no harm, and that you will treat the premises and occupants with respect. Do you agree? Think about the people you invite into your home.

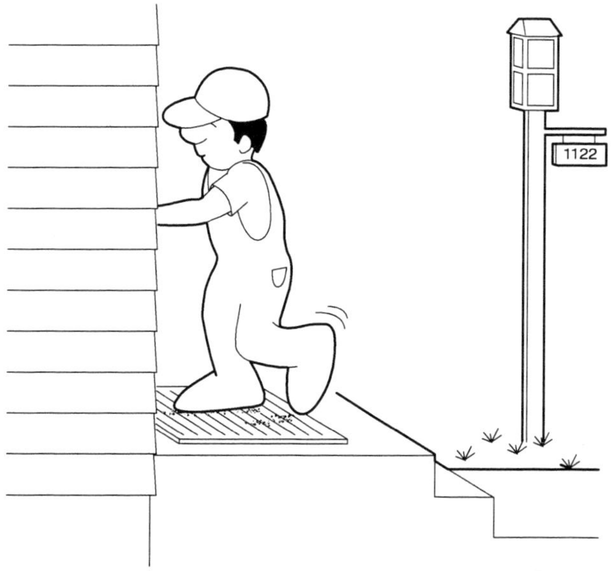

Figure 7. Treat Customers With Respect—Begin With Their Homes

Ask yourself these questions:

- Do you refrain from smoking?
- Do you remember to protect the work area?
- Do you carry rags and carefully clean up after yourself?
- Do you have dropcloths to protect hardwood floors and carpets?
- Do you respect the customer's home by not tracking in dirt?
- Do you return the home to its original condition? (Replace covers, wipe off dirty fingerprints, clean up dropcloths, etc.)

As the number of homes with more than one income has increased, the need to arrive on time has become more critical than ever, since in these cases someone has to take off from work to wait for the craftsperson. Be on time. Call if you are going to be late. Show the customer that you realize their time is valuable.

Customers notice if you arrive with a full set of tools, neatly packed. Using a wrong tool not only leaves behind poor results, it tells your customer that you are unprofessional. They notice your attitude, and whether it shows a good work ethic. They can see if you neatly repack your tools when you finish the job, too. It's part of how they judge you and your company. It's also a large contributor to whether they call your company back for more work.

Review these questions. Any "No" answers indicate areas that need work.

- Are you on time and prompt to start work?
- Do you arrive fully informed and prepared to do the job?
- Are your tools a full set, and neatly packed?
- Do you tackle the job promptly?
- Do you politely avoid general social conversation while working?

Finally, remember that your customers appreciate a positive attitude. For example, it's always best to be professional and avoid "bad mouthing" different products and/or competitors. Likewise, if for example, you are late—but it is not your fault—focus on the positive, getting the job done, rather than blaming someone else. Say something like: "I understand how you feel. I'm sorry. Please know that I will do my best to have your room painted as quickly as I can."

5.0.0 WORKING WITH OTHER TRADES

Painters working on a construction job involving other trades should treat the craftspersons in the other trades with the same courtesies given their own trade, employers, and customers. In order to maintain jobsite harmony, painters should always work closely with the other trades to coordinate and schedule their painting tasks so as not to hold up the other trades, and vice versa. A close relationship between the trades is also important to make sure that all trades *fully* understand their responsibilities on the job as defined in the job contract and specifications. This is necessary to prevent failure to complete a task, duplication of effort, and/or needless rework.

6.0.0 EMPLOYER TEAMING OR PARTNERING WITH OTHER ORGANIZATIONS

Employers often team with facility owners, other painting or general contractors, material manufacturers, or material distributors. Some benefits that can be derived through such teaming or working relationships are:

- Provides a way for supplying the manpower needed to accomplish larger jobs when an organization does not have adequate manpower to accomplish the task within the organization.
- Provides a way to acquire needed skills not possessed by one or more of the organizations.
- Provides for better coordination among all parties involved with a job, usually resulting in more efficient scheduling of the job.
- Provides for lower prices.
- Provides for timely delivery of materials.

Employees working for employers involved in such teaming relationships have a responsibility to cooperate and work with all persons from other organizations and do everything possible to help make such teaming agreements work. The main benefit derived by employees involved in such teaming relationships is more stabilized employment. This is because teaming usually results in more opportunities to work than might otherwise be available without teaming. Other benefits are the friendships and relationships developed with others both inside and outside the painting trade.

SUMMARY

Professionals in the painting trade have a long heritage. Paints and wall coverings have been used for decorating and protective purposes since prehistoric times. More advances in painting materials and methods have occurred during the latter part of the 20th century than all of prior history combined.

Because of the widespread use of painting and wallcovering products to decorate and protect structures, there are many career opportunities in the painting trade. Jobs are available with small contractors, large industrial and commercial contractors, and manufacturing and distributing firms that produce and market painting and wallcovering materials. The apprentice program provides an opportunity to learn the trade through a combination of hands-on training and related classroom study. The Youth Apprentice Program allows students to begin their training while still in high school.

The professional painter is one who is very productive. Also, the painter should always strive to make a good appearance and practice good personal hygiene. A professional is courteous and respectful towards his or her employer, fellow employees, the customer, and craftspeople in other trades.

References

For advanced study of topics covered in this task module, the following books are suggested:

Painting and Decorating Craftsman's Manual and Textbook, Painting and Decorating Contractors of America, Fairfax, Virginia.

Painting and Decorating Encyclopedia, The Goodheart-Willcox Company, Inc., Tinley Park, Illinois.

REVIEW/PRACTICE QUESTIONS

1. Paint has been used as a form of decoration since:
   a. the Roman empire.
   b. prehistoric times.
   c. the Chou Dynasty.
   d. the Ming Dynasty.

2. The Egyptians were the first to use:
   a. protective coatings.
   b. oil-based paints.
   c. white lead.
   d. lacquer.

3. The extensive use of linseed oil in paints decreased mainly as a result of:
   a. World War I.
   b. World War II.
   c. the Clean Air Act.
   d. the Korean War.

4. The Clean Air Act mandated the containment of all sources of air pollution, including:
   a. chromates.
   b. volatile organic compounds.
   c. asbestos.
   d zinc.

5. Paints containing lead pigments were commonly used in residential structures built before:
   a. 1978.
   b. 1968.
   c. 1988.
   d. both a and b.

6. The minimum number of formal training hours per year in an approved apprentice program is:
   a. 200.
   b. 180.
   c. 144.
   d. there is no minimum.

7. By participating in the Youth Apprentice Program, a student may:
   a. not receive college credit.
   b. not receive credit toward an apprenticeship, but will gain valuable experience.
   c. complete one year of the three-year apprenticeship.
   d. complete two years of the three-year apprenticeship.

8. The customer begins forming a first impression of you and your company:
   a. after you've painted their room.
   b. after you've submitted your bill.
   c. as soon as you appear at the door.
   d. as soon as you speak.

9. You should always treat customers with:
   a. kid gloves.
   b. courtesy, concern, and respect.
   c. a few good jokes to relax them.
   d. meaningful stories of past experiences.

10. Your customer asks how you feel about the "mess in Washington." You:
   a. tell him all politicians should be shot.
   b. tell him you think there isn't any "mess."
   c. politely turn your conversation to the painting/decorating task at hand.
   d. politely tell him he's wrong.

ACKNOWLEDGMENTS

Figure 4 courtesy of Painting and Decorating Contractors of America.

APPENDIX A

Ethical Principles For Members Of The Painting/Decorating Trades

Honesty: Be honest and truthful in all dealings. Conduct business according to the highest professional standards. Faithfully fulfill all contracts and commitments. Do not deliberately mislead or deceive others.

Integrity: Demonstrate personal integrity and the courage of your convictions by doing what is right even where there is great pressure to do otherwise. Do not sacrifice your principles for expediency, be hypocritical, or unscrupulous.

Loyalty: Be worthy of trust. Demonstrate fidelity and loyalty to companies, employers, fellow craftspeople, and trade institutions and organizations.

Fairness: Be fair and just in all dealings. Do not take undue advantage of another's mistakes or difficulties. Fair people display a commitment to justice, equal treatment of individuals, tolerance for and acceptance of diversity, and they are open-minded.

Respect for others: Be courteous and treat all people with equal respect and dignity regardless of sex, race, or national origin.

Law abiding: Abide by laws, rules, and regulations relating to all personal and business activities.

Commitment to excellence: Pursue excellence in performing your duties, be well-informed and prepared, and constantly endeavor to increase your proficiency by gaining new skills and knowledge.

Leadership: By your own conduct, seek to be a positive role model for others.

Painting contractor obligations: Provide the service of the industry to the fullest extent of your technical ability. Protect, preserve, and decorate structures, facilities, and all other properties by a means and with materials of the most progressive and efficient type.

Protect and build the company's good reputation and the morale of its employees by engaging in no conduct that might undermine respect and by taking whatever actions are necessary to correct or prevent inappropriate conduct of others.

Provide opportunities for training, employment, career advancement, and economic security for all those employees who would apply themselves.

NCCER CRAFT TRAINING USER UPDATES

The NCCER makes every effort to keep these manuals up-to-date and free of technical errors. We appreciate your help in this process. If you have an idea for improving this manual, or if you find an error, a typographical mistake, or an inaccuracy in the NCCER's Craft Training Manuals, please write us, using this form or a photocopy. Be sure to include the exact module number, page number, a description of the problem, and the correction, if possible. Your input will be brought to the attention of the Technical Review Committee. Thank you for your assistance.

Instructors – If you found that additional materials were necessary in order to teach this module effectively, please let us know so that we may include them in the Equipment/Materials list in the Instructor's Guide.

Write: Curriculum and Revision Department
National Center for Construction Education and Research
P.O. Box 141104
Gainesville, FL 32614-1104

Fax: 352-334-0932

Craft ____________________ Module Name ____________________

Module Number ____________________ Page Number(s) ____________________

Description of Problem

____

____

____

____

(Optional) Correction of Problem

____

____

____

(Optional) Your Name and Address

____

____

Safety

Module 07102

Painting Trainee Task Module 07102

SAFETY

PAINTING AND DECORATING CONTRACTORS OF AMERICA

NATIONAL CENTER FOR CONSTRUCTION EDUCATION AND RESEARCH

Objectives

Upon completion of this module, the trainee will be able to:

1. Explain the obligations and responsibilities of workers, employers, and the Occupational Safety and Health Administration (OSHA) with regard to safety.
2. Describe the nature of the hazard(s) and safety rules and guidelines for job site safety relating to:
   - Personal hygiene and work clothing
   - Personal protective and life saving equipment
   - Respiratory hazards
   - Fall hazards
   - Hazardous chemicals and materials
   - Lead hazards
   - Asbestos hazards
   - Confined space hazards
   - Painting tool and equipment hazards
   - Hot and cold weather hazards
   - Electrical hazards
3. Explain the purpose of Hazardous Communication (HazCom) programs including:
   - The elements or parts required of a HazCom program
   - The use of Material Safety Data Sheets (MSDSs)
   - The use of Hazardous Materials Identification System (HMIS) and National Fire Protection Association (NFPA) codes

Prerequisites

Successful completion of the following Task Modules is required before beginning study of this Task Module: Core Curricula, Painting Level 1, Module 07101.

Required Trainee Material

1. Trainee Module
2. Appropriate Personal Protective Equipment

Course Map

This course map shows all of the *Wheels of Learning* task modules in the first level of the Painting curricula. The suggested training order begins at the bottom and proceeds up. Skill levels increase as a trainee advances on the course map. The training order may be adjusted by the local Training Program Sponsor.

Course Map: Painting, Level 1

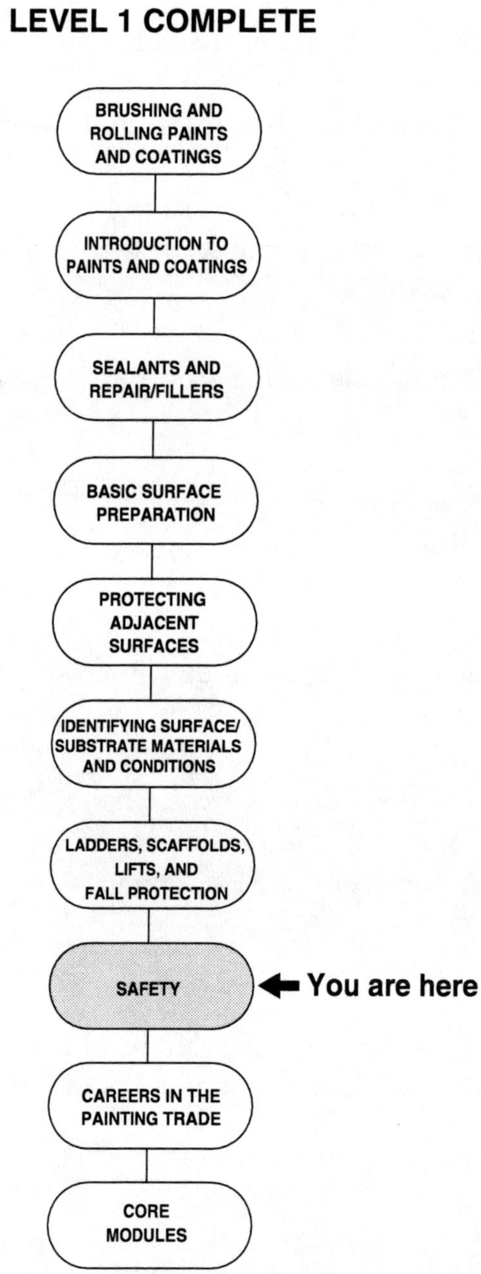

CMAP102.EPS

TABLE OF CONTENTS

Trade Terms Introduced In This Module

Acute effects of overexposure: The adverse effects that are normally evident immediately or shortly after a short term period of exposure (seconds, minutes, hours, or days) to a hazardous material without implying a degree of severity.

American National Standards Institute (ANSI): A non-profit organization that has established voluntary standards for the design, testing, and use of various products.

Chronic effects of overexposure: The effects that develop slowly over a long period of time (weeks, months, years) from exposure to a hazardous material without implying a degree of severity.

Flash point (FP): The lowest temperature of a liquid at which it gives off sufficient vapor to form an ignitable mixture with the air near the surface of the liquid or within the container used. Materials with flash points below 100°F (38°C), such as most solvents and solvent-borne coatings, are considered flammable hazards.

Hazardous Materials Identification System (HMIS): A hazard communication system of codes developed by the National Paint and Coatings Association. It uses colors, numbers, letters, and symbols that quickly identify the health hazards, flammability, and reactivity of a material. It also identifies the proper personal protection equipment and other information pertaining to the material.

High-efficiency particulate air (HEPA) filter: A high-efficiency filtering device designed to remove 99.97% of all particles larger than 0.3 micron.

Immediately dangerous to life and health (IDLH): The definition of an atmosphere that poses an immediate hazard to life or produces immediate, irreversible, and debilitating effects on health.

Material Safety Data Sheet (MSDS): A data sheet issued by a manufacturer or producer of material containing information and instructions on the chemical and physical characteristics of a substance, its hazards and risks, safe handling requirements, and actions to be taken in the event of fire, spill, overexposure, etc.

Micron: A unit of measurement equal to one-millionth of a meter or approximately 0.00003937 inch (25 microns equal about 0.001 inch).

National Institute for Occupational Safety and Health (NIOSH): An agency that performs studies, analyzes data, and publishes information, guidelines, and recommendations related to worker health and safety.

Occupational Safety and Health Administration (OSHA): An agency of the U.S. Department of Labor that is responsible for developing and enforcing regulations to support the Occupational Safety and Health act, the federal law designed to protect people from unsafe work environments.

Permissible exposure limit (PEL): The airborne concentration that has been established by OSHA as the enforceable exposure limit. PELs are expressed in parts per million (ppm) or micrograms per cubic meter (mg/m^3). Unless otherwise indicated, they are based on time-weighted average (TWA) concentrations for a normal 8-hour workday and a 40-hour workweek.

Short-term exposure limit (STEL): The airborne concentration to which workers can be exposed for up to 15 minutes without suffering ill effects.

Spontaneous combustion: The process of a material catching on fire and burning as a result of the heat generated by internal chemical action.

Threshold limit value (TLV): The airborne concentration of a substance to which a normal, healthy worker may be exposed for eight hours a day, 40 hours a week during a working lifetime without adverse health effects. TLVs are recommended limits established by the American Conference of Governmental Industrial Hygienists (ACGIH). Normally, the TLV represents a time-weighted average (TWA).

Toxic: Poisonous. Any material which can cause illness or injury.

1.0.0 INTRODUCTION

Using incorrect safety procedures and having a poor safety attitude can make painting a hazardous trade. Painters must climb on ladders, rig tall structures, and work on platforms and other scaffolding. They use high-pressure cleaning and painting equipment as well as high-speed power tools. They are often required to work around high voltages and in confined spaces. Painters use a wide variety of materials that can have negative health effects. These materials include paints, solvents, blast materials and residues, chemical strippers, cleaning agents, and etching agents.

The focus of this module is on safety issues and practices specific to the painting trade. This module supplements some of the general safety information studied earlier in *Basic Safety* and it introduces new safety topics of importance to painters.

2.0.0 EMPLOYER AND EMPLOYEE SAFETY OBLIGATIONS

It is important to understand the obligations that exist for everyone's safety. An obligation is like a promise or a contract. In exchange for the benefits of your employment, you agree to work safely. In other words, you are *obligated* to work safely. You are also obligated to make sure anyone you happen to supervise is working safely. Your employer is also obligated to maintain a safe workplace for all employees. Safety is everyone's responsibility (*Figure 1*).

Figure 1. Safety Is Everyone's Responsibility

Some employers will have safety committees. If you work for such an employer, you are then obligated to that committee to help maintain a safe working environment. This means two things:

1. Follow the safety committee's rules for proper working procedures and practices.
2. Report any unsafe equipment and conditions directly to the committee or to your supervisor.

Here is a basic rule to follow every working day:

If you see something that is not safe, REPORT IT! Don't ignore it. It won't correct itself. You have an obligation to report it.

Suppose you see a faulty electrical hook-up. You know enough to stay away from it, and you do—but then you forget about it. Why should you worry? It's not going to hurt you. Let somebody else deal with it. The next thing that happens is that your best friend accidentally grabs the live wire.

In the long run, even if you don't think an unsafe condition affects you—it does. Don't mess around. Report what isn't safe. Don't think your employer will be angry because your productivity suffers while the condition is corrected. On the contrary, your employer will be very pleased.

Your employer knows that the short time lost in making conditions safe again is nothing compared with shutting down the whole job because of a major disaster. If that happens, you are out of work anyway. So don't ignore an unsafe condition. In fact, **Occupational Safety and Health Administration (OSHA)** regulations require you to report hazardous conditions.

This applies to every part of the construction industry. Whether you work for a large contractor or a small subcontractor, you are obligated to report unsafe conditions. The easiest way to do this is to tell your supervisor. If that person is ignoring the unsafe condition, report it to the next highest supervisor. If it is the owner who is being unsafe, let that person know what you think. If nothing is done about it, report it to OSHA. If you are worried about your job being on the line, think about it in terms of your life being on the line.

The U.S. Congress passed the Occupational Safety and Health Act in 1970. The act also created OSHA. It is part of the U.S. Department of Labor. The job of OSHA is to set occupational safety and health standards for all places of employment, to enforce these standards, and to provide research and educational programs to support safe working practices.

OSHA requires each employer to provide a safe and hazard-free working environment. OSHA also requires that employees comply with OSHA rules and regulations that relate to their conduct on the job.

According to OSHA standards, you are entitled to on-the-job safety training. As a new employee, you must be:

- Shown how to do your job safely.
- Provided with the proper personal protective equipment.
- Warned about specific hazards in the work and in the surroundings.
- Supervised for safety while performing the work.

OSHA was adopted in 1970 with the stated purpose "to assure as far as possible every working man and woman in the nation safe and healthful working conditions and to preserve our human resources."

The enforcement of this act of Congress is provided by the federal and state safety inspectors who have the legal authority to make employers pay fines for safety violations. The law allows states to have their own safety regulations and agencies to enforce them, but they must first be approved by the U.S. Secretary of Labor. For states that do not develop such regulations and agencies, federal OSHA standards must be obeyed.

These standards are listed in *OSHA Safety and Health Standards for the Construction Industry* (29 CFR, Part 1926), sometimes called *OSHA Standards 1926* (*Figure 2*). Other safety standards that apply to the painting trade are published in *OSHA Safety and Health Standards for General Industry* (29 CFR, Parts 1900 to 1910).

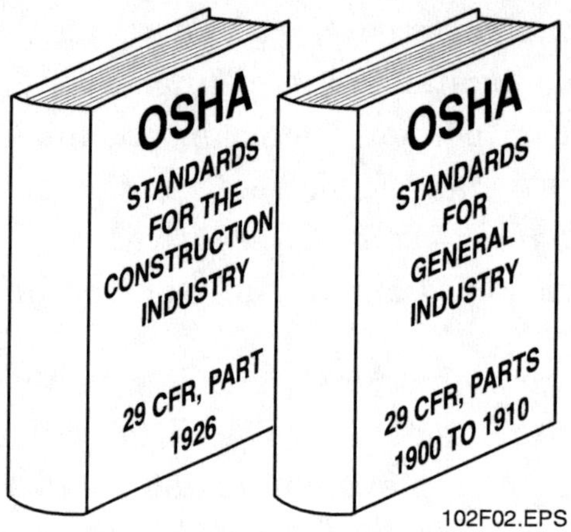

Figure 2. OSHA Standards Govern Workplace Safety

The most important general requirements that OSHA makes on employers in the construction industry are:

- The employer must perform frequent and regular job site inspections of equipment.
- The employer must instruct all employees to recognize and avoid unsafe conditions, and to know the regulations that pertain to the job so they may control or eliminate any hazards.
- No one may use any tools, equipment, machines, or materials that do not comply with *OSHA Standards 1926*.

3.0.0 PERSONAL HYGIENE AND WORK CLOTHING

3.1.0 PERSONAL HYGIENE

Personal hygiene plays a large part in avoiding painting hazards. For example, poor hygiene habits can allow chemicals to enter the body and cause harm. Depending on the type of chemical, it can be inhaled, absorbed, or ingested. This could cause acute or chronic problems later on.

The basic hygiene guidelines for all painting job sites are:

- Wash hands thoroughly before eating and/or drinking (*Figure 3*).
- Avoid sticking objects in your mouth while working.
- Do not rub your eyes. More diseases are transported by your hands than by any other means.
- Long hair must be neat and secured to avoid a vision or safety hazard.
- Mustaches and beards should be trimmed neatly. However, you must be clean shaven if required to wear a half-mask respirator.

Figure 3. Wash Hands Before Eating And/Or Drinking

Lead hazards are covered in detail later in this module, but basic hygiene guidelines for lead paint removal job sites are:

- Face, hands, and arms must be thoroughly washed prior to eating, drinking, or smoking.
- Hands must be thoroughly washed before and after using the restroom.
- Eating, drinking, or smoking is allowed only in approved areas at the lead paint removal site.
- Disposable coveralls, shoe covers, gloves, etc., must be removed only in the decontamination area. The disposable clothing is to be placed in the hazardous waste disposal container. Hazardous waste disposal must be in accordance with all local, state, and federal regulations.
- After you have washed and changed back into your normal work or street clothes, re-entry into the lead paint removal site is not allowed.

3.2.0 WORK CLOTHING

You must dress properly for construction work. The actual type of clothing worn usually depends on your company's dress code. A painter typically wears white painter's pants or overalls, a white long-sleeved shirt, and a painter's cap or a hard hat (where required). Work clothing should be in good condition, fit properly, be comfortable, and not interfere with the free movement of your body.

- Pants should not be so long as to create a tripping hazard.
- Long-sleeved shirts are recommended because they offer better protection against skin contact with hazardous materials. Long shirt sleeves should always be buttoned to prevent a sleeve from getting caught in rotating and/or moving machinery.

4.0.0 PERSONAL PROTECTIVE AND LIFE SAVING EQUIPMENT

Depending on the specific job and conditions, protective equipment is often needed. Protective equipment can include:

- Foot protection
- Hand protection
- Eye and face protection
- Head protection
- Hearing protection

4.1.0 FOOT PROTECTION

OSHA requires that protective footwear be worn when working where falling, rolling, or sharp objects pose a danger of foot injuries, and where feet are exposed to electrical hazards. Leather footwear with reinforced soles, or innersoles of flexible metal are recommended for any construction site. Steel-toed safety shoes should be worn for operations involving the handling of heavy materials. Protective footwear must comply with ANSI Z41-1991. ***Sneakers, tennis shoes, or similar types of footwear should never be worn at the job site.***

4.2.0 HAND PROTECTION

A painter's hands may be exposed to a variety of hazards (i.e., temperature extremes, abrasive materials, paints, and solvents) that may cause inflammation of the skin or dermatitis. Gloves (*Figure 4*) are the primary type of hand protection. They may be made of leather, rubber, cotton, or a variety of plastics or synthetics. There is no all-purpose glove. You must select your gloves on the basis of the hazards involved in doing the work.

- For abrasive blasting, you should wear heavy duty canvas or leather gloves.
- For water blasting, you should wear gloves that protect against chemicals (if you are using chemicals) and water.

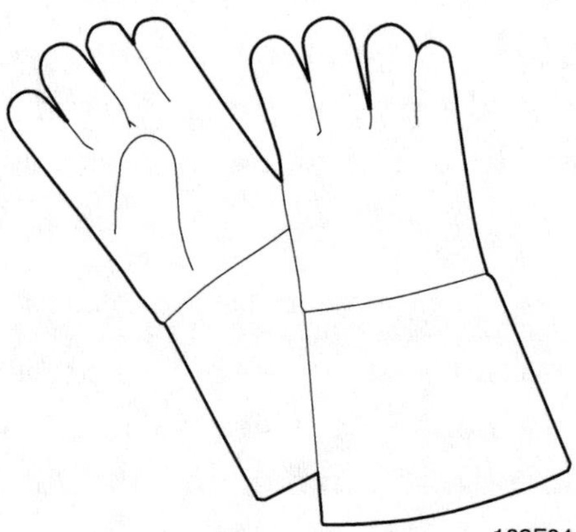

102F04.EPS

Figure 4. Gloves Are The Primary Type Of Hand Protection

- For spraying, brushing, or rolling paint, you should wear solvent-resistant gloves or use a skin barrier cream or lotion to protect and control drying of the skin.

WARNING! Never wear gloves around machinery because the moving parts can snag the gloves and pull your hands into the machinery.

4.3.0 EYE AND FACE PROTECTION

You only have one pair of eyes. Lost vision cannot be replaced. Proper eye wear and face protection (*Figure 5)* must be worn for the job at hand. There are tinted safety glasses for working out-of-doors and prescription safety glasses of all types. Have new glasses fitted properly and wear head straps when necessary.

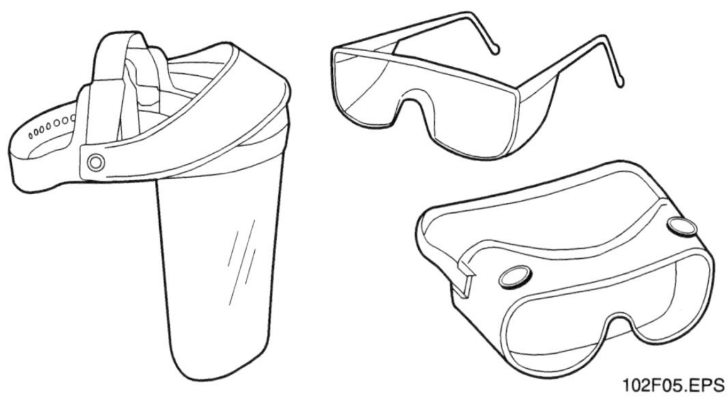

Figure 5. Use Proper Eye And/Or Face Protection For The Job At Hand

When spray painting or performing similar tasks, it is difficult to maintain clean lenses on safety glasses, goggles, shields, etc. As it is often difficult to prevent overspray from falling on your lenses, you should consider keeping an extra pair of glasses handy to exchange as necessary. Clean the lenses often using a non-abrasive cleaner and keep them in a protective case to prevent scratching.

Some tasks require the use of safety goggles, chemical-resistant goggles, dust goggles, or face shields. Always check the **Material Safety Data Sheet (MSDS)** for the product being used to find out what type of eye protection is needed. Detailed information on the types of data provided by Material Safety Data Sheets is given later in this module.

4.4.0 HEAD PROTECTION

An approved hard hat (*Figure 6*) must be worn on the job site in all posted areas and other areas where there is any danger of head injury due to falling or flying objects. Non-conductive hard hats must be worn when working in the vicinity of equipment or electrical lines that expose you to high-voltage electrical shock hazards. Hard hats used for protection against falling and/or flying objects must meet the **American National Standards Institute**

(ANSI) Z89.1-1969 requirements. Those used to protect against electrical shock must meet ANSI standard Z89.2-1971. The ANSI standard for a hard hat is stamped or labeled on the inside of the hat by the manufacturer. Only hard hats stamped or labeled with the correct ANSI classification should be used.

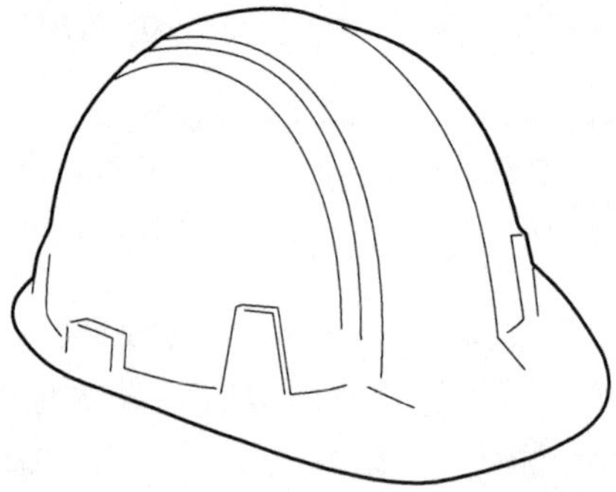

102F06.EPS

Figure 6. Wear Head Protection In All Posted Areas And When There Is A Danger Of Head Injury From Falling Or Flying Objects

The correct way to wear a hard hat is to adjust its inner suspension so that the hat is slightly raised off the head. A suspension that is too flexible will permit contact with the head upon impact and can result in a skull fracture or concussion. A suspension that is too rigid can transmit the shock impact and fracture the neck vertebrae. ***Never remove the hard hat's inner suspension for any reason or keep anything under your hard hat.*** Both actions will prevent the suspension from working as a shock absorber. Other practices that should be followed are:

- Clean your hard hat and suspension at least every 30 days.
- Immediately replace a broken or punctured hard hat.
- Never drill holes in your hard hat for ventilation.
- Never leave your hard hat in the rear window of your car or truck. The sunlight may affect its protective quality, and an emergency stop can turn your hard hat into a dangerous projectile.

4.5.0 HEARING PROTECTION

Construction site activities and/or the use of some equipment can sometimes create noise at unhealthy levels. If you find yourself saying "What?", "Pardon", or "What did you say?" because you can't hear what another person is saying, then you are probably in an area that contains a noise level that can cause permanent hearing loss.

Not all sounds have the same effect on hearing. Sound or noise has three changing factors: intensity, pitch, and length of exposure. *Intensity* means loudness of sound and is measured in decibels, or dBAs. *Pitch* refers to the frequency of sound waves. A high-frequency (high-pitched) whistle, even though it isn't any louder than a low-pitched thud, is generally more harmful than the low-frequency sound. *Length of exposure* refers to the amount of time a person is subjected to a noise. Continual exposure to certain noises can be more harmful than occasional bursts of offensive sound. Exposure levels can range from 0 dBA (the faintest sound a human ear can hear) to 140 dBA (the threshold of pain) and higher. *Table 1* shows the relative sound levels for some common noises.

Sound Level (dBA)	Source
0	Threshold of human hearing
30	Soft whisper
60	Normal conversation
60 to 80	Printers and copiers
90	Lawn mower
100	Chain saw
110	Loud discotheque
115	Portable stereo headset on high setting
140	Gunshot
180	Rocket pad during launch

Table 1. Sound Levels Of Some Common Noises

OSHA regulations allow workers to be exposed to no more than an average of 90 dBA of continuous noise (as opposed to impact noise, such as a shotgun blast) over an 8-hour period. When exposed to sound levels that exceed allowable levels (*Table 2*), OSHA requires that approved ear plugs or earmuffs (*Figure 7*) be used to provide hearing protection.

Duration Per Day (Hours)	Sound Level (dBA)
8	90
6	92
4	95
3	97
2	100
1½	102
1	105
½	110
¼ or less	115

Table 2. OSHA Permissible Noise Levels

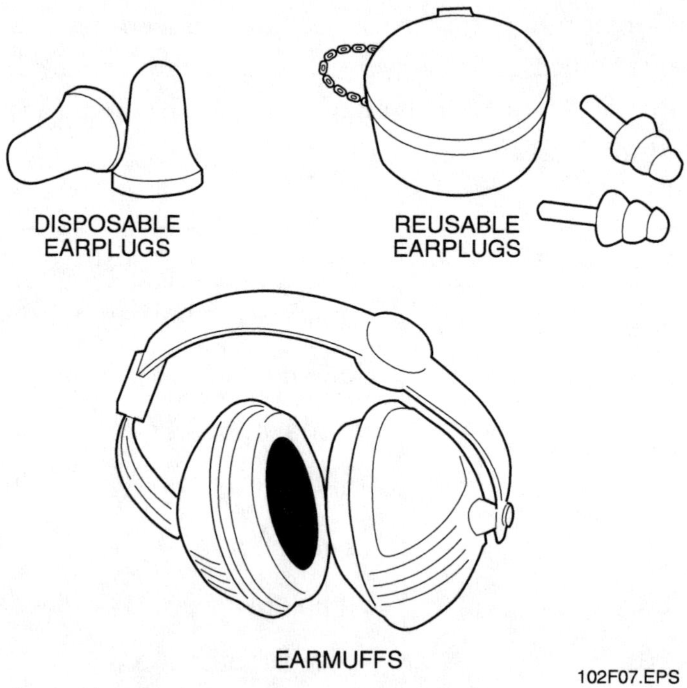

Figure 7. Use Hearing Protection When Sound Levels Exceed Allowable Limits

Disposable ear plugs are the most common form of hearing protection used in the industry today. These devices usually have an outer layer of pliable foam and a core layer of acoustical fiber which filters out harmful noise, yet allows you to hear normal conversation. To use disposable ear plugs, simply roll into a cylinder and insert the tapered end into the ear canal. The ear plugs will expand, filling the ear canal and making a proper fit. Reusable ear plugs are also common.

Earmuffs are usually worn in severe noise environments. They offer greater protection from all noise, including low frequencies. Most earmuffs have adjustable headbands that can be worn over the head, behind the neck, or under the chin. To use earmuffs, adjust the tension on the headband and ear cushion pads to obtain the best possible seal. Check the earmuff shell for cracks and the ear cushion pads for tears before each use. Any damaged, cracked, or torn part must be repaired or replaced.

5.0.0 RESPIRATORY HAZARDS AND PROTECTION

Performing painting tasks or being in an area where vapors, fumes, or particles are being released into the air can be dangerous. Painters continuously come into contact with and inhale vapors emitted from chemicals contained in paints, solvents, cleaners, strippers, and other materials. They can also inhale very fine dust particles when sanding, performing abrasive blasting, or cleaning with power tools. Dust particles finer than 10 microns in diameter are not visible, but they are respirable, which means they can enter the lungs.

A **micron** is equal to one-millionth of a meter or approximately 0.00003937 inch. OSHA has set **permissible exposure limits (PELs)** for a number of toxic materials found in the work place. *Appendix A* lists the PELs for some common materials used while painting.

5.1.0 TYPES OF RESPIRATORS

Respiratory (lung) protection is required whenever there is a danger of a respiration hazard or suffocation from lack of oxygen. There are four general types of respirators (*Figure 8*):

- *Half-mask* – A half-mask with a mechanical filter is used only in areas where dust or other solid particles can be inhaled. With the proper chemical canisters, half-mask respirators can be used for protection from vapors or fumes.
- *Full face mask* – Full face masks with chemical canisters are normally used for exposure to dangerous vapors or fumes. A distinction is made here concerning the terms *vapors* and *fumes* because it makes a difference to painters. Most painting products emit vapors. Examples of fumes are the smoke billowing from a fire or fumes generated when welding.

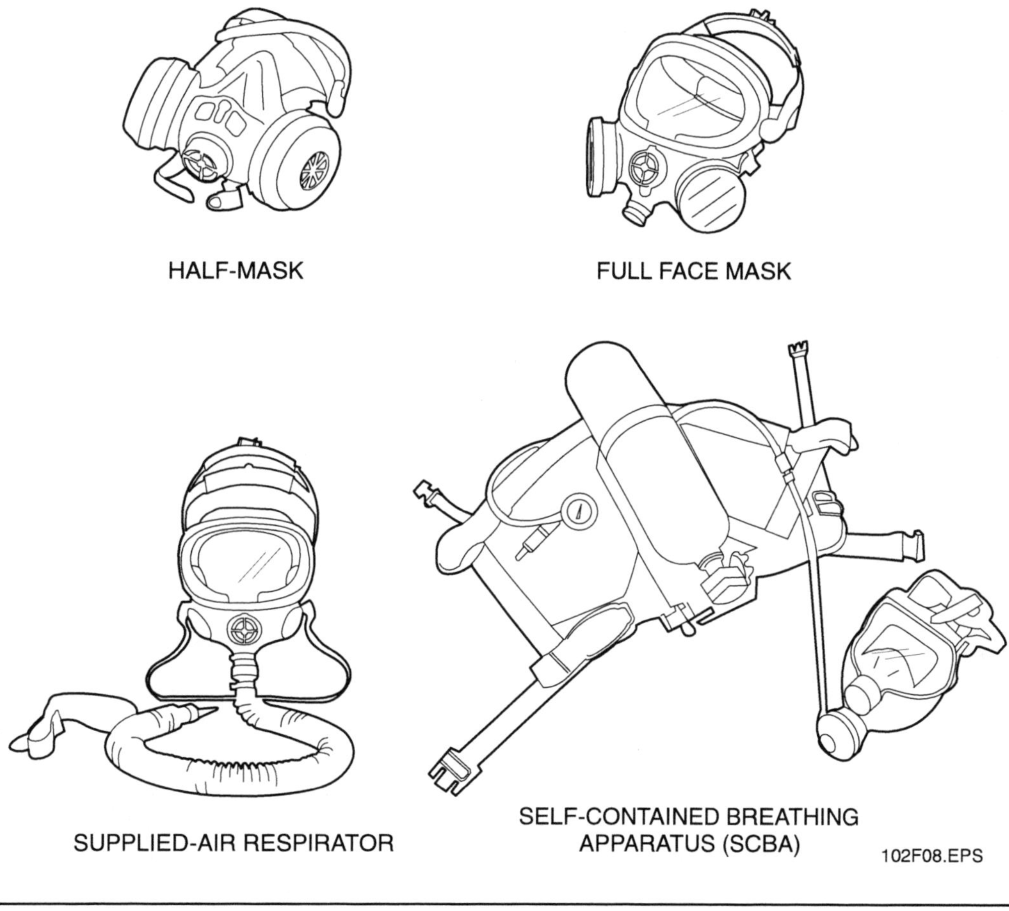

Figure 8. Use A Respirator When There Is A Danger Of A Respiration Hazard Or Suffocation

- *Supplied-air respirator* – A supplied-air respirator or hood uses a remote compressor or air tank. A hose supplies air from the tank to the mask or hood. It is used under the same conditions as half-mask and full face mask respirators. Some supplied-air respirators are approved by the **National Institute for Occupational Safety and Health (NIOSH)** to be used in atmospheres where there is a lack of oxygen or those that contain vapors, fumes, and/or particle contaminants which are immediately dangerous to life and health. An atmosphere is considered to have a lack of oxygen when there is less than 19.5% oxygen by volume at sea level. An atmosphere is considered **immediately dangerous to life and health (IDLH)** if it poses an immediate hazard to life or produces immediate, irreversible, and debilitating effects on health.
- *Self-contained breathing apparatus (SCBA)* – SCBA respirators have their own air supply carried in a compressed air tank. They may be used in areas where there is a lack of oxygen or those that contain vapors, fumes and/or particle contaminants which are immediately dangerous to life and health.

5.2.0 RESPIRATORY PROGRAM

Local and OSHA procedures must be followed when selecting the proper type of respirator for a particular job (*Figure 9*). A respirator must be properly selected (based on the contaminant present and its concentration level), properly fitted, and used in accordance with the manufacturer's instructions. It must be worn during all times of exposure. When sandblasting, make sure to use a respirator designed for that purpose, since most types of respirators are not designed for sandblasting. Regardless of the kind of respirator needed, OSHA regulations require employers to have a respirator protection program consisting of:

- Standard operating procedures for selection and use
- Employee training
- Regular cleaning and disinfecting
- Sanitary storage
- Regular inspection

Figure 9. Use The Right Respirator For The Job

As an employee, you are responsible for wearing respiratory protection when needed. When it comes to vapors or fumes, both can be eliminated (in certain concentrations) by the use of air purifying devices as long as the oxygen levels are acceptable. Always check the cartridge on your respirator to make sure it is the correct type to use for the air conditions and contaminants. Read what it says: "Chemical Cartridge for Organic Vapors." Remember that there is a difference between vapors and fumes. Look at the limitations marked on the cartridge. Does it say "Approved only for use in atmospheres containing (1) at least 19.5% oxygen, (2) less than 0.1% organic vapors by volume"? What does all this mean?

Since painters usually work in well-ventilated areas, we can assume that the oxygen level is OK. We must also check the product's MSDS to see what the "percentage by volume" is. Most water-based latex paints have a PEL of 50 ppm. The chemical cartridge for the respirator is good for 1,000 ppm (that is 20 times the PEL for this particular type of paint). To safeguard yourself, always read the product's MSDS. It should list the type of respirator and cartridge recommended for use with the product.

Limitations that apply to all half-mask (air-purifying) respirators are as follows:

- These respirators do not completely eliminate exposure to contaminants, but will reduce the level of exposure to below hazardous levels.
- These respirators do not supply oxygen and must not be used in areas where the oxygen level is below 19.5%.
- Do not use these respirators in atmospheres which are IDLH.
- Do not use these respirators in areas where chemicals have poor warning signs (no taste, odor, etc.).

If your breathing becomes difficult, if you become dizzy or nauseated, or if you smell or taste the chemical or have other noticeable effects, leave the area immediately, return to a fresh air area, and seek any necessary assistance.

5.2.1 Positive And Negative Fit Checks

All respirators are useless unless properly fit tested for each individual. To obtain the best protection from your respirator, you must perform positive and negative fit checks each time you wear it. This test fit must be done until you have obtained a good face seal.

To perform the positive fit check, do the following:

Step 1 Adjust the face piece for the best fit, then adjust the head and neck straps to assure good fit and comfort.

CAUTION Do not overtighten the head and neck straps. Tighten only enough to stop leakage. Overtightening can cause face piece distortion and dangerous leaks.

Step 2 Block the exhalation valve with your hand or other material.

Step 3 Breathe out into the mask.

Step 4 Check for air leakage around the edges of the face piece.

Step 5 If the face piece puffs out slightly for a few seconds, a good face seal has been obtained.

To perform a negative fit check, do the following:

Step 1 Block the inhalation valve with your hand or other material.

Step 2 Attempt to inhale.

Step 3 Check for air leakage around the edges of the face piece.

Step 4 If the face piece caves in slightly for a few seconds, a good face seal has been obtained.

5.2.2 Respirator Inspection, Care, And Maintenance

A respirator must be clean, in good condition, and all of its parts must be in place for it to give you proper protection. Respirators must be cleaned every day. Failure to do so will limit their effectiveness and offer little or no protection. For example, suppose you wore the respirator yesterday and did not clean it. The bacteria from breathing into the respirator, plus the airborne contaminants that managed to enter the face piece, have now made the inside of your respirator very unsanitary. Continued use may cause you more harm than good. Remember, only a clean and complete respirator will provide you with the necessary protection.

- Inspect the condition of your respirator before and after each use.
- Do not wear respirators if the face piece is distorted or if it is worn and cracked. You will not be able to get a proper face seal.
- Do not wear respirators if any part is missing. Replace worn straps or missing parts before using.
- Do not expose respirators to excessive heat or cold, chemicals, or sunlight.
- Clean and wash your respirator each day. Remove the cartridge and filter, hand wash the respirator using mild soap and a soft brush, and let it air dry overnight.
- Sanitize your respirator each week. Remove the cartridge and filter, then soak the respirator in a sanitizing solution for at least two minutes. Thoroughly rinse with warm water and let it air dry overnight.
- Store the clean and sanitized respirator in its resealable plastic bag. Do not store the respirator face down. This will cause distortion of the face piece.

6.0.0 FALL HAZARDS AND PROTECTION DEVICES

6.1.0 FALL HAZARDS

The hazard of falling enters into the daily work of almost every painter and decorator. Falls can occur even from a stepladder if the painter is careless. Working with rigging or scaffolding requires a good attitude about safety and training in the proper use of the equipment.

Equipment, work practices, electricity, and weather conditions are all factors in safely working from elevations. If rigging, scaffolding, and climbing equipment is not in good condition, it might fail and cause a fall. If it is makeshift, it might fail because it does not comply with safety standards. It may also fail if it is improperly selected.

Unsafe work practices include general carelessness, failure to anticipate hazards, poor housekeeping that increases the possibility of tripping on platforms or scaffolds, and failure to obey load limits. Electrical power lines create a hazard of electrocution when ladders or other access equipment are used nearby. Finally, weather conditions such as rain and high winds can make climbing and working from elevations especially hazardous.

6.2.0 FALL PROTECTION DEVICES

Fall protection devices come under two main categories: fall arrest devices and fall restraint devices. Fall arrest devices and systems are those that activate and catch workers after they have fallen. Fall restraint devices and methods are used to prevent a fall from happening.

The three main types of fall protection a painting and decorating craftsman is likely to use are:

- Guardrails on scaffolds and work platforms
- Body harnesses, lifelines, and lanyards
- Safety nets

Elevated work platforms and scaffolds are required to have guardrails, midrails, and toe boards on open sides and ends. OSHA outlines the specific requirements, dimensions, and construction materials for these safeguards.

Personal fall arrest equipment (*Figure 10*) includes a body harness, lanyard, and lifeline. This fall protection system, which may also include a deceleration device such as a rope grab, is designed to prevent a worker from free falling more than six feet from an elevated workplace. Each part of the system must meet OSHA minimum tensile load or break strength requirements.

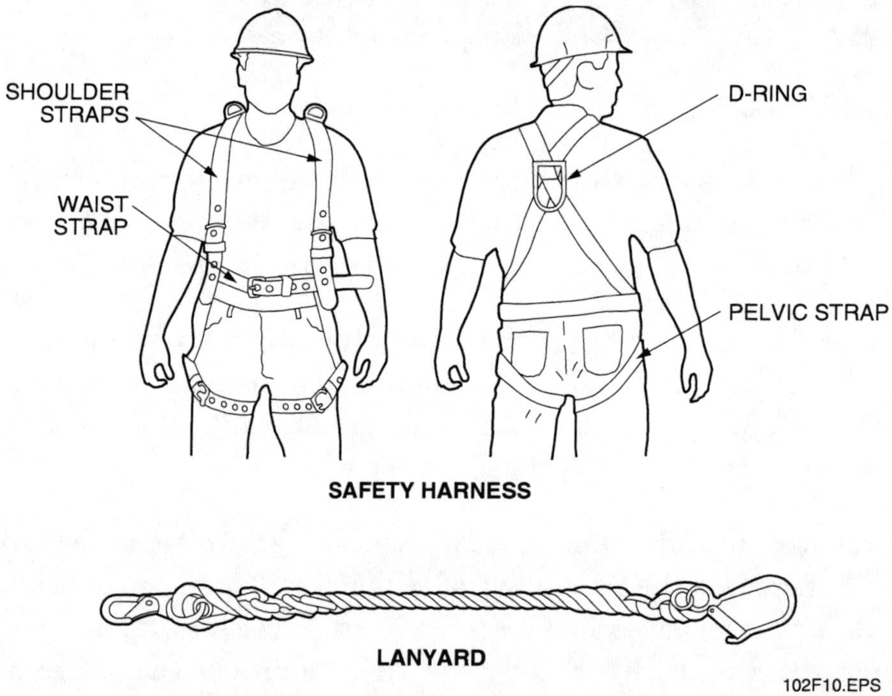

Figure 10. Use Fall Protection Equipment When Working At High Elevations

CAUTION As of January 1, 1998, body belts are prohibited by OSHA for use in personal fall arrest systems. This is because they do not provide the same protection as a body harness, which straps around the torso and absorbs the impact of a fall over a larger part of the body.

Safety nets are another device commonly used for fall protection. More information about the safe use of ladders, scaffolds, and the three types of conventional fall protection (guardrails, safety nets, and personal fall arrest systems) is given in the next module, which covers ladders, scaffolds, and lifts.

7.0.0 HAZARDOUS CHEMICALS AND MATERIALS

Most paints, coatings, and solvents will fall under one or more of the OSHA definitions of hazardous materials, as listed below:

- *Chemical* – Any element, chemical compound, or mixture of elements and/or compounds.
- *Hazardous chemical* – Any chemical which is a physical hazard or health hazard.
- *Health hazard* – Any chemical for which there is evidence to indicate acute or chronic effects may occur in exposed individuals. Health hazards include chemicals which are:
  - Carcinogens
  - Toxic or highly toxic
  - Reproductive toxins (chemicals which affect reproductive capabilities)
  - Irritants and corrosives
  - Sensitizers

- Hepatotoxins (chemicals which produce liver damage)
- Nephrotoxins (chemicals which produce kidney damage)
- Neurotoxins (chemicals which produce their primary toxic effects on the nervous system)
- Agents which act on the blood or hematopietic system (blood-forming system)
- Agents which damage the lungs, skin, eyes, or mucous membranes

- *Physical hazard* – Any chemical for which there is scientifically valid evidence that it is:
  - A combustible liquid
  - A compressed gas
  - Explosive
  - Flammable
  - An organic peroxide
  - An oxidizer
  - Pyrophoric (capable of spontaneous ignition)
  - Unstable (reactive)

7.1.0 HAZARD COMMUNICATIONS PROGRAMS

OSHA's Hazard Communication Standard (HazCom) requires that all employers make their employees aware of and knowledgeable about the hazardous chemicals and materials they may be exposed to on the job. OSHA mandates that information on the various hazardous materials be obtained by the employer, then given to employees by means of a comprehensive communications program or system which includes:

- *Evaluating the hazard* – As required by federal law, manufacturers or producers of materials conduct scientific and laboratory studies on their products, then based on the results generate an MSDS for each of their products. Copies of these MSDSs are made available to your employer or a contractor when the material is purchased. As applicable, MSDSs list the contents, hazards, precautions, recommendations for safe use, and emergency procedures used with the product.
- *Product labeling* – Manufacturers label their product containers with their name, the name of the material, and safe use and storage information. If applicable, the label also provides a hazard warning and identifies what parts of the body or internal organs could be affected.
- *Training* – As part of HazCom information system, your employer will provide training so you can learn more about materials that you work with and how you can protect yourself. You will also learn emergency and first aid procedures, how to recognize when there is a problem, and where to go or who to contact if you want more information.
- *Written program* – Your employer or contractor has taken the time to document how the system works for you. This also includes a list of all the hazardous materials and the system for keeping it up to date.

Depending on state and/or local requirements, HazCom and/or "Right To Know" systems used by your employer or contractor may only focus on hazardous chemical materials, while others may also include many of the physical problems such as noise, temperature extremes, etc. encountered on the job.

Your responsibilities under OSHA's HazCom at the job site are:

- Know the location of MSDSs.
- Spot and report hazards.
- Know the physical and health hazards of any hazardous material.
- Know and practice the actions necessary to protect yourself from any hazards.
- Know the actions necessary in an emergency.
- Know the location and content of your employer's HazCom program document.

7.2.0 UNDERSTANDING MATERIAL SAFETY DATA SHEETS

All MSDSs must contain the same basic information in at least nine required sections. However, some MSDSs may have more sections covering transportation, environmental issues, etc. as deemed necessary by the manufacturer. Depending on the manufacturer, MSDSs are often formatted in ways that make them look different from one another. *Figure 11* shows an MSDS produced by one major manufacturer. The information contained in the nine required sections of an MSDS are described below. Note that manufacturers may list information on their MSDSs in a different order than that given below.

Section 1. Material Identification – This section lists the trade name and may also contain the chemical name, chemical formula, original manufacturer, and so on.

Section 2. Ingredients – The chemical ingredients of the subject will be listed in this section, along with the relative percentage of each, either by weight or by volume.

Section 3. Physical Data – This section contains the physical data that would generally be of use to professionals with technical expertise in chemical safety, such as toxicologists, industrial hygienists, safety engineers, and so on. The data includes boiling and melting points, vapor pressure, specific gravity, and so on.

Section 4. Fire and Explosion Hazard Data – This section lists the physical parameters necessary for a fire or explosion to occur, as well as the measures that should be taken to fight fires involving this substance.

An important item listed here is the **flash point (FP)**. This is the lowest temperature at which a liquid will begin to give off vapors that will ignite when an ignition source is held close to the liquid's surface. The FP is used to classify a liquid as being flammable or not. Substances with an FP of less than 100°F (38°C) are classified as flammable liquids. Gasoline, a highly flammable substance, has an FP of 45°F (7°C). Liquids with flash points higher than the flammable classification are considered combustibles.

95 —SECTION I— PRODUCT IDENTIFICATION

MATERIAL SAFETY DATA SHEET

THE SHERWIN - WILLIAMS CO.
101 PROSPECT AVE. N.W.
CLEVELAND, OH 44115

EMERGENCY TELEPHONE NO.
(216) 566-2917
INFORMATION TELEPHONE NO.
(216) 566-2902

DATE OF PREPARATION
1 - MAY - 95

Reducers

SOL/2

CAS No.	—SECTION II— HAZARDOUS INGREDIENT (percent by weight)	ACGIH TLV <STEL>	OSHA PEL <STEL>	Units	Vapor Pressure (mm Hg)	Toluene R2 K 1	Xylene R2 K 4	H. Fl. Naphtha - 100 R2 K 5 R7K100	High Flash Naphtha - 150	Acrylic Enamel Reducer R4 K 35	Acrylic Enamel Reducer R4 K 36
						154-2364 154-2372 154-8668 (153-7547) (153-7554) (153-7737) (153-7745) (153-7851) (153-7869) (153-7976) (153-7964) (153-8222) (153-8230)	154-2380 154-2398 154-8684 (153-7562) (153-7570) (153-7752) (153-7760) (153-7877) (153-7885) (153-7992) (153-8008) (153-8248) (153-8255)	154-4576 154-4584 154-8767	154-4592 154-4600 154-8809		
64742-89-8	Lt. Aliphatic Hydrocarbons	100	100	PPM	53.0					29	
108-88-3	§ Toluene	50	100 <150>	PPM (Skin)	22.0	100				38	29
100-41-4	§ Ethylbenzene	100 <125>	100 <125>	PPM	7.1		15	1			7
1330-20-7	§ Xylene	100 <150>	100 <150>	PPM	5.9		85	5		5	42
64742-95-6	Light Aromatic Hydrocarbons	Not Established			3.8			22			
98-82-8	§ Cumene	50	50	PPM (Skin)	10.0			5			
108-67-8	1,3,5-Trimethylbenzene	25	25	PPM	10.0			27	1		
95-63-6	§ 1,2,4-Trimethylbenzene	25	25	PPM	2.0			40	2		
64742-94-5	Med. Aromatic Hydrocarbons	Not Established			0.1				84		
91-20-3	§ Naphthalene	10 <15>	10 <15>	PPM	1.0				13		
111-76-2	§ 2-Butoxyethanol	25	25	PPM (Skin)	0.6					6	
67-64-1	§ Acetone	750 <1000>	750 <1000>	PPM	180.0					15	
78-93-3	§ Methyl Ethyl Ketone	200 <300>	200 <300>	PPM	70.0						4
112-07-2	§ 2-Butoxyethyl Acetate	Not Established			1.0					3	16
	Weight per Gallon (lbs.)					7.18	7.17	7.24	7.40	6.76	7.25
	Volatile Organic Compounds (VOC) - lbs./gal.					7.18	7.17	7.24	7.40	6.76	7.25
	Photochemically Reactive					Yes	Yes	Yes	Yes	Yes	Yes
	Flash Point (°F)					40	80	105	140	10	10
	DOL Storage Category					1B	1C	2	3A	1B	1B
	Flammability Classification					Flammable	Flammable	Combustible	Combustible	Flammable	Flammable
	HMIS (NFPA) Rating (health - flammability - reactivity)					2 3 0	2 3 0	3 2 0	3 2 0	2 3 0	2 3 0

§ Ingredient subject to the reporting requirements of the Superfund Amendments and Reauthorization Act (SARA) Section 313, 40 CFR 372.65 C

→→→ *MSDS Text Page Follows* →→→

3

FRONT (SECTIONS 1 AND 2)

Figure 11. Example Of A Material Safety Data Sheet (MSDS) (Sheet 1 of 2)

Reducers

SOL/2

Section III — PHYSICAL DATA

PRODUCT WEIGHT –	See TABLE	*EVAPORATION RATE* –	Slower than Ether
SPECIFIC GRAVITY –	0.79-0.81	*VAPOR DENSITY* –	Heavier than Air
BOILING RANGE –	132-395 °F	*MELTING POINT* –	N.A.
VOLATILE VOLUME –	100 %	*SOLUBILITY IN WATER* –	N.A.

Section IV — FIRE AND EXPLOSION HAZARD DATA

FLAMMABILITY CLASSIFICATION *FLASH POINT* See TABLE *LEL* 0.5 *UEL* 36.5

See TABLE

EXTINGUISHING MEDIA

Carbon Dioxide, Dry Chemical, Foam

UNUSUAL FIRE AND EXPLOSION HAZARDS

Keep containers tightly closed. Isolate from heat, electrical equipment, sparks, and open flame. Closed containers may explode when exposed to extreme heat. Application to hot surfaces requires special precautions. During emergency conditions overexposure to decomposition products may cause a health hazard. Symptoms may not be immediately apparent. Obtain medical attention.

SPECIAL FIRE FIGHTING PROCEDURES

Full protective equipment including self-contained breathing apparatus should be used. Water spray may be ineffective. If water is used, fog nozzles are preferable. Water may be used to cool closed containers to prevent pressure build-up and possible autoignition or explosion when exposed to extreme heat.

Section V — HEALTH HAZARD DATA

ROUTES OF EXPOSURE

Exposure may be by INHALATION and/or SKIN or EYE contact, depending on conditions of use. To minimize exposure, follow recommendations for proper use, ventilation, and personal protective equipment.

ACUTE Health Hazards

EFFECTS OF OVEREXPOSURE

Irritation of eyes, skin and respiratory system. May cause nervous system depression. Extreme overexposure may result in unconsciousness and possibly death.

SIGNS AND SYMPTOMS OF OVEREXPOSURE

Headache, dizziness, nausea, and loss of coordination are indications of excessive exposure to vapors or spray mists.

Redness and itching or burning sensation may indicate eye or excessive skin exposure.

MEDICAL CONDITIONS AGGRAVATED BY EXPOSURE

None generally recognized.

EMERGENCY AND FIRST AID PROCEDURES

If INHALED: If affected, remove from exposure. Restore breathing. Keep warm and quiet.

If on SKIN: Wash affected area thoroughly with soap and water. Remove contaminated clothing and launder before re-use.

If in EYES: Flush eyes with large amounts of water for 15 minutes. Get medical attention.

If SWALLOWED: Never give anything by mouth to an unconscious person. DO NOT INDUCE VOMITING. Give several glasses of water. Seek medical attention.

CHRONIC Health Hazards

No ingredient in these products is an IARC, NTP or OSHA listed carcinogen.

Methyl Ethyl Ketone may increase the nervous system effects of other solvents.

Prolonged overexposure to solvent ingredients in Section II may cause adverse effects to the liver, urinary, blood-forming, cardiovascular, and reproductive systems.

Reports have associated repeated and prolonged overexposure to solvents with permanent brain and nervous system damage.

Section VI — REACTIVITY DATA

STABILITY – Stable

CONDITIONS TO AVOID

None known.

INCOMPATIBILITY

None known.

HAZARDOUS DECOMPOSITION PRODUCTS

By fire: Carbon Dioxide, Carbon Monoxide

HAZARDOUS POLYMERIZATION – Will Not Occur

Section VII — SPILL OR LEAK PROCEDURES

STEPS TO BE TAKEN IN CASE MATERIAL IS RELEASED OR SPILLED

Remove all sources of ignition. Ventilate and remove with inert absorbent.

WASTE DISPOSAL METHOD

Waste from these products may be hazardous as defined under the Resource Conservation and Recovery Act (RCRA) 40 CFR 261. Waste must be tested for ignitability to determine the applicable EPA hazardous waste numbers. Waste from products containing Methyl Ethyl Ketone may also require testing for extractability.

Incinerate in approved facility. Do not incinerate closed container. Dispose of in accordance with Federal, State, and Local regulations regarding pollution.

Section VIII — PROTECTION INFORMATION

PRECAUTIONS TO BE TAKEN IN USE

Use only with adequate ventilation. Avoid breathing vapor and spray mist. Avoid contact with skin and eyes. Wash hands after using.

VENTILATION

Local exhaust preferable. General exhaust acceptable if the exposure to materials in Section II is maintained below applicable exposure limits. Refer to OSHA Standards 1910.94, 1910.107, 1910.108.

RESPIRATORY PROTECTION

If personal exposure cannot be controlled below applicable limits by ventilation, wear a properly fitted organic vapor/particulate respirator approved by NIOSH/MSHA for protection against materials in Section II.

PROTECTIVE GLOVES

Wear gloves which are recommended by glove supplier for protection against materials in Section II.

EYE PROTECTION

Wear safety spectacles with unperforated sideshields.

Section IX — PRECAUTIONS

DOL STORAGE CATEGORY – See TABLE

PRECAUTIONS TO BE TAKEN IN HANDLING AND STORING

Keep away from heat, sparks, and open flame. Vapors will accumulate readily and may ignite explosively.

During use and until all vapors are gone: Keep area ventilated - Do not smoke - Extinguish all flames, pilot lights, and heaters - Turn off stoves, electric tools and appliances, and any other sources of ignition.

Consult NFPA Code. Use approved Bonding and Grounding procedures.

Keep container closed when not in use. Transfer only to approved containers with complete and appropriate labeling. Do not take internally. Keep out of the reach of children.

OTHER PRECAUTIONS

Intentional misuse by deliberately concentrating and inhaling the contents can be harmful or fatal.

Section X — OTHER REGULATORY INFORMATION

CALIFORNIA PROPOSITION 65

WARNING: These products, except High Flash Naphtha - 150, contain chemicals known to the State of California to cause cancer and birth defects or other reproductive harm.

The above information pertains to these products as currently formulated, and is based on the information available at this time. Addition of additives or other coatings materials to these products may substantially alter the composition and hazards of the products. Since conditions of use are outside our control, we make no warranties, express or implied, and assume no liability in connection with any use of this information.

BACK (SECTIONS 3 THROUGH 10)

102F11.TIF

Figure 11. Example Of A Material Safety Data Sheet (MSDS) (Sheet 2 of 2)

Section 5. Health Hazard Data – This section lists the known effects of overexposure, both acute (high exposure concentration over a short period of time—i.e., minutes or hours) and chronic (low concentrations with exposures of at least several days). Emergency and first aid procedures are also provided in the event overexposure occurs.

Section 6. Reactivity Data – This section lists those conditions, if any, that will cause the compound to become unstable and to react chemically (possibly resulting in an explosion or fire). Included is a list of incompatible materials known to react with the substance. The chemical should not be mixed, stored, or transported with any of these materials. This section also lists the hazardous decomposition products that might be given off in the event the substance is involved in a fire or is heated to extreme temperatures.

Section 7. Spill or Leak Procedures – This section lists the steps to be taken in the event the chemical is accidentally released or spilled.

Section 8. Personal Protection Information – Specifies the personal protection equipment (safety eyewear, respirators, and so on) that should be worn when handling the chemical. The equipment prescribed is usually for protection against continuous "worst case" occupational exposures that may be present in an industrial setting.

Section 9. Handling and Storage Precautions – Special precautions, if any, for handling or storage will be listed in this section.

7.3.0 HAZARDOUS MATERIALS IDENTIFICATION SYSTEM

Hazardous Materials Identification System (HMIS) and National Fire Protection Agency (NFPA) codes are used on MSDSs and product containers and packages to show the characteristics of materials. These codes use colors, numbers, letters, and symbols that allow for quick identification of a material's chemical hazards relating to health, flammability, and reactivity. They also identify the proper personal protection equipment and other information pertaining to the material.

7.3.1 HMIS Codes

The HMIS hazard and personal protection codes are summarized below.

- Health Hazard Index:
  - 4 Severe hazard
  - 3 Serious hazard
  - 2 Moderate hazard
  - 1 Slight hazard
  - 0 Minimal hazard

- Personal Protection Index:
  - A Safety glasses
  - B Safety glasses, gloves
  - C Safety glasses, gloves, synthetic apron
  - D Face shield, gloves, synthetic apron
  - E Safety glasses, gloves, dust respirator
  - F Safety glasses, gloves, synthetic apron, dust respirator
  - G Safety glasses, gloves, vapor respirator
  - H Safety goggles, gloves, synthetic apron, vapor respirator
  - I Safety glasses, gloves, dust and vapor respirator
  - J Safety goggles, gloves, synthetic apron, dust and vapor respirator
  - K Air-supplied hood or mask, gloves, full body protection (coveralls), boots
  - X Ask your supervisor for guidance

7.3.2 National Fire Protection Agency (NFPA) Codes

The NFPA hazard identification coding system uses a set of four colored diamonds (*Figure 12*) to indicate health, fire, reactivity, and specific hazard information. Each of these areas are ranked from 0 to 4 by the degree of hazard and the numbers inserted in the appropriate diamond. The NFPA codes are summarized below.

- Health Codes (Blue Diamond):
  - 4 Deadly
  - 3 Extreme danger
  - 2 Hazardous
  - 1 Slightly hazardous
  - 0 Normal material

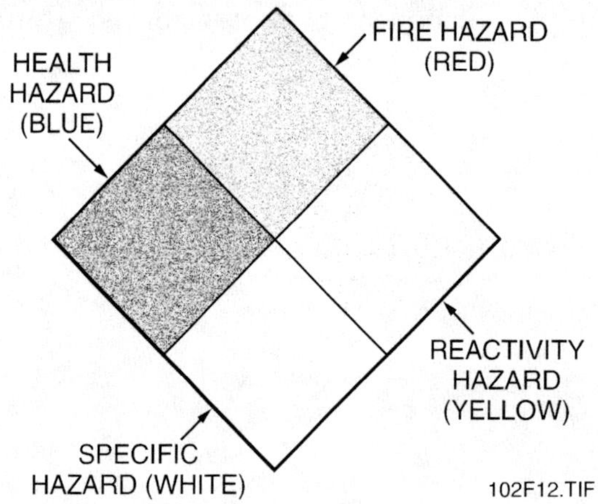

Figure 12. National Fire Protection Agency (NFPA) Hazard Identification System Diamonds

- Fire Hazard Flash Points (Red Diamond):
  - 4 Below 73°F
  - 3 Below 100°F
  - 2 Below 200°F
  - 1 Above 200°F
  - 0 Will not burn

- Reactivity (Yellow Diamond):
  - 4 May detonate
  - 3 Shock and heat may detonate
  - 2 Violent chemical change
  - 1 Unstable if heated
  - 0 Stable

- Specific Hazard (White Diamond):

Oxidizer	OX
Corrosive	COR
Acid	ACID
Alkali	ALK
Use No Water	~~W~~
Radioactive	☢

7.4.0 CHEMICAL HAZARDS

There are four major hazards associated with chemicals:

- Flammables
- Corrosives
- Toxic materials
- Reactives

7.4.1 Flammables

Flammable materials include those which will burn when ignited at, or below, room temperature. Combustibles (also included in this category) must be heated before they will burn. All flammable materials burn only when there is a sufficient concentration of the material in the air. Solvent, fuel materials, and fuel gases are the most common flammables. The best way to find out if a material is flammable is to check the label (*Figure 13*). Flammable materials may become more hazardous when mixed with other kinds of chemicals. Corrosives, when stored with flammables, can deteriorate the containers and release the material into the atmosphere.

Figure 13. The Best Way To Find Out If A Material Is Flammable Is To Check The Label

To use flammables safely:

- Don't smoke. Eliminate all sources of flame or ignition.
- Keep containers of flammable materials as small as possible.
- Reduce surface area of all containers.
- Use adequate ventilation.
- Clean up spills promptly.
- Cover containers when not in use.
- Store flammable-soaked rags in covered protective containers.
- Bond and ground all containers when dispensing.
- Use explosion-proof wiring and equipment.

Safe handling of flammables often requires personal protective equipment. Repeated skin contact can remove the oily protective layer of the skin and lead to irritation. Some flammables are also toxic and may require the use of a respirator. Also, take care to avoid splashes onto the skin and into the eyes.

Flammables, especially solvents, should be stored in an unbreakable container designed for this purpose. It should have flame arresters and a spring-loaded cover. If storage is indoors, a small amount can be kept in a special cabinet. Larger amounts should be stored in special storage rooms which have devices and controls to minimize the risk of fire or explosion.

Care should be taken to provide storage of flammables away from oxidizers and corrosives. Oxidizers may ignite an otherwise non-flammable mixture. Corrosives may destroy the container and allow the release of flammable vapors into the atmosphere. Concentrated vapors may sink to the floor and travel some distance to a source of ignition, with the flashback traveling back to the source container.

Small leaks should be cleaned up quickly. If it is possible to limit the leak by closing a valve, shutting down the equipment, or moving the container, it should be done. Turn off electrical equipment which may provide a source of ignition.

If the leak is large or your skin, eyes, or clothes are contaminated, leave the area immediately. Wash eyes, skin, and clothes with lots of water. Get to fresh air. Notify your supervisor as soon as possible. Unless you have special training and the proper protective equipment, do not try to clean up a large leak.

Fires can result from **spontaneous combustion** of solvent or oil-soaked rags that may generate heat by chemical reaction. For this reason, it is important to store used rags in a suitable container or to soak them in water until they can be properly disposed of.

7.4.2 Corrosives

Corrosive materials can cause damage on contact with the skin or eyes, or when inhaled. They can come in many forms and many strengths. Weak corrosives are sometimes called *irritants*. Acids can range from mild citric acid in fruit to strong mineral acids, such as sulfuric or nitric. Even strong acids and bases can be diluted so that they are less dangerous. Caustics (or bases or alkalis) can be weak, like baking soda, or strong, like potassium hydroxide.

Strong acids are usually liquid. The shipping containers are marked "corrosive" and have the Department of Transportation (DOT) sign for corrosives. The shipping containers are usually plastic or a combination of plastic and stainless steel. Bases can come in powdered form or strong solutions. Some acids are also found as powders.

Be sure to read the label to see if a material is a corrosive (*Figure 14*). Acids and bases come in different forms and are used for many purposes. In construction, most corrosive materials are found in cleaners. Examples of acids are phosphoric, nitric, sulfuric, hydrochloride, acetic, and chromic. Even mild concentrations can cause burns which go on long after the splash is washed off.

Caustics are used to clean metal and to treat water. Examples of caustics include ammonium, potassium and sodium hydroxide, sodium bicarbonate, trisodium phosphate, and lime.

Acids react violently when combined with bases or water. If combining acids with water, always add acid to water. Keep caustics and acids stored separately. Use appropriate personal protective equipment when working with acids and bases. Gloves, aprons, and goggles are usually required. In some cases, boots or other foot protection is also required. Always use ventilation. Contact lenses may aggravate eye injury from corrosives and should not be used when working with them.

Figure 14. Read The Labels And Signs to Find Out If A Material Is Corrosive

Acids and bases should be stored in separate places. Keep all corrosives away from flammables. They may cause leaks in the flammable containers and lead to a fire. Some corrosives, such as nitric acid, are also oxidizers and may contribute to any combustion processes. Again, check the label or MSDS for storage instructions.

If a corrosive material is splashed in the eyes or on your skin and clothes, immediately wash the affected area with water for 15 minutes. Use safety showers and eyewashes if available. If not, get water from a pail or hose. Remove any contaminated clothing. Get to fresh air if you feel burning in the nose or lungs. Do not vomit if you swallow a corrosive; drink large quantities of water and see a doctor.

Spills or leaks should be stopped or contained as soon as possible. Proper protective equipment must be used to deal with leaks and spills. Turn off any leaking valve or process, then dike or dam the spill. If the spill is too large, contact emergency authorities. Ventilate the area before resuming activities.

7.4.3 Toxic Materials

Any material can be hazardous under the wrong conditions. The degree of the hazard depends on the dose. Small amounts of most materials can cause mild symptoms which disappear once the person is removed from the exposure. Larger doses can cause more severe illness, and extremely large doses can cause irreversible illness and even death.

Each person responds differently to hazardous materials. OSHA permissible exposure limits are intended to protect the average person from the harmful effects of chemical exposures over a working lifetime. OSHA's definition of toxic or highly toxic material applies to only a very few potent poisons seldom used by construction trades. We will use the more common definition of **toxic** as any material which can cause illness or injury.

Toxic materials can poison the body and cause harmful effects (*Figure 15*). There is a difference between the acute effect and the chronic effect of exposure to toxic chemical hazards. Acute effects are usually due to a sudden overexposure to large quantities or concentrations of a material. The acute effects usually disappear after the exposure ends. Sometimes, supportive medical treatment is needed, but the body usually returns to normal.

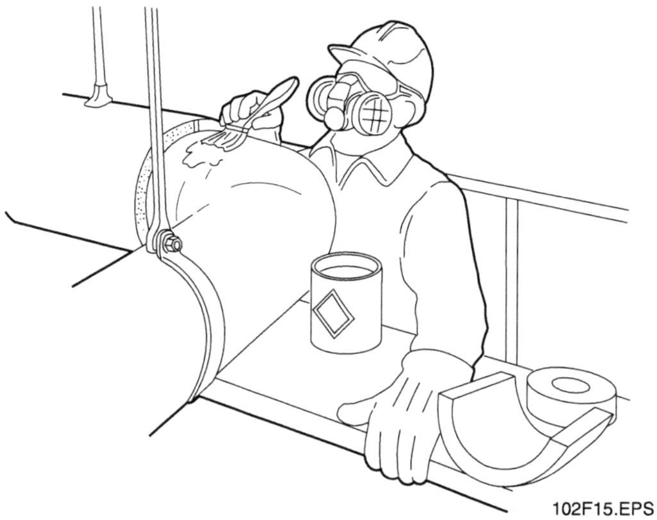

Figure 15. Toxic Materials Can Poison The Body And Cause Harmful Effects

Chronic toxic effects are not as easy to recognize. They are often the result of lower levels of exposure over a long period of time. Typically, they affect one or more of the body's organ systems. If the problem is identified, the effect can often be reduced. If the exposure to a toxic material is stopped, healing of the organ or organ system can return the body to normal. Because of the slow nature of some diseases caused by exposure, irreversible damage can be caused by long-term exposure to a chemical hazard. Asbestosis is a debilitating, chronic disease common to older insulation workers. It results from long-term exposure to asbestos fibers in the air.

Chemicals can enter your body in three ways:

- *Inhalation* – Whenever you are doing a job that uses a toxic material, you need to be careful not to breathe too much of that material. This is the most common way that chemicals get into the bloodstream. As we breathe the material, it enters our lungs. It is then easily transferred into our blood and taken throughout the body. To prevent this from happening, good ventilation is very important. Open doors and windows or set up a fan that directs the air away from you. Respirators may also be necessary to protect you. Be sure you choose the correct one and know how to use it. Air filtration respirators take the toxic material out of the air you breathe, while air-supplied respirators provide you with clean air from a tank or other source. If you think you need a respirator, check with your supervisor and read the label or MSDS.

- *Ingestion* – Some chemicals can hurt you if you accidentally eat or swallow them. Good personal habits can stop this route of entry. Washing hands before you eat and keeping your clothes clean are good practices. Check the MSDS or label for emergency and first aid procedures, and see a doctor, if necessary.
- *Skin absorption* – Some materials have the ability to pass through unprotected skin into the bloodstream. Wearing proper gloves and other skin or face protection will reduce the chances of this happening. The label and MSDS both tell you if gloves or other equipment is recommended. Remember that not all gloves are alike, nor will they protect you from all materials. Use the right ones.

Solvents are among the most common toxic materials in the workplace. Many processes (mixing and cleaning) use or give off solvent vapors. They are also used as thinners in paints and adhesives. Solvents vary in their toxicity from practically non-toxic materials such as freons and acetones, through moderately toxic materials such as alcohols, ketones, and halogenated solvents, to the very toxic such as dimethyl acetamide, methyl acrylate, and other materials. Some solvents are also flammable or reactive.

The following list provides a rough guide to the toxicity of solvents or other toxic materials you may work with. The values listed represent the **threshold limit value (TLV)**, which is the airborne concentration of a substance to which a normal, healthy worker may be exposed for eight hours a day, 40 hours a week during a working lifetime without adverse health effects. TLVs are recommended limits established by the American Conference of Governmental Industrial Hygienists (ACGIH). Normally, the TLV represents a time-weighted average (TWA). Check the glues, solvents, or cleaners you use against this list. So you don't get confused, the more hazardous chemicals have the *lowest* TLVs, while the safer ones usually have *high* values listed.

- Mildly toxic: TLV 500 to 1,000 ppm
- Moderately toxic: TLV 50 to 500 ppm
- Toxic: TLV 1 to 50 ppm
- Highly toxic: TLV less than 1 ppm

Another method for rating toxicity is known as the **short-term exposure limit (STEL)**. This represents the airborne concentration to which workers can be exposed for up to 15 minutes without suffering ill effects.

All solvents may irritate the eyes and skin in high concentrations. Most will dissolve the protective layer of oils on the skin and leave it looking white in the small cracks. Solvents should never be used to clean the skin; if there is a problem with contamination, some form of glove or barrier cream should be used to protect the skin. The early signs of overexposure often include headaches, dizziness, and nausea, but there are many other causes of these symptoms.

In general, minimizing contact with toxic materials will minimize their toxic effect. Use controls, such as ventilation, to draw contaminants away from the workplace air. Use respiratory protection to minimize the inhaled dose. Use goggles, gloves, aprons, and other protective gear to keep the material off the skin, out of the eyes, and away from the body. Although the body can get rid of a certain amount of most toxic chemicals and the standards are there to maintain the level below that point, you can minimize your exposure by understanding the routes of entry and methods of control.

Storage methods which minimize the release of volatile materials should be used. Keep solvents tightly capped and stored in ventilated areas whenever possible. Keep the number of cleaning tanks and trays to a minimum. Use local exhaust ventilation to capture any vapors released during storage. Be sure to keep flammable and corrosive materials apart during storage.

For small spills (usually under a gallon), immediately clean using wipes or other absorbent material. For larger spills, or spills of moderately to highly toxic material, leave the area and notify your supervisor. If you are unsure how to deal with a problem, leave the area, alert others in the area to the problem, and let trained personnel deal with the situation.

7.4.4 Reactives

Reactives are materials which can change violently when combined with certain other materials or conditions. There are very few reactives in use by the construction trades. However, knowing the hazard can help you when you do come in contact with a reactive material.

Oxidizers add oxygen to any situation where burning is occurring and make the fire more intense and more difficult to put out. Some reactives explode or give off gas and heat in air or on contact with water.

Many materials can act like reactives when mixed with incompatible materials. Acids and bases react strongly with each other, giving off heat, often enough to cause boiling and splattering of the mixture. The MSDS should tell you what materials may be incompatible with the chemical you are working with and what other materials to avoid.

The most common reactive mixture in construction is found in gas welding or brazing. Acetylene gas mixes with oxygen gas to provide an extremely powerful reaction in the form of a very intense flame (*Figure 16*).

Oxidizers are the most commonly found reactive class. Most oxidizers are also corrosive, so keep them away from the skin and eyes.

Figure 16. The Most Common Reactive Mixture In Construction Is Found In Gas Welding And Brazing

Reactives should be stored away from other types of materials. Read the MSDS carefully when you see the words *oxidizer* or *reactive*. Note what chemicals are compatible with the materials that you are using and avoid situations where they might become mixed. Be sure to use any protective clothing or respiratory protection required by the MSDS or the process documentation. Many reactives are toxic, corrosive, or both. Protect yourself against the health hazards as well as the physical hazards of reactives.

In an emergency, if it is practical, shut down any electrical equipment. If possible, stop the spill or leak from continuing, but if there is any doubt, leave the area and notify your supervisor. Do not try to neutralize the material or clean up the spill unless you have appropriate protective equipment and have been properly trained.

8.0.0 LEAD HAZARDS AND PROTECTION MEASURES

The use of lead-based paint has been banned in all residential buildings since 1978. When preparing surfaces in older buildings previously painted with lead-based paint, painting contractors and painters must take precautions to prevent lead poisoning. These precautions are as set forth in the *Occupational Safety and Health Standards for the Construction Industry* (29 CFR 1926.62).

8.1.0 LEAD HAZARDS

Lead is a toxic metal that can enter your body through inhalation or ingestion. It is not absorbed through your skin. Lead-based paint becomes dangerous once it is disturbed by a removal process, normally through sanding, power tool cleaning, or abrasive blasting. The smaller the lead particles, the more dangerous they become because they are more easily ingested when eating, drinking, or smoking. At 10 microns or smaller, lead particles in the air become respirable. Inhalation or ingestion of lead allows it to get into the bloodstream. It is then circulated throughout the body and stored in various organs and tissues and can result in damage to the blood, nervous, digestive, kidney, and reproductive systems.

The adverse effects that are normally evident immediately or shortly after a short-term period of exposure (seconds, minutes, hours, or days) to a hazardous material are called the **acute effects of overexposure**. For lead these can be: weakness, vomiting, loss of appetite, uncoordinated body movements, convulsions, stupor, bloody stools and possibly coma. The effects that develop slowly over a long period of time (weeks, months, or years) from exposure to a hazardous material are called the **chronic effects of overexposure**. For lead these can be: weakness, insomnia, hypertension, metallic taste in the mouth, anemia, headache, and muscle and joint pain. There is also a risk to developing fetuses and the reproductive systems in both men and women.

8.2.0 PROTECTION MEASURES

Currently, OSHA has established a permissible exposure limit of 50 mg/m^3, averaged over an 8-hour period, for workers exposed to lead. It also established an action level of 30 mg/m^3 over an 8-hour period. If the action level is reached, your employer must conduct exposure monitoring, medical surveillance, and employee training. If any lead is present in the workplace, your employer must monitor workers' exposure.

Methods used to control exposure to lead include the use of other paint removal methods, such as chemical stripping instead of abrasive blasting, or the use of special ventilated containment structures to reduce exposure levels when abrasive blasting is used. Containment structures to protect the surrounding environment are often built to enclose a lead hazard work area. However, this can seriously endanger the health of the worker inside the containment structure because it increases the level of exposure to lead dust unless strict protection measures are put into place. Respirators must be worn when other forms of control cannot lower the exposure below the permissible exposure limit. Workers who wear respirators in lead exposure zones must comply with a rigorous respirator protection plan and a demanding hygiene plan in order to keep exposures below the permissible limit.

To control lead hazards when working in lead exposure zones, OSHA requires that the actions shown in *Figure 17* be implemented. In addition, as applicable, you must strictly comply with the following guidelines:

- Wear your respirator at all times.
- Clean your respirator daily and sanitize it weekly. Properly store it when not in use.
- Always perform a positive and negative fit check each time you put the respirator on; if you cannot obtain a good face seal, you will not be permitted to enter the work area.
- Change HEPA cartridges and filters as needed to assure maximum respiratory protection. **High-efficiency particulate air (HEPA)** filter cartridges and filters are those designed to remove 99.97% of all particles larger than 0.3 micron.
- Wash hands thoroughly before and after using the restroom.
- Wash hands thoroughly before eating, drinking, or smoking.
- Do not bring food or beverages into the lead-based paint removal work area.

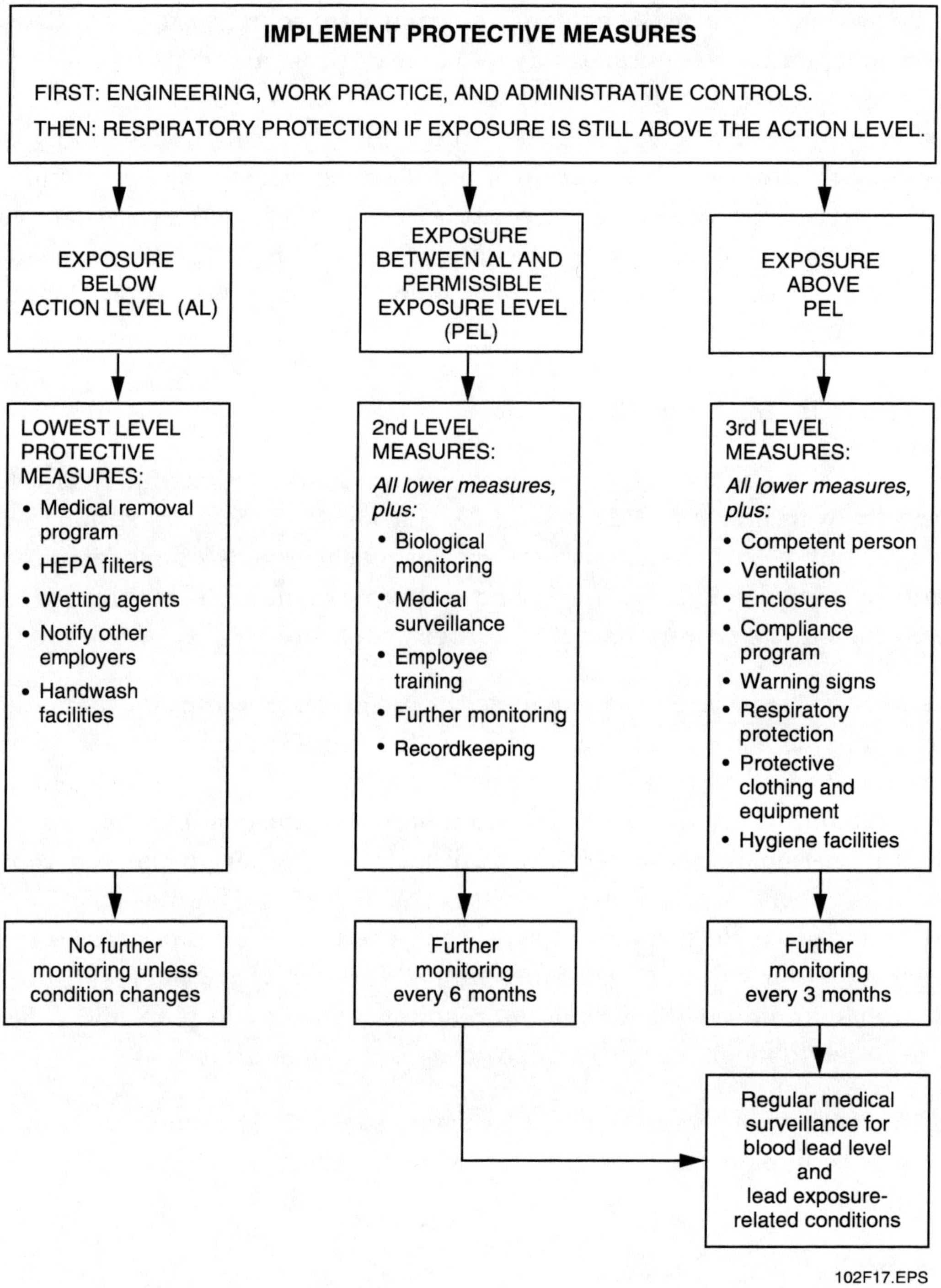

Figure 17. Protective Measures Required By OSHA To Control Lead Hazards At A Job Site

- Do not enter the lead paint removal site wearing your normal work or street clothes. Wear the disposable coveralls, shoe covers, head covers, and gloves that are provided for you.
- At the end of your shift, you must remove the contaminated disposable clothing in the decontamination area only, and place the contaminated clothing in the hazardous waste container. Then thoroughly wash your face, neck, hands, and arms before leaving the area.
- Strong oxidizers or acids are incompatible with lead paints. If these chemicals come in contact with each other, it creates a reactivity hazard which may result in the release of hydrogen gas.

9.0.0 ASBESTOS HAZARDS AND PROTECTION MEASURES

Airborne asbestos fibers can be inhaled, allowing them to penetrate into the lungs. Inhaling asbestos fibers usually shows no immediate symptoms, but can cause a long-term risk of a lung disease called *asbestosis* and a cancer that attacks the lining of the chest cavity or abdomen. Smokers have an increased risk of developing asbestos-induced lung cancer.

Asbestos can occur in many forms, including amosite, chrysotile, and crocidolite. In construction materials, it can be found both indoors and outdoors, although as a result of asbestos abatement programs, its use has been severely limited in recent years. Materials containing asbestos were commonly used in duct and pipe thermal insulation, fireproofing materials, insulation cladding for structural members, acoustical materials, floor tiles, and roofing materials. Where asbestos material was used and is in good condition, exposure is normally insignificant. However, elevated levels of airborne asbestos fibers can occur when asbestos-containing materials are deteriorating or have been disturbed or abraded, releasing its fibers. It is best to leave asbestos material alone if it is not damaged.

For work areas that contain hazards from airborne asbestos fibers, OSHA regulations require that your employer monitor the air to make sure that no worker is exposed to a concentration in excess of 0.2 fibers per cubic centimeter (cc) of air (8-hour TWA) or exposed to a concentration in excess of 0.1 fiber per cc of air over a sampling period of 30 minutes. Where possible, your employer should use negative-pressure enclosures or other engineering solutions to reduce exposure in the work area before starting any removal and/or renovation of asbestos materials. Your responsibility when working in such an area is to strictly follow your employer's asbestos containment plan. This includes wearing the proper respirator and protective clothing, such as coveralls (or similar whole body clothing), head covering, gloves, and foot coverings.

10.0.0 CONFINED SPACE HAZARDS AND ENTRY

10.1.0 HAZARDS

A confined space is any space that has limited or restricted openings for entry or exit that would make escape difficult in an emergency (*Figure 18*). It is also any space that could contain moving or rotating machinery, dangerous levels of toxic air contaminants, flammable or explosive atmospheres, a lack of oxygen (less than 19.6% by volume), or excess amounts of oxygen (more than 23.5% by volume).

When oxygen in the air decreases from the normal 21%, a person suffers increasingly severe effects, beginning with accelerated breathing and heartbeat at about 16% to rapid death at 6%. Many processes inside a confined space can deplete the atmosphere of oxygen. These include flash rusting of metal, fermenting of matter, and rotting of debris.

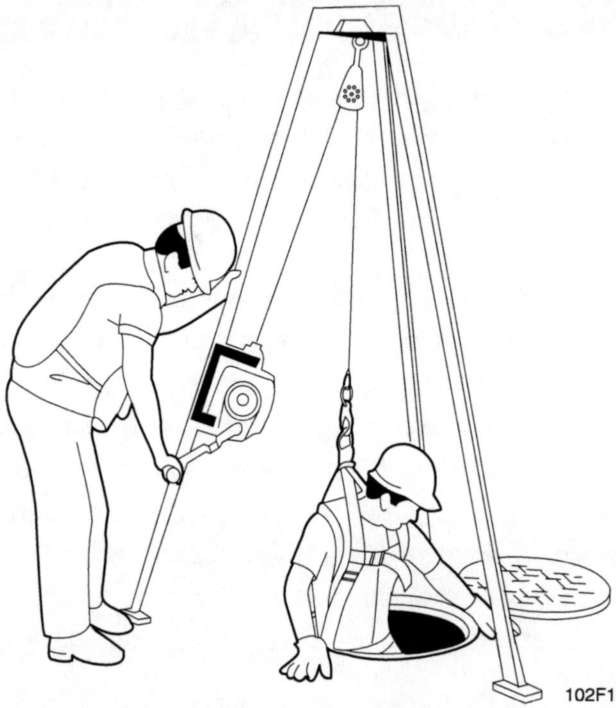

Figure 18. Worker Entering A Confined Space With A Restricted Opening For Entry And Exit

Fires and explosions can occur in confined spaces as a result of the presence of residual chemicals, the accumulation of vapors from paint and solvents, or the release of trapped gases from disturbed sludge or scale. At explosive concentrations, gases or vapors can be ignited by sparks from tools, from static electricity created by air movers or other equipment, or by lights. Under these conditions, explosion-proof lights, motors, exhaust fans, etc. should always be used to prevent fires and/or explosions.

While not all gases and vapors are explosive, they are often toxic. In poorly ventilated spaces and at high levels of exposure, some paints and solvents can be very toxic. Even in lower doses, some chemicals can adversely affect the respiratory or nervous systems. Many toxic gases, such as carbon monoxide, are deadly but cannot be detected by sight or smell.

In some confined spaces, there can also be physical hazards. For example, moving or rotating gears, augers, etc. might accidentally be energized and injure a worker. In tanks, there is the hazard of valves being opened accidentally and allowing liquids to pour in. Where solids such as grain are stored, there is the hazard of engulfment, which may lead to crushing or suffocation. Noise, excessive heat, and the potential for falls are also hazards of confined spaces.

10.2.0 CONFINED SPACE ENTRY

To overcome the hazards of confined spaces, OSHA requires a program of strict compliance with regulations on confined space entry that apply to all persons involved with the entry into confined spaces. These include the owners of facilities with confined spaces, contractors who place workers in confined spaces, the worker(s) who enter confined spaces, and entry attendants and supervisors who monitor the access into confined spaces. As part of any compliance program, OSHA also requires training for the entrants, attendants, and supervisors.

By OSHA requirements, the owner of the facility where the confined space is located has the following responsibilities:

- Identify confined spaces in the facility and put warning signs at the entrances; also, establish policies preventing unauthorized entry.
- Identify hazards associated with each confined space, and establish test procedures for monitoring atmospheres.
- Establish a written permit system for entering confined spaces.
- Train and certify the proficiency of entrants, supervisors, and attendants on how to test, enter, and work safely inside confined spaces.
- Develop an emergency action plan and establish rapid response rescue teams and train them in confined space rescue. The elements of an emergency action plan are described later in this module.
- Provide specific equipment necessary for safe confined space entry, including equipment for monitoring, communicating, ventilating, rescuing, locking and tagging out, breathing, and warnings.

Contractors who place workers in confined spaces must comply with the confined space program of the facility owner. As part of this process, the facility owner must inform the contractor about:

- The location of all confined spaces.
- The required entry permit system.
- Details of the hazards to be encountered.
- How workers can be protected from the hazards.

Authorized entrants to a confined space have the following responsibilities:

- Know the hazards to be encountered.
- Recognize the symptoms of asphyxiation or toxic exposure; understand the result of such exposure. Make sure the confined space has been isolated, ventilated, and tested for hazardous atmosphere before entering and during the operations conducted there if a hazard is being generated.
- Maintain contact with the attendant; know how to escape when necessary.
- Understand the need to exit immediately upon hearing an evacuation alarm.
- Wear a harness and lifeline for aid in evacuation.
- Be trained in the use of retrieval lines, barriers, signs, respirators, and protective clothing. Know when to use explosion-proof equipment.

Attendants have the following responsibilities:

- Maintain a station at the exit of the confined space; keep a count of personnel inside; know what work is allowed or disallowed in the space.
- Recognize hazards; know how to monitor for safety; recognize safe and unsafe amounts registered on air testing and other monitoring equipment.
- Test the atmosphere remotely using a probe inserted into a small hole.
- Maintain contact with entrants; order evacuation when problems occur; recognize symptoms of physical distress.
- Know how to summon rescuers and use the alarm system.
- Refuse entry to unauthorized personnel.
- Do not enter the space for rescue attempts.

Entry supervisors have the following responsibilities:

- Authorize and oversee entry; be highly trained in entry procedures.
- Know the hazards of the confined space; assure the correctness and completeness of the entry permit.
- Assure that proper lockout/tagout procedures are performed to deactivate all equipment.
- Assure that procedures and equipment for entry are available.
- Assure that workers understand the scope of the work and safe work practices.
- Assure the availability of rescue plans and personnel.
- Be responsible for canceling entry authorization and terminating entry when unacceptable conditions are present, and then for evaluating the problem and recommending the point at which reentry is allowed.
- Conclude the entry operation and transfer control of the confined space to the facility owner.

11.0.0 PAINTING TOOL AND EQUIPMENT HAZARDS

Operation of tools and equipment without the proper training can be hazardous to yourself and others on the job. The dangers of any tool or piece of equipment should be well known by every worker before using it. Only trained and qualified painters should operate air abrasive and water blast equipment, spraying equipment, or climbing or lifting equipment. When untrained persons use such equipment for the first time, it must always be under the guidance and direct supervision of an experienced operator so that proper safety and skills training are received in the use of the equipment. To prevent accidents or injury, always read and fully understand the manufacturer's operating literature for the equipment before using it for the first time. Pay particular attention to and follow the instructions given in all Warnings, Cautions, and Notes contained in the manufacturer's literature. If asked to use equipment that you are unfamiliar with, don't be afraid to ask your supervisor to provide you with proper instruction in its use before doing so. This can prevent injury to you, others, or both. Information about the operation and safety procedures that pertain to specific tools and equipment used in the painting trade is given where appropriate in the modules you will be studying throughout your training.

12.0.0 HOT AND COLD WEATHER PRECAUTIONS

Performing work outdoors in extremely hot or cold weather requires that appropriate precautions be taken to prevent an abnormal gain or loss of body heat.

12.1.0 HOT WEATHER PRECAUTIONS

Heat stroke, heat exhaustion, or heat cramps can result from heavy exertion while working in high temperature or high humidity conditions. When working in these conditions, you should dress in loose, cool cotton clothing. You should also take periodic cool-down breaks and make sure not to overexert yourself.

12.1.1 Heat Stroke

Heat stroke can be life-threatening. It happens because the body's heat-regulating mechanism stops working. This can cause convulsions, unconsciousness, and even death if the body is not cooled quickly.

WARNING! If heat stroke occurs, get immediate medical attention.

To help reduce the body temperature, move the person to a cool place and remove as much clothing as possible. Douse with water or wrap in a wet sheet. Symptoms of heat stroke are:

- Sudden onset
- Dry, hot, and flushed skin
- Dilated pupils
- Early loss of consciousness
- Full and fast pulse
- Deep breathing at first, later shallow or almost absent
- Muscle twitching, growing into convulsions
- Body temperature reaching 105°F or higher

12.1.2 Heat Exhaustion

Heat exhaustion happens when body fluids are lost through heavy sweating but the sweat cannot evaporate fast enough from the body to cool it properly. It may be accompanied by heat cramps. To help reduce the body temperature, the affected person should be moved to a cool place and as much clothing removed as possible. The person should be given sips of salted drinking water (about one teaspoon of salt to one quart of water). Following this, get medical attention or qualified medical advice.

Symptoms of heat exhaustion are:

- Weak pulse
- Rapid and unusually shallow breathing
- General weakness
- Pale, clammy skin
- Heavy sweating
- Dizziness, disorientation
- Temperature slightly above normal body temperature

Heat cramps are muscular pains and spasms. They can result from heavy exertion and the related loss of water and salt by heavy sweating. The affected person should be moved to a cool area and given sips of salted drinking water. The cramped muscles can also be gently stretched and massaged. Symptoms of heat cramps are:

- Painful muscle cramps
- Sweaty skin
- Faintness

12.2.0 COLD WEATHER PRECAUTIONS

Working in extremely cold weather, especially outdoors, requires that you dress adequately to protect against the loss of body heat that can result in possible hypothermia or frostbite.

You should dress in layers in order to allow you to adjust to the prevailing temperature conditions. Cotton or lightweight wool should be worn next to the skin with wool layers over the undergarments. Outer garments should be made of waterproof, wind-resistant fabrics such as nylon. Since a great deal of body heat is lost through the head, a hat with ear protection should be worn. Water chills the body far more rapidly than air or wind. For this reason, it is recommended that a dry set of clothing be taken along whenever working outdoors. Waterproof boots should be worn in damp or snowy weather.

12.2.1 Hypothermia

The effects of hypothermia are gradual and can often go unnoticed until it is too late. It is recommended that a buddy system be used when working in extreme cold for an extended period of time, where each person checks on the other for overexposure to the cold. If working alone, let someone know where you will be and what time you expect to return.

WARNING! Get immediate medical attention for a person affected by hypothermia.

Move the person indoors as soon as possible to warm up. If necessary, remove any wet clothing. If the person is conscious, they can be given hot liquids and/or a hot bath to help speed up the warming process. Symptoms and the progression of hypothermia are:

- Shivering
- Slurred speech
- Mental confusion
- Drowsiness and weakness
- Glassy stare
- Respiration and pulse rate become slower and slower
- Extremities freeze
- Death

12.2.2 Frostbite

The ears, nose, hands, and feet can be affected by frostbite as a result of exposure to intensely cold air or liquids.

WARNING! Cover the frostbitten area to protect it and get immediate medical attention.

Symptoms and progression of frostbite are:

- Exposed skin reddens
- Skin takes on a gray or blotchy appearance, especially the ear lobes, cheeks, and nose
- Exposed skin surface becomes numb
- All sensation is lost and the skin becomes dead white

13.0.0 ELECTRICAL HAZARDS

There are three possible results when electrical current shocks a worker. One is bodily harm from the shock itself that can range from minor to severe; another is injury or death resulting from an accident caused by involuntary muscle response to shock, such as falling from a scaffold; and the third is death from electrocution.

The amount of current that passes through the human body determines the outcome of an electrical shock. The higher the voltage, the greater the chance for a fatal shock. Electrical current flows along the path of least resistance to return to its source. If you come in contact with a live conductor, you become a load. *Figure 19* shows how much resistance the human body presents under various circumstances and how this converts to amps or milliamps when the voltage is 110V. Note that the potential for shock increases dramatically if the skin is damp. A cut will also reduce your resistance. Currents of less than 1 amp can severely injure and even kill a person (*Table 3*).

WARNING!	High voltage, defined as 600 volts or more, is almost ten times as likely to kill as low voltage. On the job, you spend most of your time working on or near lower voltages. However, lower voltages can also kill.

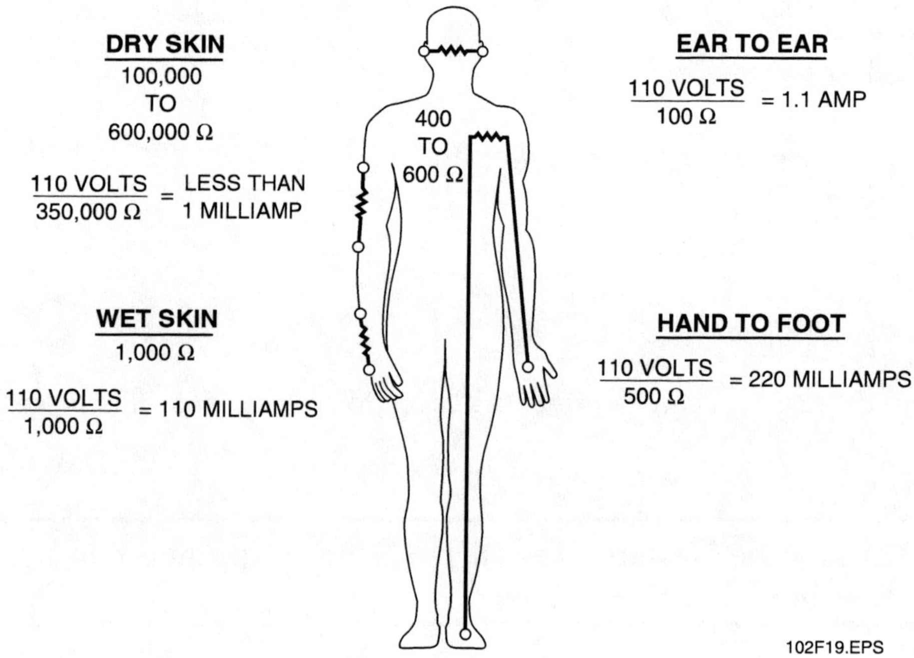

Figure 19. Typical Body Resistances And Current

CURRENT VALUE	TYPICAL EFFECTS
Less than 1 milliamp	No sensation.
1 to 20 milliamps	Sensation of shock, possibly painful. May lose some muscular control between 10 and 20 milliamps.
20 to 50 milliamps	Painful shock, severe muscular contractions, breathing difficulties.
50 to 200 milliamps	Up to 100 milliamps same symptoms as above, only more severe. Between 100 and 200 milliamps, ventricular fibrillation may occur. This typically results in almost immediate death unless special medical equipment and treatment are available.
Over 200 milliamps	Severe burns and muscular contractions. The chest muscles contract and stop the heart for the duration of the shock.

Table 3. Current Effects On The Human Body

In addition to shock, electricity poses the hazard of fire from short circuiting or overheating, and the hazard of explosion from arcing in an atmosphere that contains combustible dust or flammable vapors. Even the low energy discharge of static electricity from the operation of electrical equipment can cause a major disaster. The severity of these hazards makes it necessary to take careful precautions to avoid them.

There are two areas where electrical hazards must be considered. The first is general electrical tools and equipment; the second is high voltages, which are encountered most often in industrial plants and on transmission, substation, and power generation equipment.

The following precautions can be taken to minimize the hazards of electricity when using power tools and equipment:

- When available, use double-insulated power tools (ones that are encased entirely in a nonconductive, shatterproof material and have a nonconductive switch).
- Do not use electrical tools in a wet environment.
- Inspect the cords and plugs of tools before use to make sure they are serviceable.
- Do not trim the larger, polarized prong of a two-pronged plug to make it fit into an outlet. This prong is for safety.
- When using an adapter for a three-pronged plug, secure the grounding wire on the adapter to the faceplate of a grounded outlet box.
- Wherever possible, use an outlet box with a ground fault circuit interrupter (GFCI) which is designed to protect workers when current leakage is detected.
- Use properly grounded, three-wire extension cords.
- Use jacketed extension cords with power tools, and specially rated cords with heating devices that require large amounts of current.
- Protect extension cords from sharp surfaces. Tape them down in areas of foot traffic.

When working around high voltages, workers must have special training, use standard operating procedures, and follow a site safety plan. Safety measures must be taken to avoid high voltages, which can lead to death. Chief among these safety measures are locking and tagging out equipment to avoid accidental energizing. Lockout and tagout procedures involve placing locks and warning tags on each means used to de-energize circuits or equipment to prevent their use.

14.0.0 SAFETY PLANNING AND EMERGENCY ACTION PLANS

Planning is critical to safety. Accidents occur if you let them, but if you anticipate and plan ways to avoid them, accidents can be reduced or eliminated. Hazards in painting can arise from conditions at the work site and from the tools, equipment, and materials being used. Before work begins, the site should be surveyed for hazardous materials and unsafe conditions. If the job only involves painting a room in a house, the survey will be less extensive and time consuming than if it involves painting structures such as an industrial tank or bridge. Regardless of the size of the job, the site should always be evaluated to reveal any hazard such as confined spaces, unguarded openings, possible electrical hazards, etc.

Once all potential hazards have been identified, a comprehensive emergency action plan should be developed to deal with all emergencies. Depending on prevailing OSHA standards and the number of people involved, the plan may be required to be in writing or it may be communicated orally. As appropriate, an emergency action plan must include the following elements.

- Emergency escape procedures and emergency route assignments.
- Procedures to be followed by employees who remain to perform or shut down critical equipment before evacuating the area or remain to take corrective actions.
- Procedures that account for all employees after any emergency evacuation has been completed.
- Rescue and medical duties for those individuals who are to perform them.
- The preferred means for reporting fires and other emergencies including all phone numbers, etc.
- Names and/or job titles of persons or departments to be contacted for further information or explanation of duties under the plan.

SUMMARY

Accidents are very harmful for employees and employers both, and they are often caused by poor behavior and unsafe conditions. However, most accidents can be prevented. By knowing and avoiding the behaviors that cause accidents and keeping working conditions safe, it is possible to avoid injuries and reduce hazards. The most important governmental agency concerned with accident prevention is OSHA, which has imposed requirements on trade workers designed to keep job sites and personnel safe from harm. OSHA rules apply to protective clothing and equipment, housekeeping, electrical safety, and all types of tool and machine operations. Particular hazards are presented by chemicals, ladders, scaffolds, and tools. Developing an attitude of safety is an excellent way for every worker to avoid or reduce all of these hazards.

Safety is the responsibility of each and every one of us. No one person can constantly watch and guide every operation that is going on at the job site. You should know how to do your job safely. The training you receive, complying with your employer's safety program, and the use of everyday common sense will prevent you from being involved in an accident. An employee trying to bluff his way through a job he does not understand is just asking for trouble. Even if you think you know the correct procedures, a review may bring out an important part of the job that you may have forgotten. Don't be afraid to ask questions. The responses you receive may help a new or less experienced coworker get answers to questions they may be too bashful to ask.

Practicing good safety attitudes means that you:

- Report all unsafe conditions and acts immediately.
- Keep work areas clean and orderly at all times.
- Immediately report all accidents and injuries, no matter how minor.
- Be certain you completely understand the instructions given before starting work.
- Know how and where needed medical help may be obtained.
- Wear the required protective devices when working in a hazardous operation area.
- Do not use alcohol or drugs. If you are ill and must take prescribed medication, notify your supervisor immediately.

References

For advanced study of topics covered in this task module, the following books are suggested:

Basic Safety, Core Curricula.

Compliance Manual for OSHA Lead in Construction Regulation, Painting and Decorating Contractors of America, Fairfax, Virginia.

Employees Safety Handbook, Painting and Decorating Contractors of America, Fairfax, Virginia.

Occupational Safety and Health Standards for the Construction Industry - 29 CFR, Part 1926, U.S. Government Printing Office, Washington, DC.

Occupational Safety and Health Standards for General Industry - 29 CFR, Parts 1900 through 1910, U.S. Government Printing Office, Washington, DC.

Painting and Coatings Industry Hazard Communication Employee Handbook, Painting and Decorating Contractors of America, Fairfax, Virginia.

Painting and Decorating Craftsman's Manual and Textbook, Painting and Decorating Contractors of America, Fairfax, Virginia.

Supervisor's Safety Manual, Painting and Decorating Contractors of America, Fairfax, Virginia.

REVIEW/PRACTICE QUESTIONS

1. If your employer has a safety committee, you are obligated to:
   a. help maintain a safe working environment.
   b. volunteer to serve on the committee at least once a year.
   c. report any unsafe equipment and/or conditions to a committee member.
   d. both a and c.

2. OSHA is responsible for setting occupational safety and health standards for:
   a. the construction industry.
   b. general industry.
   c. the chemical industry.
   d. both the construction and general industries.

3. Tools, equipment, machines, or materials that do not comply with *OSHA Standards 1926*:
   a. may not be used.
   b. may be used if supplied by your employer.
   c. may be used if owned by the worker.
   d. may be used with specific restrictions.

4. When water blasting with a water and chemical solution, painters should protect their hands with:
   a. a good pair of general-purpose gloves.
   b. chemical-resistant gloves.
   c. heavy duty canvas gloves.
   d. heavy duty leather gloves.

5. Sound or noise intensity refers to:
   a. the sound frequency measured in Hertz.
   b. the duration or length of time exposed to the sound.
   c. sound loudness as measured in decibels.
   d. offensive bursts of sound.

6. When you are working in an area with either a lack of oxygen or high levels of toxic materials in the air, which type of respirator provides the most protection?
   a. Half mask
   b. Full mask
   c. Supplied air
   d. SCBA

7. To make sure your respirator provides proper protection, before using you must:
   a. perform a positive fit check.
   b. perform both positive and negative fit checks.
   c. perform a negative fit check.
   d. tighten the respirator head and neck straps as much as possible.

8. One device that should *not* be used by painters and decorators in a fall arrest system is a:
   a. body belt.
   b. body harness.
   c. lifeline.
   d. lanyard.

9. Hazardous materials can be found as:
   a. solids.
   b. liquids.
   c. gases.
   d. all of the above.

10. Depending on the prevailing OSHA, state, and local regulations, the scope of an employer's hazard communications (HazCom) and/or "Right To Know" programs may focus on:
    a. chemical hazards only.
    b. general employee safety training without specific reference to on-the-job hazards.
    c. physical hazards only.
    d. both a and c.

11. An MSDS is a form used to:
    a. file a workers' compensation claim with an insurance company.
    b. list the contents, hazards, precautions, etc. that pertain to a chemical or material.
    c. record unsafe conditions that exist at a job site.
    d. recommend changes to your employer's safety program.

12. The lowest temperature at which a flammable liquid gives off enough vapor to burn is called the:
    a. ignition temperature.
    b. combustion point.
    c. vaporization point.
    d. flash point.

13. Acute effects of overexposure to a concentration of a hazardous material refers to the:
    a. short-term exposure time (seconds, minutes, hours, days).
    b. long-term exposure time (weeks, months, years).
    c. degree of severity for a short-term exposure.
    d. degree of severity for a long-term exposure.

14. A material labeled with the health hazard index number of "4" has a:
    a. minimal hazard.
    b. moderate hazard.
    c. slight hazard.
    d. severe hazard.

15. The blue diamond used with the NFPA hazard code applies to:
    a. a specific hazard.
    b. health.
    c. reactivity.
    d. fire hazard flash points.

16. An MSDS lists the threshold limit value (TLV) of a solvent as less than 1 ppm. This means that the solvent is:
    a. toxic.
    b. mildly toxic.
    c. highly toxic.
    d. moderately toxic.

17. A reactive material is one that:
    a. can change violently when combined with certain other materials or conditions.
    b. will burn when ignited at or below room temperatures.
    c. can cause damage on contact with the body or eyes, or when inhaled.
    d. can be poisonous under the wrong conditions.

18. When preparing the wall surfaces for painting inside a 1950s vintage house, the chance that the walls have previously been painted with lead paint is:
    a. high.
    b. moderate.
    c. low.
    d. none.

19. A confined space is one that:
    a. has restricted openings for entry and exit.
    b. has <19.6% or >23.5% oxygen by volume.
    c. has a flammable atmosphere.
    d. both a and b.

20. The potential for electrical shock:
    a. increases when the skin is damp.
    b. decreases when the skin is cut.
    c. decreases when the skin is damp.
    d. remains the same when the skin is damp.

ACKNOWLEDGMENTS

Figure 11 courtesy of Sherwin-Williams Company.

Figures 13 - 16 courtesy of Painting and Decorating Contractors of America.

APPENDIX A

OSHA Construction PELs For Substances Commonly Found In Blasting And Coating Operations (1994)

Substance	Materials Found In	PEL
Cyclohexanone	Solvent for degreasing; resins, shellacs	50 ppm TWA
Epichlorohydrin	Solvents, resins for coatings	5 ppm TWA
Hexone (methyl isobutyl ketone)	Solvent for coatings	100 ppm TWA
Hydrochloric acid	Pickling solution; concrete etching solution	5 ppm
Iron oxide (fumes and dust)	Pigment in paints; abrasive	10 mg/m^3 TWA
Lead, inorganic	Pigment in coatings	50 $\mu g/m^3$ TWA
Mercury	Catalysts and pigments	0.1 mg/m^3 TWA
Methyl alcohol	Solvent in paints	200 ppm TWA
Methyl chloroform (1,1,1-trichloroethane)	Solvent in degreasing; solvent in coatings	350 ppm TWA
Methyl ethyl ketone	Solvent in resins, paint removers, coatings	200 ppm TWA
Methylene bisphenyl isocyanate	Resins for coatings	0.02 ppm
Muriatic acid	Concrete etching solution	5 ppm
n-butyl acetate	Lacquers	150 ppm TWA
2-nitropropane	Solvent; compound in paint	25 ppm TWA
Oxalic acid	Bleaching agent	1 mg/m^3 TWA
Phosphoric acid	Wash primer	1 mg/m^3 TWA
Silica, crystalline cristobalite respirable dust	Abrasives	5 $mg/m^3 \div (\%SiO_2 + 2)$
Silica, crystalline cristobalite total dust	Abrasives	15 $mg/m^3 \div (\%SiO_2 + 2)$
Silica, crystalline quartz respirable dust	Abrasives	10 $mg/m^3 \div (\%SiO_2 + 2)$
Silica, crystalline quartz total dust	Abrasives	30 $mg/m^3 \div (\%SiO_2 + 2)$
Sodium hydroxide	Paint strippers	2 mg/m^3 TWA
Titanium dioxide total dust	Pigment for coatings	15 mg/m^3 TWA
Toluene	Solvent for coatings	200 ppm TWA
Toluene diisocyanate (TDI)	Resins for coatings	0.02 ppm
Xylene	Solvent for coatings	100 ppm TWA

TWA = Time-weighted average

102APPXA.EPS

NCCER CRAFT TRAINING USER UPDATES

The NCCER makes every effort to keep these manuals up-to-date and free of technical errors. We appreciate your help in this process. If you have an idea for improving this manual, or if you find an error, a typographical mistake, or an inaccuracy in the NCCER's Craft Training Manuals, please write us, using this form or a photocopy. Be sure to include the exact module number, page number, a description of the problem, and the correction, if possible. Your input will be brought to the attention of the Technical Review Committee. Thank you for your assistance.

Instructors – If you found that additional materials were necessary in order to teach this module effectively, please let us know so that we may include them in the Equipment/Materials list in the Instructor's Guide.

Write: Curriculum and Revision Department
National Center for Construction Education and Research
P.O. Box 141104
Gainesville, FL 32614-1104

Fax: 352-334-0932

Craft ____________________ Module Name ____________________

Module Number ____________________ Page Number(s) ____________________

Description of Problem ____

____

____

____

(Optional) Correction of Problem ____

____

____

(Optional) Your Name and Address ____

____

Ladders, Scaffolds, Lifts and Fall Protection

Module 07103

Painting Trainee Task Module 07103

PAINTING AND DECORATING CONTRACTORS OF AMERICA

NATIONAL CENTER FOR CONSTRUCTION EDUCATION AND RESEARCH

LADDERS, SCAFFOLDS, LIFTS, AND FALL PROTECTION

Objectives

Upon completion of this module, the trainee will be able to:

1. Recognize, erect, and safely use the different kinds of ladders, including:
   - Stepladders
   - Single ladders
   - Extension ladders
   - Trestle and extension trestle ladders
2. Recognize, erect, and safely use the different kinds of scaffolds, including:
   - Built-up scaffolds
   - Swing scaffolds
   - Beam-suspended scaffolds
3. Recognize and safely use aerial work platforms and scissor lifts.
4. Recognize and safely use fall arresting and other fall protection equipment, including:
   - Body harnesses
   - Lanyards
   - Deceleration devices
   - Lifelines
   - Anchoring devices and equipment connectors

Prerequisites

Successful completion of the following Task Modules is required before beginning study of this Task Module: Core Curricula, Painting Level 1, Modules 07101 and 07102.

Required Trainee Material

1. Trainee Module
2. Appropriate Personal Protective Equipment

Course Map

This course map shows all of the *Wheels of Learning* task modules in the first level of the Painting curricula. The suggested training order begins at the bottom and proceeds up. Skill levels increase as a trainee advances on the course map. The training order may be adjusted by the local Training Program Sponsor.

Course Map: Painting, Level 1

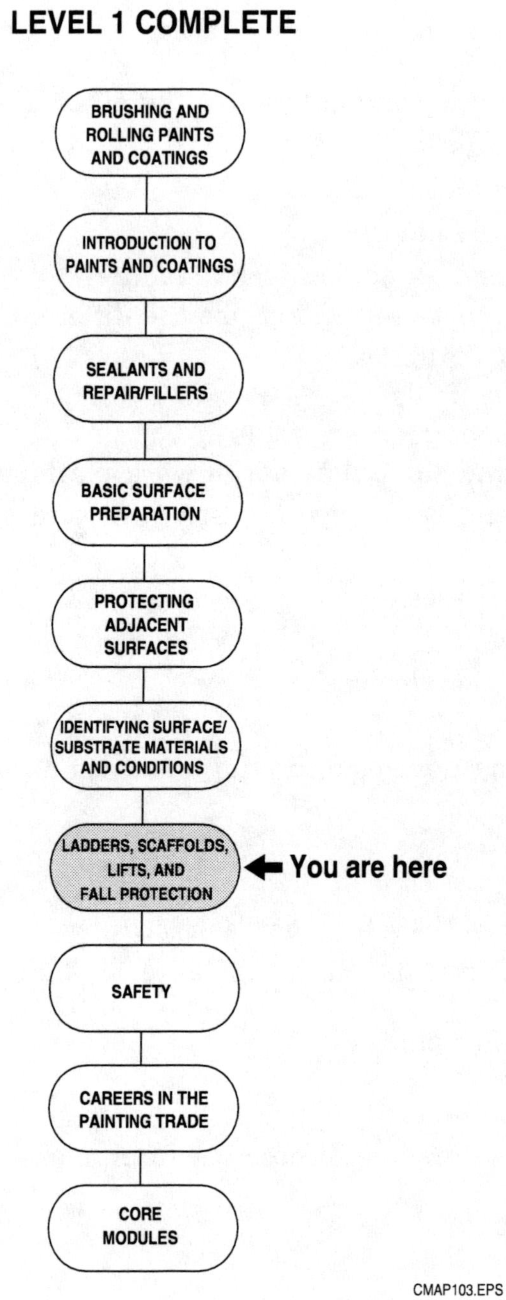

TABLE OF CONTENTS

Trade Terms Introduced In This Module

Aerial lifts: Portable, heavy-duty equipment used to raise or lower workers to and from an elevated job site.

Arresting force: The force needed to stop a person from falling. The greater the free fall distance, the greater the force needed to stop or arrest the fall.

Deceleration device: A device, such as a shock-absorbing lanyard or self-retracting lifeline, that brings a falling person to a stop without injury.

Deceleration distance: The distance it takes before a person comes to a stop. The required deceleration distance for a fall arrest system is a maximum of 3½ feet (1.07 meters).

Duty rating: American National Standards Institute (ANSI) rating assigned to ladders. It indicates the amount of use the ladder is designed to handle (industrial, commercial, or household) and the maximum working load limit (weight capacity) of the ladder. The working load limit is the maximum combined weight of the user, tools, and any materials bearing down on the rungs of a ladder.

Free fall distance: The vertical distance a worker moves during a fall before a fall protection deceleration device activates.

Ladder: A wood, metal, or fiberglass framework consisting of two parallel sidepieces (rails) connected by rungs on which a person steps when climbing up or down. Ladders may either be of a fixed-length that are permanently attached to a building or structure or portable. Portable ladders have either fixed or adjustable length, and are either self-supporting or not self-supporting.

Personal fall arrest systems: Safety systems that activate and catch workers after they have fallen.

Personal positioning systems: Safety systems that allow workers to hold themselves in place, keeping their hands free to accomplish a task. Whenever the worker leans back, the system is activated.

Scaffold: A temporary built-up framework or suspended platform or work area designed to support workers, materials, and equipment at elevated or otherwise inaccessible job sites.

Single-point suspension scaffold: A manually- or power-operated platform designed for light-duty use supported by one cable or rope from an overhead support. It allows the raising and lowering of the platform to the working position.

Two-point suspension scaffold: A manually- or power-operated platform which is supported by hangers at two points suspended from overhead supports in a way that allows it to be raised or lowered to the working position.

1.0.0 INTRODUCTION

Ladders, scaffolds, and/or lifts are used daily by painters and decorators to gain access to high places in order to perform surface preparation and application of paint or wallcovering. This module describes each type of equipment, including how to properly erect, use, and maintain it. Because falls from ladders, scaffolds, or lifts can cause injury or even death, this module also covers the related fall protection equipment that must be used and the safety practices that must be followed when working on ladders, scaffolds, or lifts.

2.0.0 LADDERS

Ladders are a framework consisting of two parallel sidepieces (rails) connected by rungs on which a person steps when climbing up or down. There are many types of ladders, each designed for a specific purpose and climbing conditions. *OSHA Standards 1926,* Subpart X defines and governs the use of portable, fixed, and job-made ladders as well as stairways. Portable ladders are defined as ladders that can be moved or carried; fixed ladders are permanently attached to a structure and cannot be moved; job-made ladders are ladders built by workers at a job site for a specific purpose. The remainder of this section focuses on portable ladders which are the type most often used for painting and decorating. Some general OSHA regulations pertaining to the use of ladders and stairways are:

- A ladder or stairway must be provided where levels of a work area are separated by 19 inches (48 cm) or more and where there is no other means of access, such as a ramp, runway, sloped embankment, or hoist.
- Two or more ladders are needed when there is no other way into and out of a work area for 25 or more workers, or when a ladder must handle two-way traffic at the same time.
- Employers must provide training and retraining as needed for workers who use ladders so that they recognize the hazards involved and learn how to avoid them.

2.1.0 TYPES OF LADDERS

Portable ladders can be grouped into five basic types:

- Stepladders
- Single ladders
- Extension ladders
- Trestle and extension trestle ladders
- Other ladders

For most types of portable ladders, OSHA requires that they support at least four times their maximum intended load without failure. Ladder rungs and steps must be shaped or treated to be slip-resistant and they must be level, parallel, and evenly spaced. Ladders can be made of fiberglass, aluminum, or wood. Fiberglass and clean, dry wooden ladders are nonconductive, allowing them to be used when working around electricity. Aluminum ladders are lightweight and easy to move, but should never be used where contact with electrical equipment or lines is possible.

Manufacturers attach labels (*Figure 1*) to ladders to inform users about the safe selection, use, and care of the ladder, and to warn them of hazards. You should always read and follow these labels whenever using any ladder. To prevent accidents, OSHA-approved ladders have a **duty rating** assigned in accordance with American National Standards Institute (ANSI) standards. This rating indicates the amount of use the ladder is designed to handle and a maximum working load limit. The ladder working load limit includes the combined weight of the user, tools, and any materials bearing down on the ladder rungs. The ANSI duty ratings assigned to ladders are listed below. Painters should use Type IA and Type I industrial ladders because these ladders are built to carry heavy loads and withstand continuous use.

- *Type IA* – Extra heavy-duty industrial use with a limit of 300 pounds (136 kilograms).
- *Type I* – Heavy-duty industrial use with a limit of 250 pounds (113 kilograms).
- *Type II* – Medium-duty commercial use with a limit of 225 pounds (102 kilograms).
- *Type III* – Light-duty household use with a limit of 200 pounds (91 kilograms).

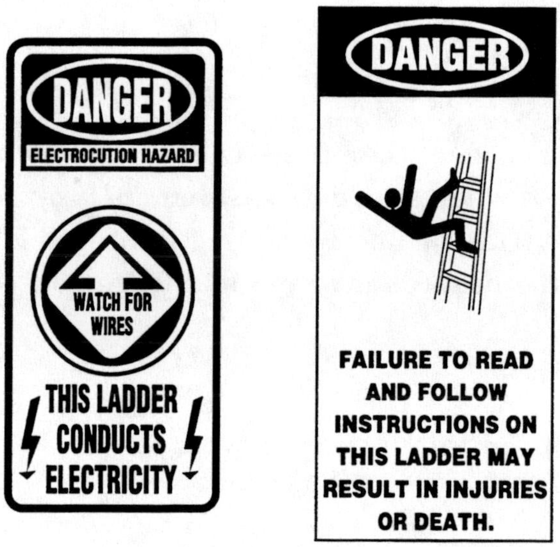

103F01.EPS

Figure 1. Typical Labels Found On Ladders

2.1.1 Stepladders

Stepladders (*Figure 2*) are self-supporting, non-adjustable ladders made in sizes ranging from 4 to 16 feet (1.2 to 4.9 meters). They usually have flat, level steps, a movable back frame hinged to the front legs, and spreaders that hold and lock the ladder open in the safe position.

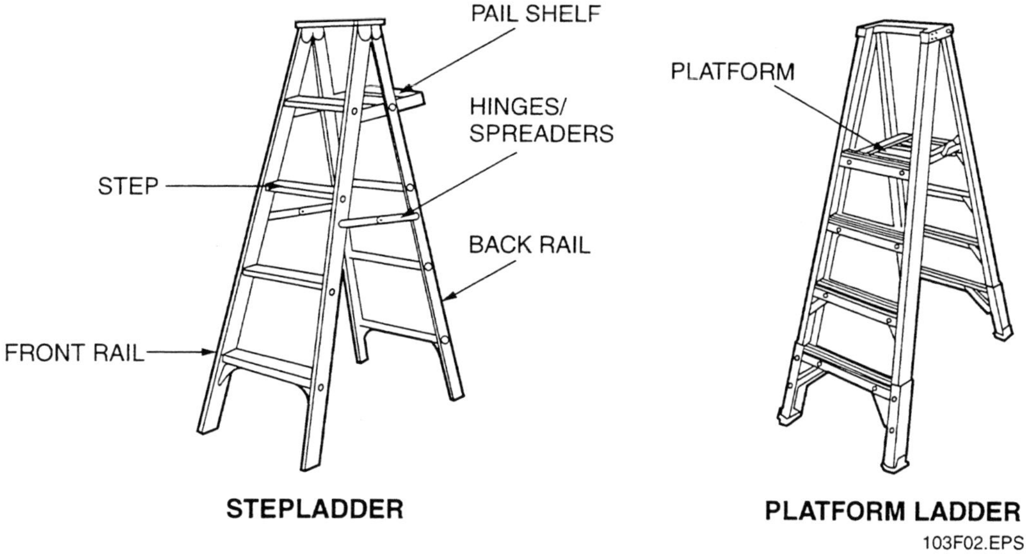

Figure 2. Step And Platform Ladders

They may have a flat top cap or a pail shelf on the back side. Another type of stepladder, called a *double stepladder*, has steps on both sides that can be climbed by one person at a time. A platform ladder is like a stepladder but has a platform at the highest standing level.

2.1.2 Single And Extension Ladders

Single ladders (*Figure 3*), also called *straight ladders*, are not self-supporting. This means that they must be leaned against a structure or surface for support. Straight ladders are made in sizes ranging from 6 to 20 feet (1.8 to 6.1 meters). They usually have two parallel sides, or rails, of fixed length that are joined by rungs.

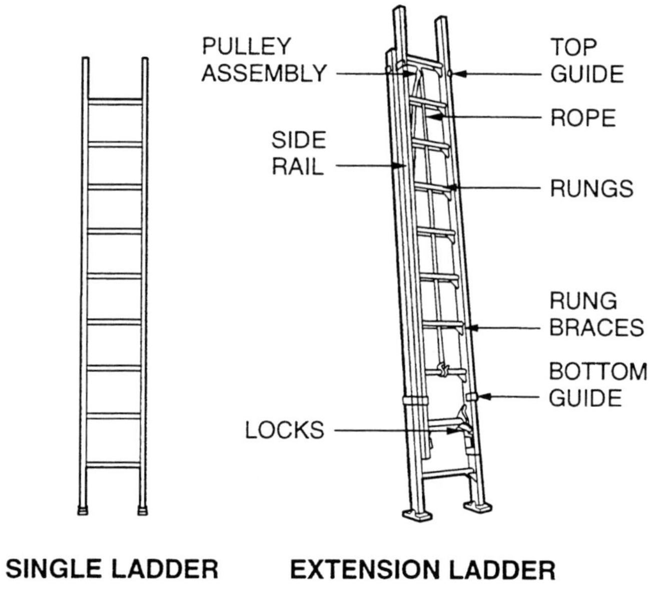

Figure 3. Single And Extension Ladders

Extension ladders are also not self-supporting. They are typically made in sizes from 16 to 60 feet (4.9 to 18.3 meters). Extension ladders are formed by two simple ladders that slide within one another to obtain the height capacity. Do not confuse the term *simple ladder* with *single ladder*, since a simple ladder is not built to be used alone. Most extension ladders are equipped with a rope and pulley system to help raise and lower the upper ladder section. Self-locking rung latches attached to one of the sections support and secure the raised section.

2.1.3 Trestle And Extension Trestle Ladders

Common trestle ladders (*Figure 4*) are self-supporting and have two fixed-length sections of rungs joined with hinges and a spreader to hold the sections open in the safe position. Extension trestle ladders have a vertical section that is adjustable in height and extends up between the two sections of the base. The base and extension should overlap at least 3 to 6 feet, depending on the size of the base section. The base sections of common and extension trestle ladders are typically made in sizes ranging from 6 to 16 feet (1.8 to 4.9 meters). The extended length for extension trestle ladders with the same base sizes typically range from about 8 to 24 feet (2.4 to 7.3 meters). A common practice is to suspend a plank between two extension trestle ladders to form a work platform.

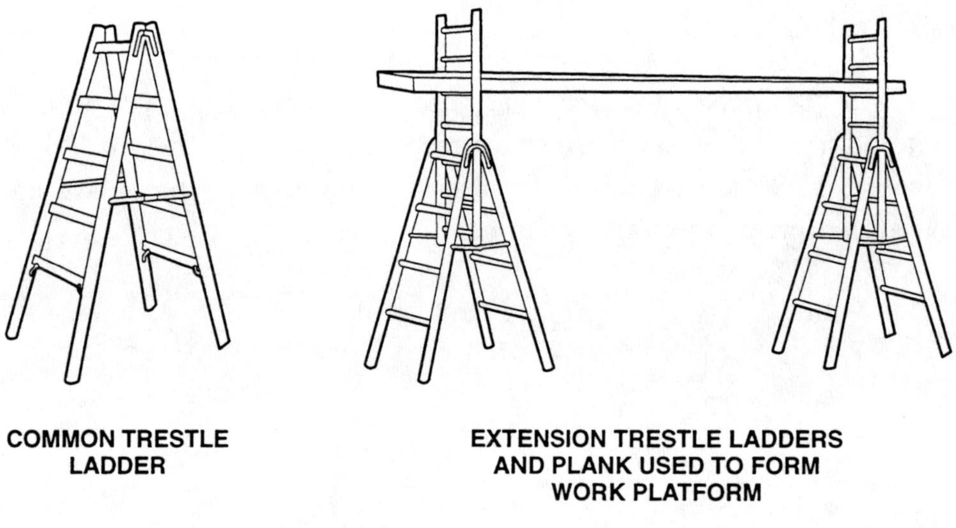

Figure 4. Trestle Ladders

2.1.4 Other Ladders

In addition to the ladders described above, there are many other kinds of ladders designed for specific uses. Three kinds commonly used by painters and decorators are sectional ladders, combination ladders, and articulated ladders. Sectional ladders (*Figure 5*) are made of two or more sections that are designed to be locked together to form a single, fixed-length ladder.

103F05.EPS

Figure 5. Sectional Ladders

The maximum assembled length of a typical three-section sectional ladder is between 21 and 26 feet (6.4 to 7.9 meters), depending on the length of the individual sections.

Combination ladders (*Figure 6*) and articulated ladders can be easily reconfigured to form different kinds of ladders. Combination ladders can typically be used as a stepladder, double stepladder, two single ladders, an extension ladder, or a stairway ladder. Articulated ladders can typically be used as a stepladder, single ladder, or placed in a scaffold position.

103F06.EPS

Figure 6. A Combination Ladder Can Be Reconfigured For Many Uses

2.1.5 Ladder Accessories

Numerous accessories (*Figure 7*) are made for use with ladders to make them more efficient and/or safer. Some common accessories are described below.

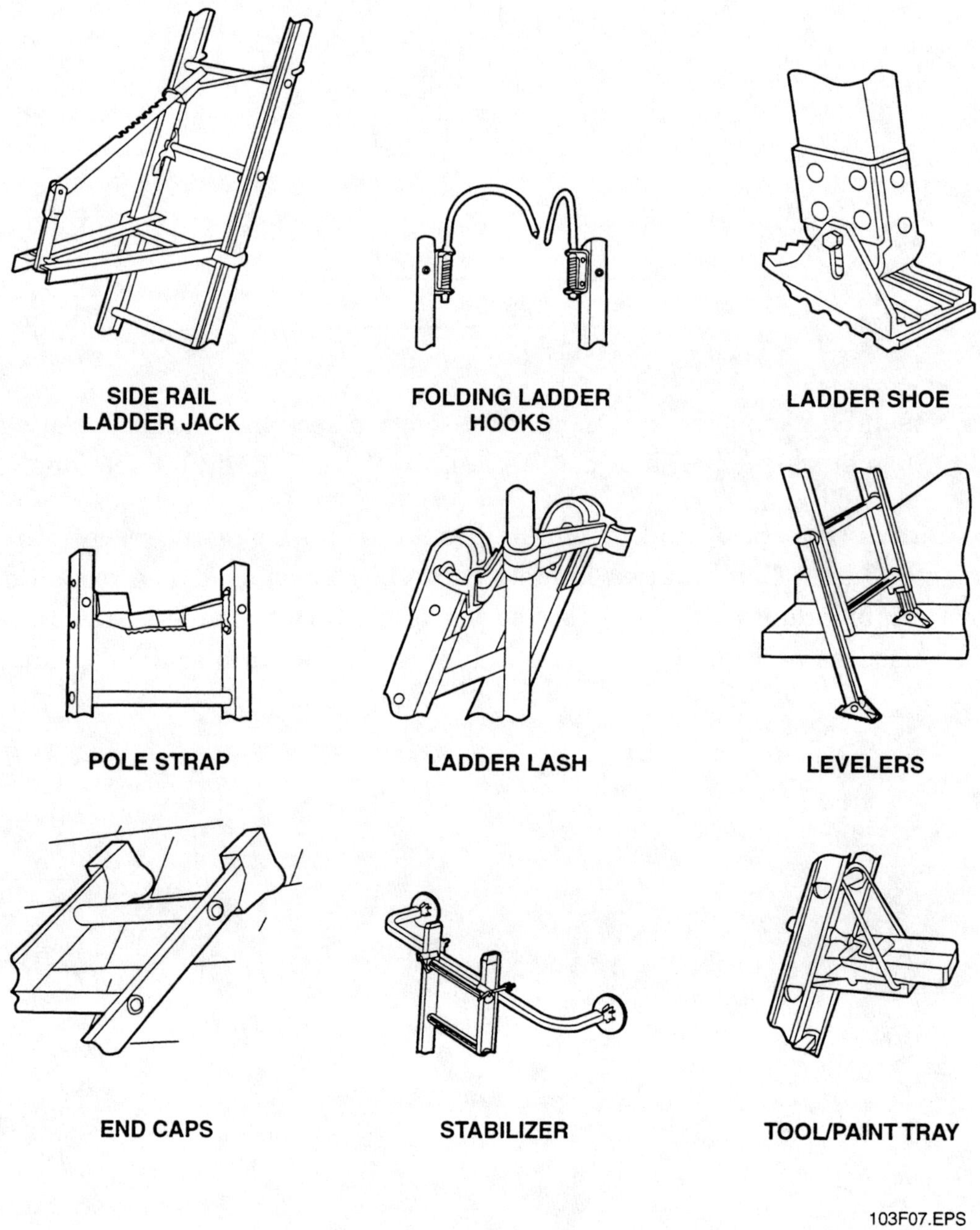

Figure 7. Ladder Accessories

- *Ladder jacks* – Ladder jacks are metal supports that attach to either the inside or the outside of a straight ladder or extension ladder. When used in pairs, ladder jacks can support a plank or lightweight stage placed across them to form a light-duty scaffold.

Some ladder jacks bear on the side rails and rungs of the ladder. If they attach to the rungs only, a bearing of at least 10 inches on each rung must be maintained. Only heavy-duty ladders (Type IA or Type I) should be used as supports, and they should be secured at both ends to prevent slipping. Ladder jacks are adjustable so that the platform is level regardless of the pitch of the ladder. The platform should be no more than 20 feet off the ground.

- *Ladder hooks* – Ladder hooks are devices that attach to the ladder and connect to a roof, pipe, or similar anchorage for support. Folding ladder hooks bolt to the ladder and fold out of the way when not in use.
- *Ladder shoes* – Ladder shoes mount onto the base of the ladder side rails and may be made of rubber or vinyl; others are made of metal with a rubber foot pad. Some have spurs or short metal spikes that can be flipped into position. Some ladder shoes are adjustable to the level of the surface. All ladder shoes are designed to give the ladder stability and resistance to slipping.
- *V-rungs* – V-rungs are bars that mount across the top of a ladder and have a V-shaped groove to give the ladder better support against poles and corners. Similar devices, called *pole lashes* and *pole straps*, attach to a ladder and wrap around a pole to give the ladder stability.
- *Levelers* – Levelers are side rail attachments that extend to a desired length and make the ladder level when the surface between the side rails is uneven, such as on a staircase.
- *End caps* – End caps cover the tops of side rails to prevent them from marring the supporting surface and to help make the ladder resistant to slipping.
- *Stabilizers* – Stabilizers attach to the upper end of a ladder to give it a wider base for support against a surface.
- *Tool and paint trays* – Tool and paint trays mount onto ladder rungs to hold tools and cans or trays of paint.

2.2.0 SELECTING AND ERECTING A LADDER FOR USE

Selection of the right ladder for the job at hand is important to complete a job as safely and efficiently as possible. When selecting a ladder, consider its features and how it meets the job. Always consider the highest duty rating and weight limit needed, as well as the height requirements. A ladder that is too long or too short does not allow the work surface to be reached easily or comfortably.

2.2.1 Step, Platform, And Trestle Ladders

To determine the height of a stepladder, platform ladder, or trestle ladder needed for a particular job, consider the ladder's highest standing level as marked on the ladder plus your reach. Next, choose a ladder at least one foot higher than you need. This gives you a wider, more stable base and allows you to place the pail shelf at a convenient working height. *Table 1* shows the ladder size and maximum standing height for common stepladders.

Ladder Size (Feet)	Maximum Standing Height (Feet)
4	2
6	4
8	6
10	8
12	10

Table 1. Stepladder Size Versus Maximum Standing Height

Step, platform, and trestle ladders should always be opened and set level on all four feet, with the spreaders locked in place (*Figure 8*). Never use a stepladder like a straight ladder. When climbing, never stand on the top or top two steps. For balance, lean your body into the ladder. When you can no longer reach your work comfortably, get down and move the ladder.

Figure 8. Position Stepladder So Feet Are Level And Spreaders Are Locked

2.2.2 Extension And Single Ladders

When selecting the size of an extension ladder, the ladder's maximum extended length and highest recommended standing level, usually the fourth rung from the top, must be considered. Also, OSHA requires that when a ladder is used to reach a roof or landing, the side rails must extend three feet above the eaves of the roof or edge of the landing. *Table 2* shows the maximum working length for 16-foot through 40-foot extension ladders. The sizes given in the table account for the proper overlap of ladder sections, the ladder angle, and a 3-foot extension above the eaves of a roof or landing.

Size (Feet)	Maximum Extended Length (Feet)	Maximum Working Height (Feet)
16	13	9
20	17	13
24	21	17
28	25	21
32	29	25
36	32	28
40	35	31

Table 2. Extension Ladder Size, Maximum Extended Length, And Maximum Working Height

Before use, ladders should always be inspected to make sure the rails, rungs, safety latches, and feet are not missing, broken, damaged, or loose.

Before climbing single or extension ladders, it is important that they be raised, placed at the proper angle, and secured firmly as outlined below.

Step 1 Lay the unextended extension or single ladder on the ground with its feet against a wall or other firm vertical surface (*Figure 9*). Starting at its top end, grasp the rungs and walk it up, hand over hand, until it is resting against the wall.

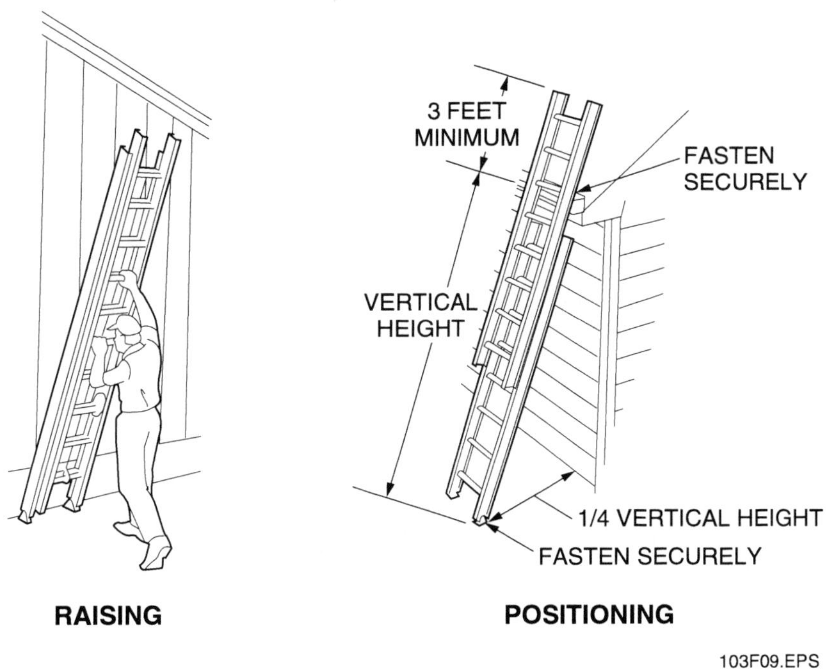

Figure 9. Raising And Positioning An Extension Ladder

Step 2 Once upright, raise the extended sections of an extension ladder to the desired height. Make sure that the safety latches for all sections securely engage the rungs of the base section or the section below.

Step 3 Position the bottom of the ladder so that the horizontal distance between its feet and the support point at the top is about 1/4 of the ladder's working length. If you are going to step off the ladder onto a platform or roof, remember to make sure the top of the ladder extends at least three feet beyond the support point. Also make sure there is solid contact of the top rails at the support point. Do not lean the rails against loose, unsafe, or movable objects.

Step 4 Make sure that the feet of the ladder are at an equal distance from the wall and have secure footing on a solid, level base. Never set a ladder on soft or uneven ground or movable objects. Ladders placed on slippery surfaces should be secured or equipped with slip-resistant feet to prevent them from moving.

Step 5 Once the ladder is in position, tie off the top and bottom to make it secure.

2.3.0 LADDER SAFETY

To prevent falls when climbing ladders, be sure to comply with the following safety rules:

- One person should hold the ladder while the other ties off and secures the ladder.
- Grasp the ladder firmly and always face the ladder when climbing up or down. Follow the three-point rule: at least three parts of your body—either two feet and a hand or two hands and a foot—should be in contact with the ladder at all times. Keep your body centered on the ladder and do not overreach. Climb down and reposition the ladder, if necessary.
- Test every rung or step before putting your full weight on it.
- Do not carry anything on a ladder that may cause loss of balance or a fall. Haul job materials or tools with a rope hoist.
- Clean grease, oil, mud, wet paint, or any slippery material from the rungs or steps of a ladder and from the bottom of work boots before climbing.
- Never climb higher than the highest standing level marked on a ladder, generally the third or fourth rung from the top of a straight or extension ladder or the second step from the top of a stepladder.
- Do not skip rungs or steps when climbing up or down.
- Never allow more than one person on a ladder at a time unless it is made for two-person use.
- Never stand or sit on the top cap or the paint shelf of a stepladder.
- Always move or extend a ladder from the ground—never the roof—and never while it is occupied.

In addition to the previously described safety rules, the following general safety rules concerning ladders should also be followed to prevent falls and other accidents:

- Keep the area around the top and bottom of a ladder free from clutter and debris.
- Never use a ladder in a manner for which it was not designed. For example, do not use a stepladder as a straight ladder.
- Do not use an extension ladder upside down.
- Avoid using a ladder in windy or stormy weather.
- Never leave a ladder in position and unattended, especially near children.
- Never use pallets or skids as ladders.

2.4.0 LADDER MAINTENANCE

Ladders must be properly maintained to provide for safe use and a long life. To avoid premature wear or damage, they should not be dropped, thrown, or exposed to excessive heat or corrosive materials. Before every use, a ladder should be thoroughly inspected for:

- Rungs or steps that are missing, broken, or loose
- Rails that are split, cracked, or dented
- Rails that form loose connections with the steps or rungs
- Hardware that is missing or loose
- Locks or spreaders that are missing, broken, worn, or do not engage properly
- Ropes that are missing or worn
- Pulleys that are missing, broken, or loose
- Accessories, such as nonslip devices, that are missing, broken, loose, or worn
- Excessive weathering of wooden and fiberglass ladders, such as degradation or fading from ultraviolet sunlight
- Corrosion of metal ladders

Note Clear wood preservative will help extend the life of a wooden ladder. Do not use an opaque coating because it will hide defects, such as cracks or splits, which are signs of trouble.

Ladders should be stored in a warm, dry place protected from the weather and from contact with the ground. They should also be kept from sagging during storage.

Defective ladders should be tagged with a warning that they are not to be used. They should then be either repaired or replaced. Repairs should be made only by a qualified person.

3.0.0 SCAFFOLDS

Scaffolds are temporary built-up frameworks or suspended platforms designed to support workers, materials, and equipment at elevated or otherwise inaccessible work locations. All scaffolds must be used and assembled in accordance with all local, state, and federal/OSHA laws and codes. They may be built up from the ground or suspended from a structure. Depending on their design, scaffolds are classified for light-duty, medium-duty, or heavy-duty use, meaning they can safely support loads up to 25, 50, or 75 pounds per square foot (psf) or 1.05, 2.1, or 3.15 kilograms per square meter, respectively. *OSHA Standards 1926*, Subpart

L defines and governs the use of scaffolds. OSHA requires that the building, moving, or dismantling of all scaffolding be supervised by a *competent person*. A competent person is one who has the training, knowledge, and experience to identify hazards on the job site and the authority to eliminate them. Some general OSHA regulations pertaining to the use of scaffolds are:

- All scaffolds must be able to support at least four times their maximum intended load; scaffold cable or rope must be able to support at least six times the rated load.
- Poles, legs, and uprights must be plumb and securely braced. Footings must be sound and able to support the maximum intended load. Planking must be secured to the scaffold. It must also be of the proper grade and size for the load.
- A ladder or other safe means must be available for accessing the scaffold.
- Any scaffold that is 10 feet (3.04 meters) or more off the ground or surface (less in some states) must have guardrails, midrails, and toeboards (*Figure 10*) made of lumber or other suitable material on all open sides and ends. The guardrail should be placed about 42 inches (106.7 cm) high. The midrail should be halfway between the guardrail and the toeboard, which should run along the base perimeter of the scaffold. The prescribed dimensions are 2 x 4 inches for guardrails, 1 x 6 inches for midrails, and at least 4 inches (10.16 cm) high for toeboards. If anyone must work or pass under the scaffold, the space between the guardrail and toeboard should be covered with wire mesh to prevent tools or materials from falling below. Any scaffold between 4 and 10 feet (1.22 and 3.05 m) high and less than 45 inches (114.3 cm) long or wide must have guardrails on all open sides and ends.
- Any damaged scaffold must be repaired or replaced immediately.

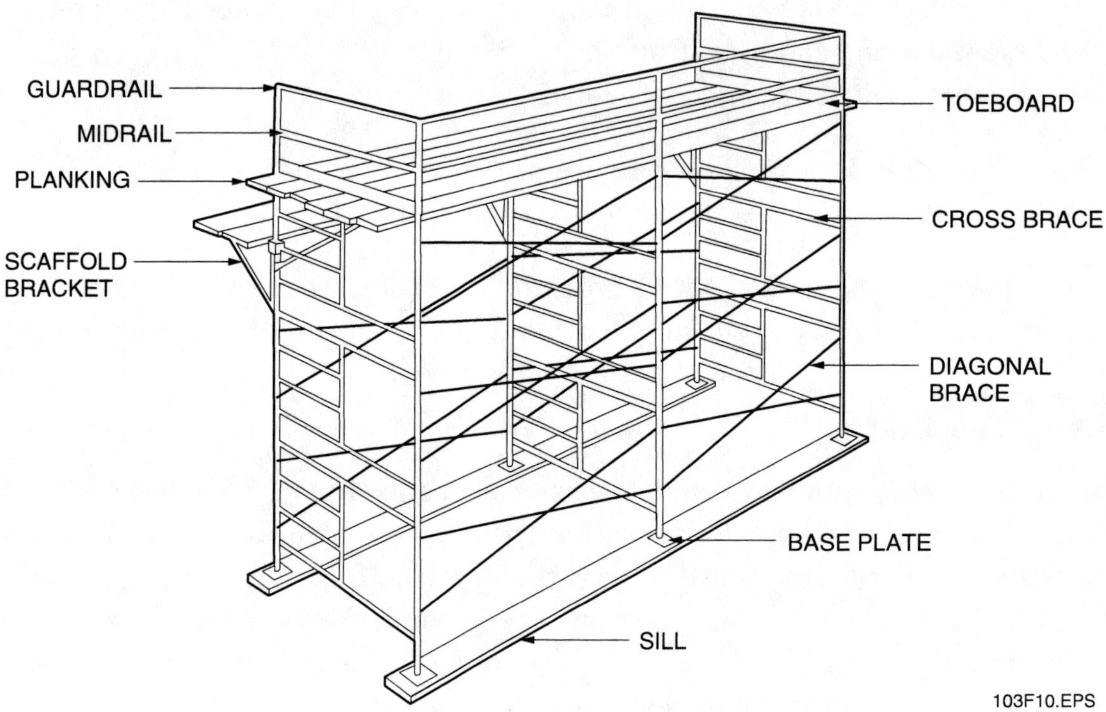

Figure 10. Parts Of A Typical Built-Up Scaffold

3.1.0 TYPES OF SCAFFOLDS

There are many kinds of scaffolds, each designed for a specific use and job site condition. Some common scaffolds include:

- Built-up scaffolds
- Swing scaffolds
- Beam-suspended scaffolds

3.1.1 Built-Up Scaffolds

Built-up scaffolds are assembled from the ground up at the job site. OSHA has a height limit of 125 feet for built-up scaffolds. Scaffolds higher than 125 feet must be designed by a professional engineer and approved by OSHA before use. Most built-up scaffolds are manufactured scaffolds made from tubular steel that can be fitted together to form complete sections. For light-duty use, scaffolds are sometimes made from aluminum or wood. There are two main types of manufactured steel built-up scaffolds: tubular welded frame scaffolds and tube and coupler scaffolds.

Tubular welded frame scaffolds (*Figure 11*) are used in accessible places with fairly level ground conditions. They use manufactured, welded-end frame sections made in various heights and widths that are joined with horizontal and diagonal cross braces and secured by pins. The braces have a fixed length that automatically squares and vertically aligns vertical members so that the erected scaffold is always plumb, square, and rigid. The scaffold is extended by adding braces and frames until the desired length is reached. Scaffold height is increased by

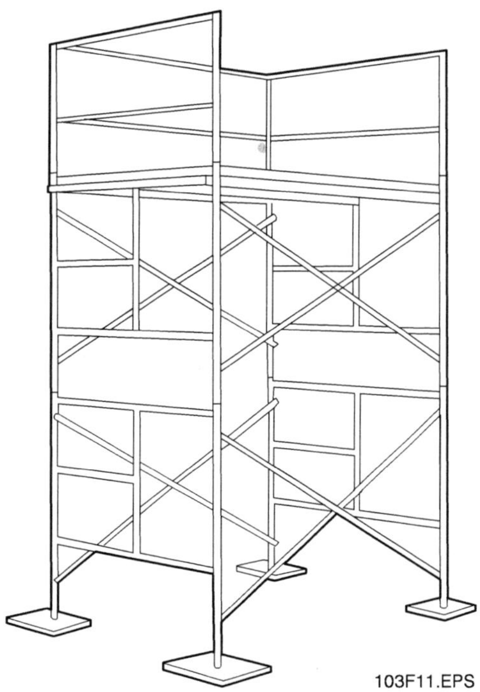

103F11.EPS

Figure 11. Tubular Welded Frame Scaffold System

stacking end frames on top of each other. The bottoms of the legs of the upper end frames slide into the tops of the legs of the lower end frames and are joined together with drop locks and coupling pins.

Tube and coupler scaffolds (*Figure 12*) are typically used in difficult or inaccessible places. The various tubes that form the scaffold are attached together as required with right-angle or swivel clamps to build up a scaffold to go over uneven ground and unusual shapes. End fittings are used to connect the end of one tube with another.

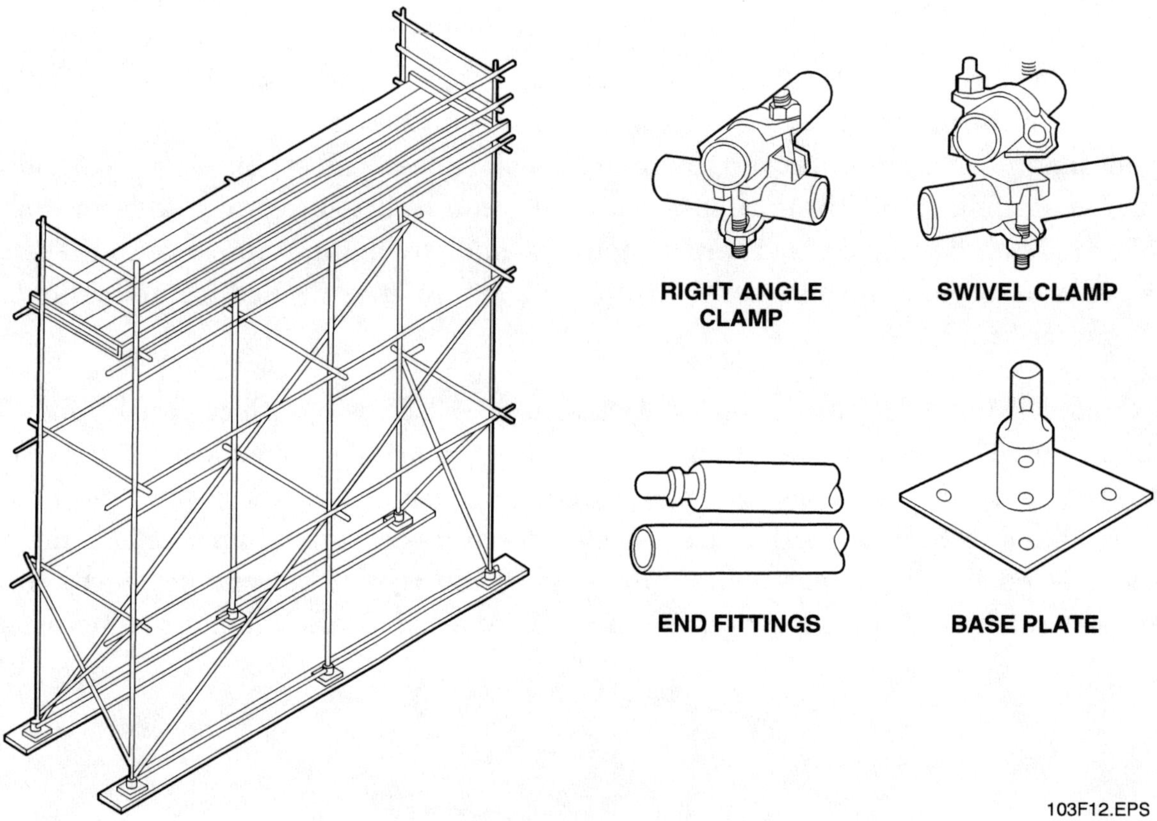

Figure 12. Tube And Coupler Scaffold System

Metal base plates attached to the tubular legs or posts of both tube and coupler and tubular welded frame scaffolds serve as a foundation for stationary scaffolds. On dirt and similar surfaces, planks called *sills* or *mud sills* must be placed beneath the base plates to provide a secure and level footing for the scaffold. Leveling jacks are frequently used with the base plates to aid in leveling the scaffold. To prevent movement, built-up scaffolds must be secured to the structure at intervals not to exceed 26 feet (7.62 meters) vertically and 30 feet (9.14 meters) horizontally.

Tubular welded frame scaffolds are often equipped with casters on the bottom of the frame legs to form a freestanding mobile scaffold tower. Note that mobile scaffolds should only be used on level, smooth surfaces that are free of obstructions and openings. Regulations require that mobile scaffold casters support four times the maximum intended load. They must also have a positive locking device to hold the scaffold in place. The height of mobile towers must not exceed four times the smaller base dimension. For example, if the scaffold's base is 3 feet (0.91 meter) by 6 feet (1.82 meters), its height should be no more than 12 feet (3.66 meters). Outriggers, which are supports that attach to both sides of the scaffold base (*Figure 13*), can be used to increase the size of the base dimension. When moving a mobile scaffold, force should be applied as close to the base as possible to avoid tipping it over.

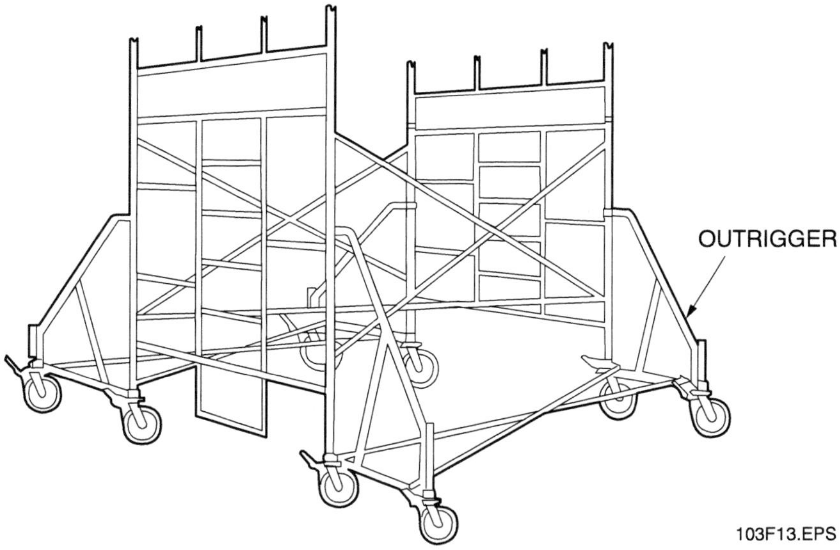

Figure 13. Outrigger And Casters Used On Mobile Scaffold

3.1.2 Swing Scaffolds

A swing scaffold (*Figure 14*), sometimes called a **two-point suspension scaffold**, is suspended from a building or structure by two ropes or cables in a manner that allows it to be raised or lowered as needed. It consists of a platform or stage, hangers, hoisting mechanism, ropes or cables, and rigging devices. The platform or stage provides for a work area of 24 to 36 inches wide that is securely attached to the hangers.

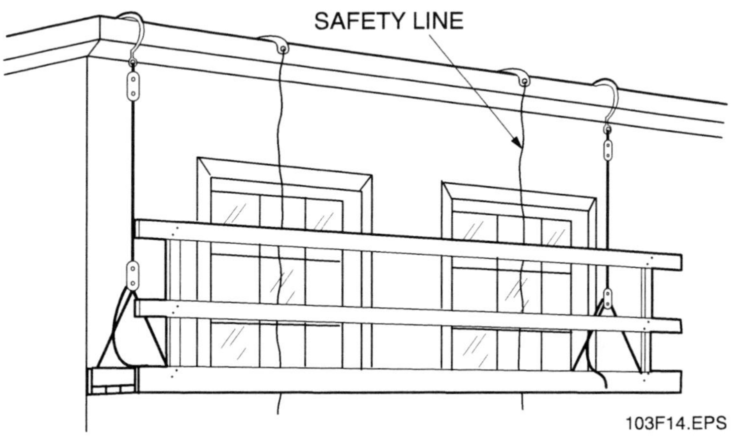

Figure 14. Swing Scaffold

The platform may be metal or wooden. Platform types are described in more detail later in this section. Suspended scaffold platforms are required to have guardrails, midrails, and toeboards on all open sides and ends.

The platform is supported at each end by steel hangers, often called *stirrups* or *triangles*. These are attached to wire cables or ropes, which may be made of wire, fiber, or synthetic material, suspended from rigging devices that are secured to the building or structure. The suspension system and rigging devices (including their support on the structure) must be able to hold four times the rated load of the hoist. However, OSHA requires that the ropes or cables used be able to support at least six times the rated load. Safety lines must be used when working on the swing scaffold. These lines should always be tied off to different anchor points than those used to support the scaffold.

Cornice hooks, parapet clamps, or outrigger beams (*Figure 15*) are rigging devices commonly used to support swing scaffolds. A cornice hook is a steel device that hooks to a roof, parapet, or structural support. A parapet clamp fits over and clamps onto a parapet that runs along the perimeter of a roof. An outrigger beam is an arm that extends beyond the edge of the roof. Some outrigging devices have wheels so that they can roll along the roof as the work progresses. Rigging devices are anchored and secured to the building or structure with tieback lines installed perpendicular (at right angles) to the face of the structure. Tieback lines must be as strong as the suspension ropes or cables.

Cornice hooks or parapet clamps are typically used when working on older buildings. For this reason, it is extremely important to make sure that the cornice or parapet to which they attach is structurally sound and capable of supporting the combined weight of the scaffold and workers. If there is any doubt, always have an engineer or other competent person check the structural integrity of the cornice or parapet before using it to support the scaffold.

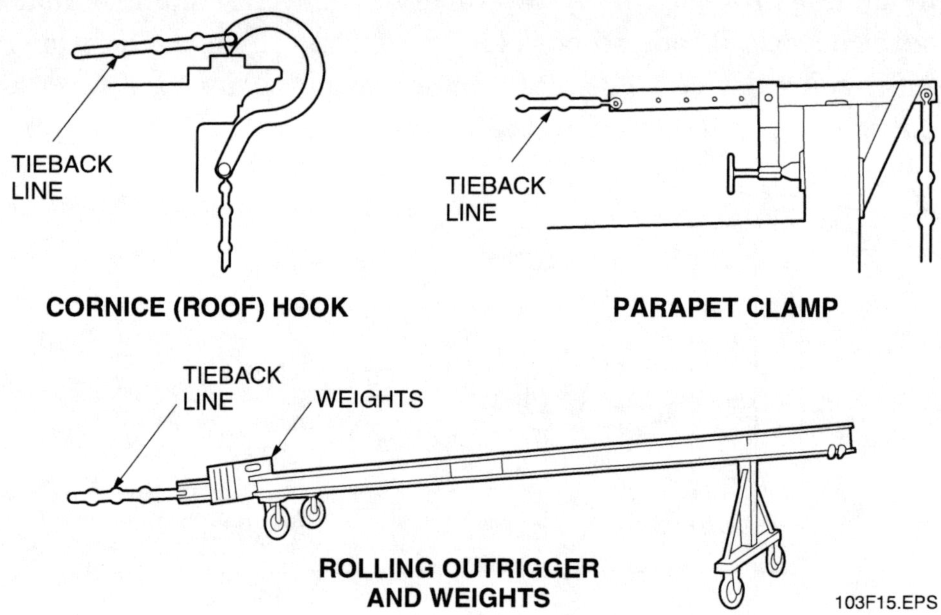

Figure 15. Typical Rigging Devices Used With Swing Scaffolds

The scaffold's hoisting mechanism enables the worker to raise or lower the scaffold. For manually-operated scaffolds, the mechanism consists of ropes or cables rigged through tackle blocks and pulleys. Many scaffolds have hoist mechanisms powered by electricity (*Figure 16*) or compressed air. Per ANSI standards, these powered hoists are limited to a maximum speed of 35 feet per minute. They must also have speed reducers and both primary and secondary brakes. Safety devices are also built into powered hoist systems to prevent operation of the scaffold if it is overloaded and to lower the unit without power if necessary.

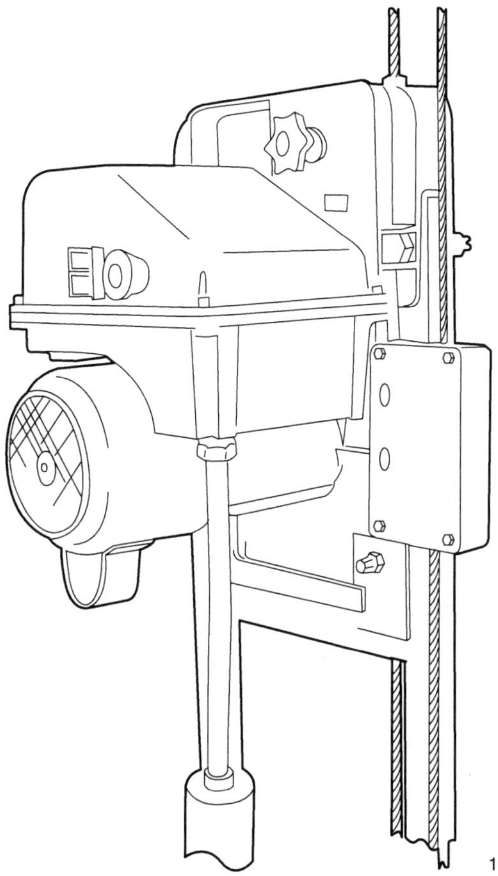

103F16.EPS

Figure 16. Electric-Powered Hoist

Regulations mandate that no more than two workers are permitted to work at one time on a swing or other suspended scaffold designed for a working load of 500 pounds (225 kilograms). With a working load of 750 pounds (337.5 kilograms), no more than three workers are permitted to work at one time. Also, all workers must wear the proper fall protection equipment when working on suspended scaffolds.

3.1.3 Beam-Suspended Scaffolds

Beam-suspended scaffolds (*Figure 17*) are typically used when working on bridges or other industrial steel structures. They are suspended with clamp or roller-type rigging devices that hook onto the lower flanges of two parallel I-beams, each having a bar or bracket to which the scaffold is attached. The roller-type device has wheels that allow the scaffold to be moved along

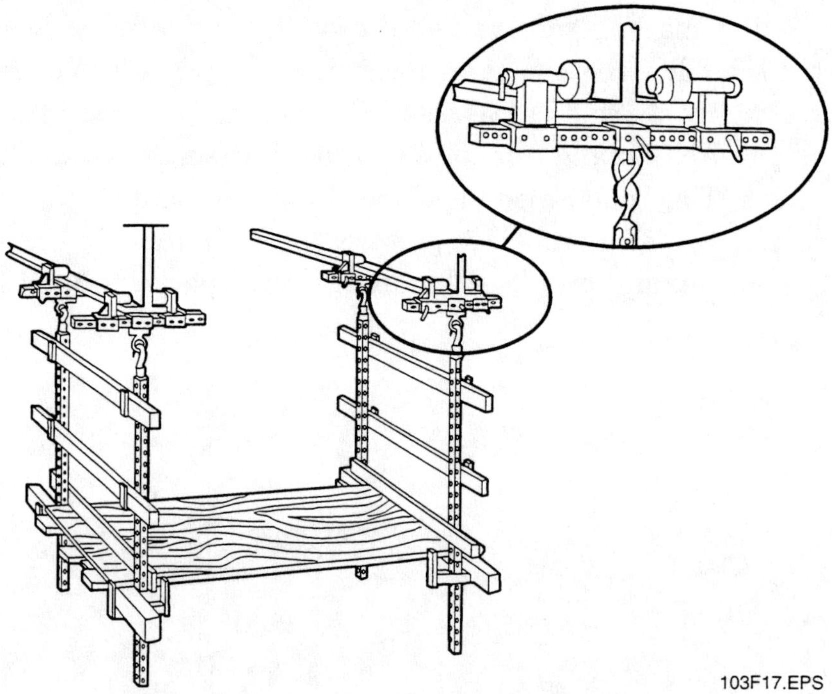

Figure 17. Beam-Suspended Scaffold

the I-beam flanges as work progresses. This allows the job to continue without having to stop to lower the scaffold and reposition the rigging.

Some important guidelines that pertain to the installation and use of beam-suspended scaffolds are given below:

- Make sure the I-beam is level, or else the rollers may move. If the I-beam is not level, secure the rigging devices to prevent accidental movement.
- Never use this type of equipment on open-ended beams that could allow the scaffold to roll off the supports.
- Install stops if the flange is interrupted anywhere along the beam.
- Check that the width of the flange is uniform along the entire length of the beam. A roller set to a particular width could derail if the width of the flange narrows unexpectedly.

3.1.4 Other Types Of Scaffolds

There are many types of scaffolds used by painters and decorators and covered by OSHA regulations or ANSI standards. Some common scaffolds are listed below:

- *Ladder-type scaffold* – A scaffold made with a plank or platform placed between two heavy-duty, self-supporting ladders (see *Figure 4*).
- *Ladder jack scaffold* – A light-duty scaffold with a platform supported by a pair of brackets called *ladder jacks*, which should be attached to heavy-duty ladders according to proper positioning requirements. A ladder jack scaffold should be no more than 20 feet high.

- *Multiple-level suspension scaffold* – A scaffold with several levels of work platforms supported by the same suspension system, which allows it to move up or down.
- *Multiple-point suspension scaffold* – A scaffold suspended by a rope or cable at each corner. It may have just one platform or multiple platforms joined together to form a larger work area.
- *Outrigger scaffold* – A scaffold for light-duty or medium-duty use that is suspended from outrigger beams anchored inside a building or structure.
- *Pump jack scaffold* – A platform placed between metal brackets that can be positioned on vertical poles that are braced to the wall of a building or structure (*Figure 18*). The poles should be no more than 30 feet high.
- *Window jack scaffold* – A one-person platform supported by a bracket anchored inside a window opening.

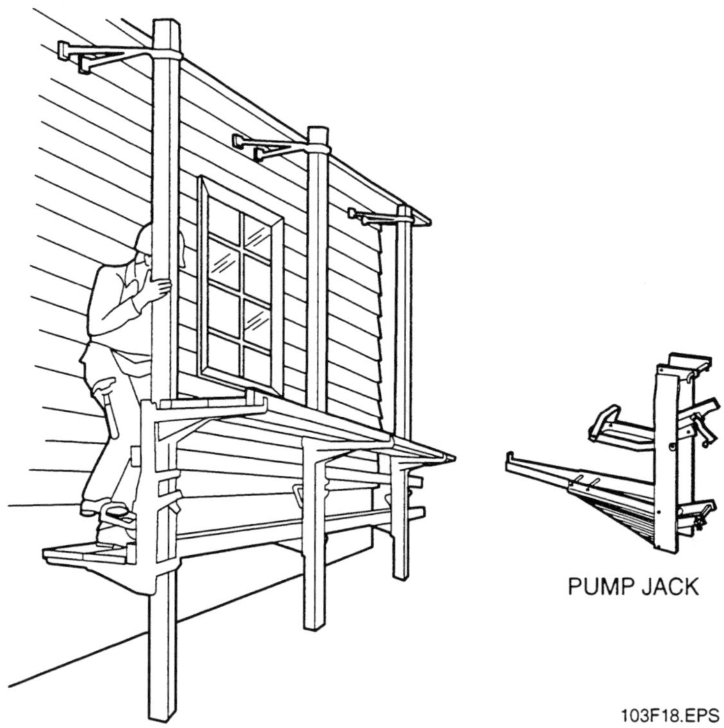

Figure 18. Pump Jack Scaffold

3.1.5 Scaffold Platforms

The platform or working area of a scaffold may be a plank, deck, or stage (*Figure 19*). Planks may be made of solid wood or assembled out of wood or metal parts. Planks can have a fixed length, or they can be expandable. An expandable plank, also called an *extension plank*, is made of two or three interlocking wood or metal sections that can be extended to lengthen the plank. For example, an 8-foot (2.44-meter) extension plank typically can be extended to 13 feet (3.96 meters). OSHA regulations specify the type and grade of wood used in wooden planks and the required lengths for different types of lumber planks depending on use. Regulations

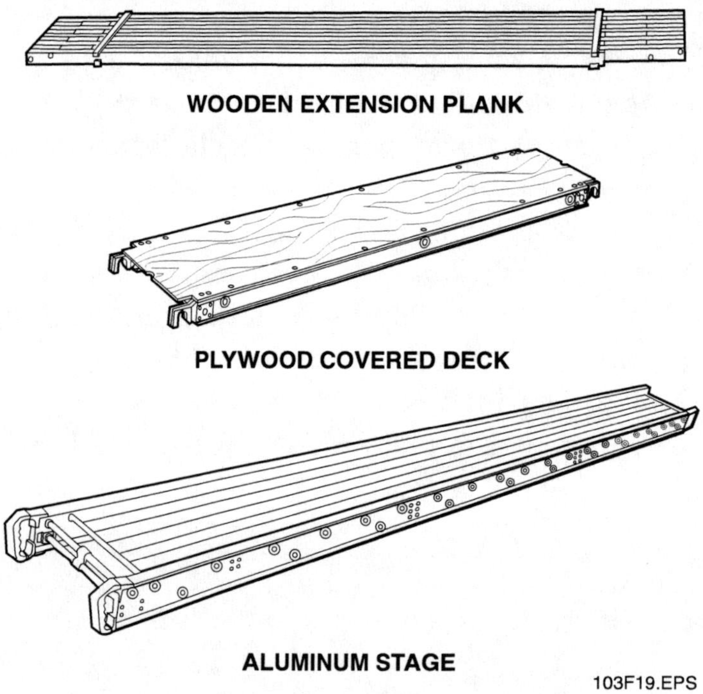

Figure 19. Scaffold Platforms

also require that the planking must be overlapped by at least one foot or be secured from moving. Planks must also extend 6 to 12 inches over the end supports.

Deck and stage platforms are similar in that both are made with two parallel side rails that are connected with rungs or cross members. The base of a deck is typically covered with plywood or aluminum sheeting. Its ends may be equipped with metal hooks used to attach the deck to the horizontal supports of tubular scaffolding. The base and covering of stage platforms are generally made from heavy-duty aluminum for use with swing or other suspended scaffolds. Many come equipped with aluminum guardrails and toeboards. Deck and stage platforms come in various lengths and widths.

3.1.6 Boatswain's (Bosun's) Chair And Work Cages

Boatswain's chairs and work cages are **single-point suspension scaffolds**, meaning they are suspended from an overhead support by a single cable or rope. Normally, they are used by one worker at a time. Both types provide for good maneuverability and accessibility for light-duty work in areas not accessible by other types of scaffolds. Depending on the design, the operator can raise or lower the platform manually with a block and tackle or use an electric or compressed air-powered hoist.

A boatswain's chair (*Figure 20*) is basically a seat attached to a suspended cable or rope. The chair is typically made of nylon rigging with a plywood seat. The worker is restricted to a sitting position with little or no room for tools or equipment on the chair. If necessary, these items can be tied to the chair.

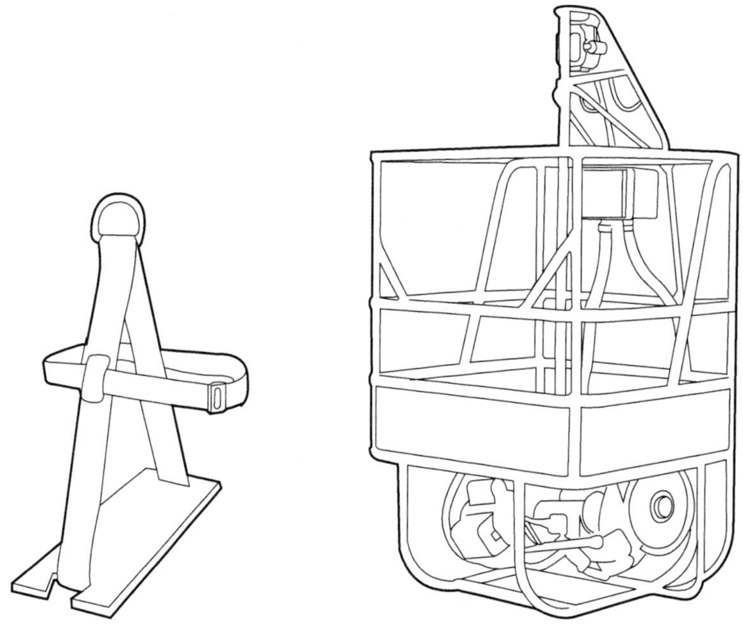

Figure 20. Boatswain's Chair And Work Cage Scaffold

A work cage (*Figure 20*) is a single-point adjustable suspension scaffold with the platform enclosed with guardrails and toeboards. It is large enough to allow the operator to work standing up. Two work cage units are sometimes rigged to form a two-point suspension (swing) scaffold.

3.2.0 SCAFFOLD SAFETY

Improper or careless use of scaffolds can result in accidents, injury, or death. Everyone who works on scaffolds can minimize their risks by following the safety guidelines given below.

3.2.1 General Safety Guidelines

Safety begins by first getting training in the proper use of scaffolds. Equally important is to always use the right safety equipment, including a hard hat and personal fall arrest equipment. Detailed information about the types and use of personal fall equipment is given later in this module. Never work on a scaffold if you:

- Are subject to seizures
- Become dizzy or lightheaded when working at an elevation
- Take medication that might affect your stability or performance
- Are under the influence of drugs or alcohol

Always erect and use scaffolds according to the manufacturer's instructions. They must also be erected and used in accordance with all local, state, and federal/OSHA requirements. Never interchange parts of scaffold systems made by different manufacturers. As appropriate, attach a green, red, or yellow tag (*Figure 21*) to any scaffold that is assembled and erected to alert users of its current mechanical and safety status. Do not rely solely on the tag. Always inspect all parts of a scaffold before each use.

Figure 21. Typical Scaffold Tags

3.2.2 Safety Guidelines For Built-Up Scaffolds

Comply with the following guidelines when erecting and using tubular built-up scaffolds:

- Inspect all scaffolding parts before assembly. Never use parts that are broken, damaged, or deteriorated. Be cautious of rusted materials. Follow the manufacturer's recommendations for the proper way to erect and use scaffolding. Do not interchange parts from different manufacturers. Do not force braces or other parts to fit. Adjust the level of the scaffold until the connections can be made easily.
- Provide adequate sills or underpinnings for all scaffolds built on filled or soft ground. Compensate for uneven ground by using adjusting screws or leveling jacks. Do not use boxes, concrete blocks, bricks, or other similar objects to support a scaffold.
- Be sure scaffolds are plumb and level at all times. Follow the prescribed spacing and positioning requirements for the parts of the scaffold. Anchor or "tie-in" scaffolding to the building at prescribed intervals.
- Use ladders rather than cross braces to climb the scaffold. Position ladders with caution to prevent the scaffold tower from tipping.
- Keep scaffolds free of clutter and any slippery material.

- Do not work on a scaffold that is more than 10 feet high without guardrails, midrails, and toeboards on open sides and ends.
- Lock the casters of a mobile scaffold when it is positioned for use.
- Do not ride a mobile scaffold.
- Avoid building scaffolds near power lines.

3.2.3 Safety Guidelines For Swing And Other Suspended Scaffolds

Comply with the following guidelines when erecting and using swing and other suspended scaffolds:

- Follow the manufacturer's recommendations for installation, use, and maintenance of the equipment. Before installation, inspect all parts of a structure to which rigging and tieback lines will be secured to be sure that they can support the load.
- Be sure rigging devices are of the proper size and design to support the scaffold and that they are installed properly. If counterweights are used to secure the inner end of outrigger beams, they should be fastened to the outrigger. Roofing materials, sand bags, or any "flowable" materials are not appropriate counterweights. Check that tieback lines are installed perpendicular to the face of the structure and secured to a solid support.
- Check for power lines or electric service wires on the job site. If they pose a hazard, contact the utility company about having them temporarily deactivated.
- Inspect all scaffold equipment each day. Check ropes and cables thoroughly for wear, fraying, corrosion, brittleness, damage, or other conditions that may weaken them. Have them replaced as necessary by qualified personnel.
- Observe the scaffold's load capacity; never overload the equipment.
- Keep suspension ropes and cables straight and perpendicular to the platform during use. Do not affix them to anything to change the line of travel.
- Lash the scaffold to the building or structure to prevent it from swaying.
- Stay off scaffolds during storms or high winds; watch for icy or slippery platforms.
- Guarantee safe access to the swing stage at all times.
- Use two-way radios for communication between workers on the scaffold and on the ground.
- Do not combine two or more two-point swing scaffolds to form one unit. (However, two single-point suspension scaffolds can be combined to form a two-point suspension scaffold.)
- Raise and lash the scaffold into a safe position when not in use.

4.0.0 AERIAL LIFTS

Aerial work platforms and scissor power lifts are portable **aerial lifts** used to raise or lower workers to and from elevated job sites. Both types are made in various models. Some are transported on a vehicle to the job site where they are unloaded. Others are trailer-mounted and towed to the job site by a vehicle. Still others are permanently mounted onto a vehicle. Depending on their design, they can be used for indoor work, outdoor work, or both. *Figure 22* shows an aerial work platform and scissor lift designed for indoor use. Boom lifts, often called *cherry pickers*, are truck or trailer-mounted lifts designed for both indoor and outdoor use. Boom lifts have a single arm that extends a work platform/enclosure capable of holding one or two workers. Some models have a jointed (articulated) arm that allows the work platform to be positioned both horizontally and vertically. Scissor lifts raise a work enclosure vertically by means of crisscrossed supports. Most aerial work platforms and scissor lifts are powered either by electric motors or hydraulically.

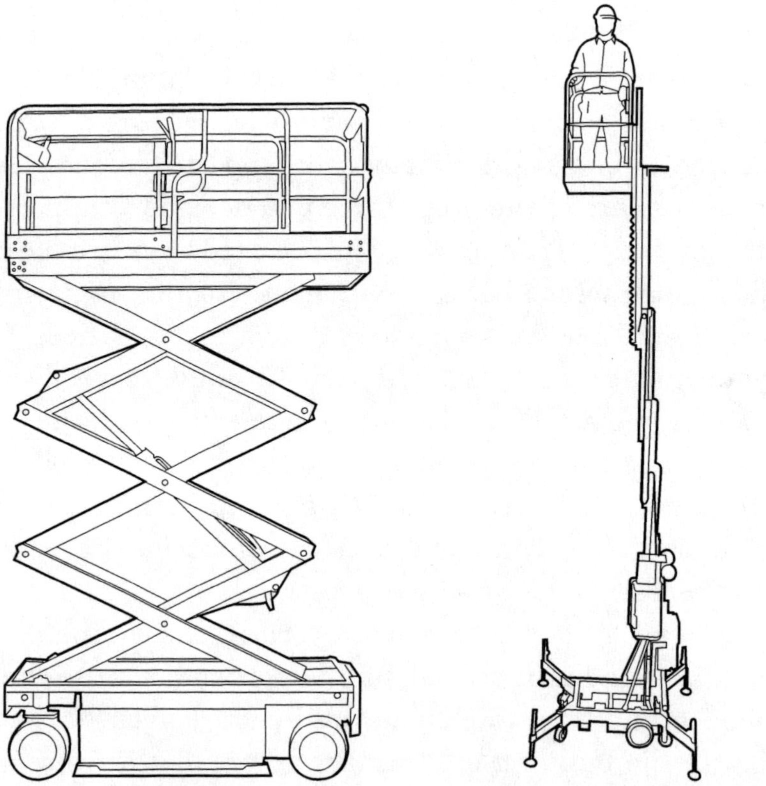

Figure 22. Aerial Lifts

OSHA Standards 1926, Subpart N defines and governs the use of aerial lifts. Some guidelines for using lifts safely are given below.

- Know the capacity and operating characteristics of the aerial platform. Do not overload the platform. Use a lift only on a solid, stable surface. Lock the wheels, especially on an incline. Avoid using a lift outdoors in stormy weather or in strong winds.
- Inspect the equipment before each use, as specified by the manufacturer, to make sure everything is in proper working order. Have any defects repaired before using the lift. Never modify or remove any part of the equipment unless authorized to do so by the manufacturer.

- Check the job site for any hazards that may cause the lift to overturn. Check for hazards above, below, and all around the path of travel. Make sure to maintain a safe distance from electrical power lines and other electrical hazards.
- Prevent people from walking beneath the work area of the platform.
- Use personal fall arrest equipment (body harness, lanyard, and lifeline), if required, for the type of lift being used.
- Lower the lift and lock it into place before moving the equipment. Also, lower the lift, shut off the engine, set the parking brake, and remove the key before leaving it unattended.
- Keep the lift free from oil, grease, wet paint, mud, or any slippery material.
- Do not lean over the guardrails of the platform, and never stand on the guardrails.

5.0.0 FALL PROTECTION EQUIPMENT

Painters and decorators spend a major part of their time working on ladders, scaffolds, and lifts in order to reach elevated work areas. Falls from such high places can cause serious injury or death when the wrong kinds of fall protection equipment are used, or when the right equipment is used improperly.

There are three common types of fall protection equipment: guardrails, personal fall arrest systems, and/or safety nets. This section will focus on personal fall arrest systems and safety nets. These devices and their use are governed by *OSHA Standards 1926*, Subpart M. Guardrails used with scaffolds were described earlier in this module. It should be pointed out that *OSHA Standards 1926*, Subpart M also covers guardrails but the rules given in this subpart do not apply to guardrails on scaffolds. The rules covering guardrails on scaffolds are contained in *OSHA Standards 1926*, Subpart L. Basically, OSHA requires that all workers use guardrail systems, safety net systems, or personal fall arrest systems to protect themselves from free falling more than 6 feet (1.8 meters) and hitting the ground or a lower work level.

5.1.0 PERSONAL FALL ARREST SYSTEMS

Personal fall arrest systems catch workers after they have fallen. They are designed to activate only if a fall occurs. As stated previously, they are designed and rigged to prevent a worker from free falling a distance of more than 6 feet (1.8 meters) and hitting the ground or a lower work area. When describing personal fall arrest systems, the terms defined below must be understood.

- **Free fall distance** – The vertical distance a worker moves after a fall before a deceleration device is activated.
- **Deceleration device** – A device such as a shock-absorbing lanyard or self-retracting lifeline that brings a falling person to a stop without injury.
- **Deceleration distance** – The distance it takes before a person comes to a stop. The required deceleration distance for a fall arrest system is a maximum of $3\frac{1}{2}$ feet (1.07 meters).

- **Arresting force** – The force needed to stop a person from falling. The greater the free fall distance, the more force is needed to stop or arrest the fall.

Equipment used in personal fall arrest systems includes:

- Body harnesses and belts
- Lanyards
- Deceleration devices
- Lifelines
- Anchoring devices and equipment connectors

5.1.1 Body Harnesses

Full-body harnesses (*Figure 23*) with sliding back D-rings are used in personal fall arrest systems. They are made of straps that are designed to be worn securely around the user's body. This allows the arresting force to be distributed via the harness straps throughout the body, including the shoulders, legs, torso, and buttocks. This distribution decreases the chance of injury. When a fall occurs, the sliding D-ring moves to the nape of the neck, keeping the worker in an upright position and helping to distribute the arresting force. This keeps the worker in a relatively comfortable position while waiting for rescue.

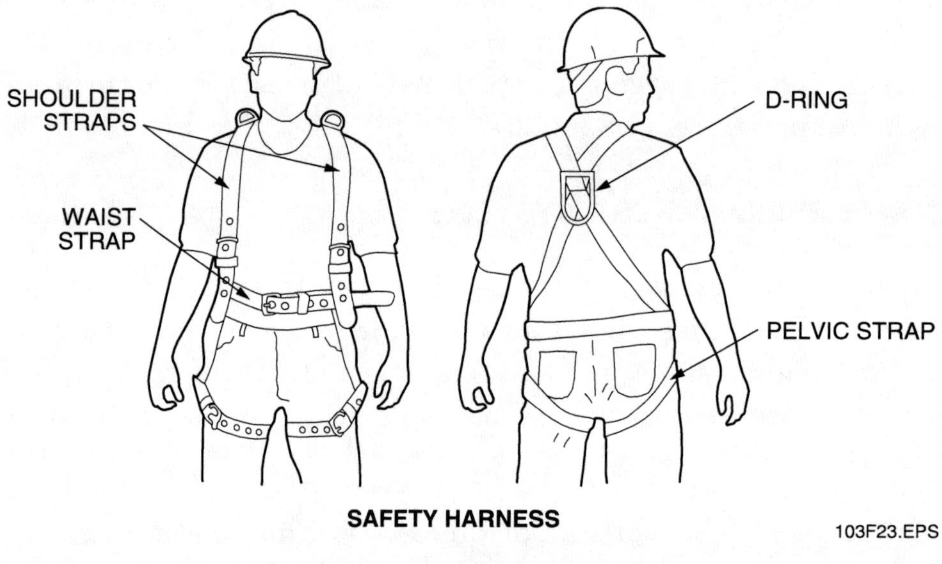

Figure 23. Full-Body Harness

Selecting the right full-body harness depends on a combination of job requirements and personal preference. Harness manufacturers normally provide selection guidelines in their product literature. Other types of full-body harnesses can be equipped with front chest

D-rings, side D-rings, or shoulder D-rings. Harnesses with front chest D-rings are typically used in ladder climbing and personal positioning systems. Those with side D-rings are also used in personal positioning systems. **Personal positioning systems** are systems that allow workers to hold themselves in place, keeping their hands free to accomplish a task. Per OSHA regulations, a personal positioning system should not allow a worker to free fall more than 2 feet (0.61 meter), and the anchorage to which it is attached should be able to support at least twice the impact load of a worker's fall or 3,000 pounds (13.3 kilonewtons), whichever is greater. Harnesses equipped with shoulder D-rings are typically used with a spreader bar or rope yoke for entry into and retrieval from confined spaces.

Note that in the past, body belts (*Figure 24*) were frequently used instead of a full-body harness as part of a fall arrest system. As of January 1, 1998, they are banned from such use. This is because body belts concentrate all of the arresting force in the abdominal area. Also, after a fall, the worker hangs in an uncomfortable and potentially dangerous position while awaiting rescue. Today, the use of body belts is limited to personal positioning systems.

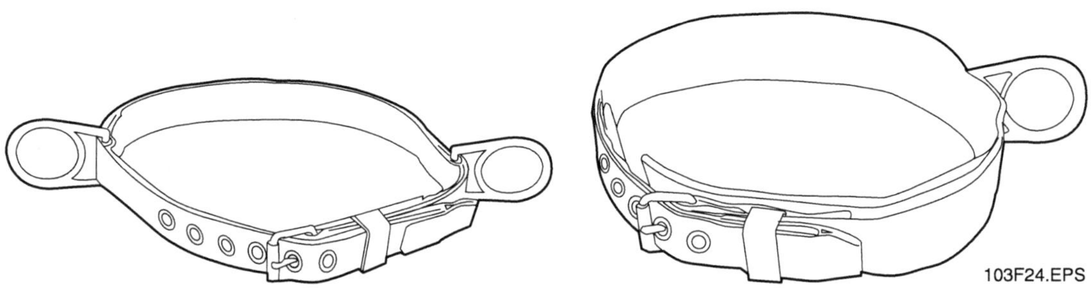

103F24.EPS

Figure 24. Body Belts

5.1.2 Lanyards

Lanyards are short, flexible lines with connectors on each end. They are used to connect a body harness or body belt to a lifeline, deceleration device, or anchorage point. There are many kinds of lanyards made for different uses and climbing situations. All must have a minimum breaking strength of 5,000 pounds (22.2 kilonewtons). They come in both fixed lengths and adjustable lengths and are made out of steel, rope, or nylon webbing. Some have a shock absorber (*Figure 25*) which absorbs up to 80% of the arresting force when a fall is being stopped. When choosing a lanyard for a particular job, it is a good idea to always follow the manufacturer's recommendations.

CAUTION When activated during the fall arresting process, a shock-absorbing lanyard stretches as it acts to reduce the fall arresting force. This potential increase in length must always be taken into consideration when determining the total free fall distance from an anchor point.

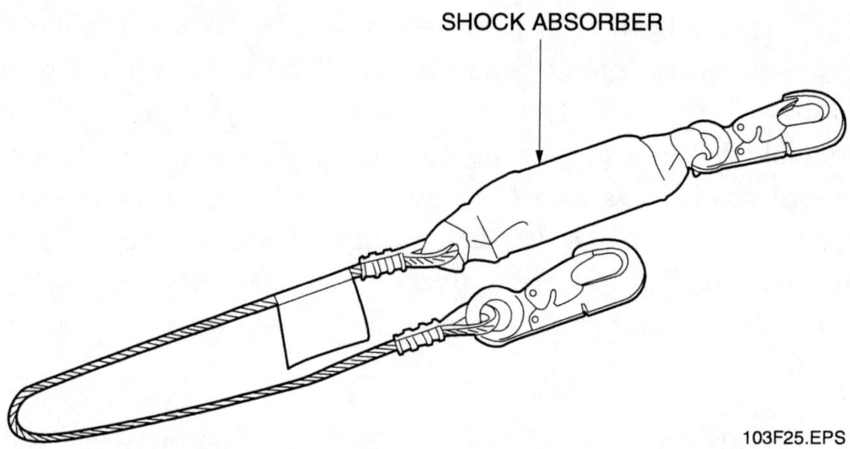

Figure 25. Typical Shock-Absorbing Lanyard

5.1.3 Deceleration Devices

Deceleration devices limit the arresting force that a worker is subjected to when the fall is stopped suddenly. Rope grabs and self-retracting lifelines are two common deceleration devices. A rope grab (*Figure 26*) connects to a lanyard and attaches to a lifeline. In the event of a fall, the rope grab is pulled down by the attached lanyard, causing it to grip the lifeline and lock in place. Some rope grabs have a mechanism that allows the worker to unlock the device and slowly descend down the lifeline to the ground or surface below.

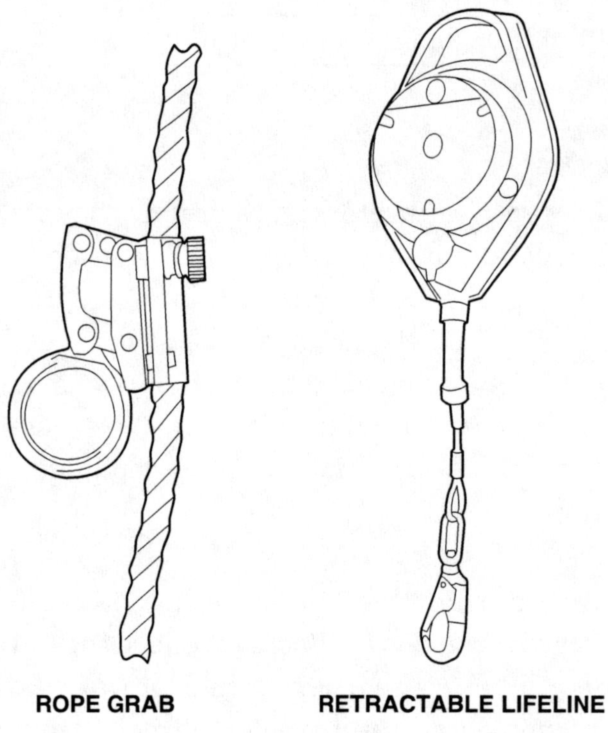

Figure 26. Deceleration Devices

Self-retracting lifelines (*Figure 26*) provide for unrestricted movement and fall protection while climbing and descending ladders, etc., or when working on multiple levels. Typically, they have a 25-foot to 100-foot galvanized steel cable that automatically takes up the slack in the attached lanyard, keeping the lanyard out of the worker's way. In the event of a fall, a centrifugal braking mechanism engages to limit the worker's free fall. Per OSHA requirements, self-retracting lifelines and lanyards that limit the free fall distance to 2 feet (0.61 meter) or less must be able to support a minimum tensile load of 3,000 pounds (13.3 kilonewtons). Those that do not limit the free fall to 2 feet (0.61 meter) or less must be able to hold a tensile load of at least 5,000 pounds (22.2 kilonewtons).

5.1.4 Lifelines

Lifelines are normally flexible steel cables that are attached to an anchorage. They provide a means for tying off personal fall protection equipment. Vertical lifelines are suspended vertically from a fixed anchorage at the upper end to which a fall arrest device such as a rope grab is attached. Vertical lifelines must have a minimum breaking strength of 5,000 pounds (22.2 kilonewtons). Each worker must use his or her own line. This is because if one worker falls, the movement of the lifeline during the fall arrest may also cause the other workers to fall. Vertical lifelines must be terminated in a way that will keep the worker from moving past its end, or must extend to the ground or the next lower working level.

Horizontal lifelines are connected horizontally between two fixed anchorages to which a fall arrest device is attached. Horizontal lifelines must be designed, installed, and used under the supervision of a qualified and competent person. The more workers tied off to a single horizontal line, the stronger the line and anchors must be.

5.1.5 Anchoring Devices And Equipment Connectors

Anchoring devices, commonly called *tie-off points*, support the entire weight of the fall arrest system. The anchorage must be capable of supporting 5,000 pounds (22.2 kilonewtons) for each worker attached. Eye bolts, overhead beams, and integral parts of building structures are all types of anchorage points.

The D-rings, buckles, snaphooks, etc. that fasten and/or connect the parts of a personal fall arrest system are called *connectors*. OSHA regulations specify how they are to be made, and require D-rings and snaphooks to have a minimum tensile strength of 5,000 pounds (22.2 kilonewtons). All such components should be designed for use with the attached hardware. As of January 1, 1998, only locking-type snaphooks are permitted for use in personal fall arrest systems.

5.2.0 PROCEDURES FOR SAFELY USING PERSONAL FALL ARREST EQUIPMENT

Before using fall protection equipment on the job, your employer should provide you with training in the basics of fall protection and the proper use of the equipment. All equipment supplied by your employer must meet OSHA standards for strength. Before each use, always read the instructions and warnings on any fall protection equipment. Inspect the equipment using the following guidelines:

- Examine harnesses and lanyards for mildew, wear, damage, and deterioration.
- Make sure no straps are cut, broken, torn, or scraped.
- Check for damage due to fire, chemicals, or corrosives.
- Hardware should be free of cracks, sharp edges, or burrs.
- Snaphooks should close and lock tightly. Buckles should work properly.
- Check ropes for wear, broken fibers, pulled stitches, and discoloration.
- Make sure lifeline anchors and mountings are not loose or damaged.

Do not mix or match equipment from different manufacturers. All substitutions must be approved by your supervisor. All damaged or defective parts must be taken out of service immediately and tagged as unusable or destroyed. If the equipment is subjected to impact from a fall, remove it from service until it can be inspected by a qualified person.

5.2.1 Wearing A Full-Body Harness

General guidelines for wearing a sliding back D-ring full-body harness are given below:

Step 1 Hold the harness by the back D-ring, then shake the harness, allowing all the straps to fall into place.

Step 2 Unbuckle and release the waist and/or leg straps.

Step 3 Slip straps over the shoulders so that the D-ring is located in the middle of the back (*Figure 27*).

Step 4 Fasten the waist strap. It should be tight but not binding.

Step 5 At each leg, pull the straps between the legs and buckle.

Step 6 After all the straps have been buckled, tighten all friction buckles so that the harness fits snugly but allows full range of movement.

Step 7 Pull the chest strap around the shoulder straps and fasten in the mid-chest area. Tighten enough to pull the shoulder straps taut.

5.2.2 Selecting An Anchor Point And Tying Off

Once the full-body harness has been put on, the next step is to connect it either directly or indirectly to a secure anchor point. This is called *tying off*. Tying off is always done before you get into a position from which you can fall. Good practice is to follow the manufacturer's

103F27.EPS

Figure 27. Slip Straps Over Shoulders So D-Ring Is Located In The Middle Of The Back

instructions on the best tie-off methods for your equipment. Use the following general guidelines when selecting an anchorage and tying off.

When selecting an anchorage point, it should be:

- Directly above the worker
- Easily accessible
- Damage free and capable of supporting 5,000 pounds (22.2 kilonewtons) per worker
- High enough so that no lower level is struck should a fall occur

Be sure to check the manufacturer's equipment labels and allow for any equipment stretch and deceleration distance.

When tying off, consider the following:

- Tie-offs that use knots are weaker than other methods of attachment. Knots can reduce the lifeline or lanyard strength by 50% or more. A stronger lifeline or lanyard should be used to compensate for this effect.
- To protect equipment from cuts, do not tie off around rough or sharp surfaces. Tying off around H-beams or I-beams can weaken the line because of the cutting action of the beam's edge. This can be prevented by using a webbing-type lanyard or wire-core lifeline.
- Never tie off in a way that would allow you to free fall more than 6 feet (1.83 meters).
- A shorter free fall can reduce your chances of falling into obstacles, being injured by the arresting force, and damaging your equipment. To limit your free fall, a shorter lanyard can be used between the lifeline and your harness. Also, the amount of slack in your lanyard can be reduced by raising your tie-off point on the lifeline. The tie-off to the lifeline or anchor must always be higher than the connection to your harness.

5.3.0 SAFETY NET SYSTEMS

Safety nets are used for fall protection on bridges and similar projects. They must be installed as close as practical beneath the work area, but not more than 30 feet (9.14 meters). OSHA requires enough clearance under a safety net to prevent a worker who falls into it from hitting the surface below. Also, there must be no obstruction between the work area and the net. Depending on the actual vertical distance between the net and the work area, the net must extend at least 8 to 13 feet (2.44 to 3.96 meters) beyond the edge of the work area. Mesh openings in the net are limited to 36 square inches (91.44 square centimeters) and 6 inches (15.24 centimeters) on the side. The border rope must have a 5,000-pound (22.2-kilonewton) minimum breaking strength, and connections between net panels must be as strong as the nets themselves. Safety nets must be inspected at least once a week and after any event that might have damaged or weakened them. Worn or damaged nets must be removed from service.

6.0.0 RESCUE AFTER A FALL

Every elevated job site should have a rescue and retrieval plan in case it is necessary to rescue a fallen worker. Planning is especially important in remote areas that are not readily accessible to a telephone. Before there is a risk of a fall, make sure that you know what your employer's rescue plan calls for you to do. Find out what rescue equipment is available and where it is located. Learn how to use equipment for self-rescue and the rescue of others.

If a fall occurs, any employee hanging from the fall arrest system must be rescued safely and quickly. Your employer should have previously determined the method of rescue for fall victims, which may include equipment that lets the victim rescue himself or herself, a system of rescue by coworkers, or a way to alert a trained rescue squad. If fall rescue depends on calling for outside help such as the fire department, rescue squad, etc., all the needed phone numbers must be posted in plain view at the work site. In the event a coworker falls, follow your employer's rescue plan. Call any special rescue service needed. Communicate with the victim and monitor him or her constantly during the rescue.

7.0.0 TESTING FALL PROTECTION SYSTEMS AND EQUIPMENT

The testing of fall arrest equipment should be performed regularly to make sure it is in compliance with OSHA requirements. Depending on the prevailing state and local laws, the tests may be either mandatory or voluntary. Guidelines for testing personal fall arrest equipment and systems are given in *OSHA Standards 1926*, Appendices C and D to Subpart M. A good practice is to tag or label all items of fall protection equipment with the date when the equipment was last tested and the date it is due for the next test.

Safety nets should be drop-tested at the job site after the initial installation, whenever relocated, after a repair, and at least every six months if left in one place. The drop test consists of a 400-pound (180-kilogram) bag of sand of 29 to 31 inches (71 to 81 cm) in diameter that is dropped into the net from at least 42 inches (1 meter) above the highest walking/working surface to which workers are exposed to fall hazards.

SUMMARY

Ladders, scaffolds, and/or lifts are used daily by painters and decorators to gain access to elevated locations in order to perform surface preparation and application of paint or wallcovering. *OSHA Standards 1926*, Subpart X defines and governs the use of ladders. There are many types of ladders, each designed for a specific purpose and climbing condition. Portable ladders are ladders that can be moved or carried around the job site. Common portable ladders include:

- Stepladders
- Single ladders
- Extension ladders
- Trestle and extension trestle ladders

Scaffolds are temporary built-up frameworks or suspended platforms designed to support workers, materials, and equipment at elevated or otherwise inaccessible locations. *OSHA Standards 1926*, Subpart L defines and governs the use of scaffolds. There are many kinds of scaffolds, each designed for a specific use and job site condition. Common scaffolds include:

- Built-up scaffolds
- Swing scaffolds
- Beam-suspended scaffolds

Aerial work platforms and scissor power lifts are portable aerial lifts used to raise or lower workers to and from elevated job sites. *OSHA Standards 1926*, Subpart N defines and governs the use of aerial lifts.

Falls from ladders, scaffolds, and lifts can cause serious injury or death when no fall protection equipment or the wrong kind of protection equipment is used, or when the right equipment is used improperly. Three types of fall protection equipment are commonly used while working at elevated levels: guardrails, personal fall arrest systems, and/or safety nets. The types and uses of personal fall arrest systems and safety nets are governed by *OSHA Standards 1926*, Subpart M. The rules covering guardrails on scaffolds are contained in *OSHA Standards 1926*, Subpart L. Basically, these OSHA standards require that all workers use guardrail systems, safety net systems, or personal fall arrest systems to protect themselves from free falling more than 6 feet (1.8 meters) and hitting the ground or a lower work level.

References

For advanced study of topics covered in this task module, the following books are suggested:

Basic Safety, Core Curricula.

Occupational Safety and Health Standards for the Construction Industry - 29 CFR, Part 1926, U.S. Government Printing Office, Washington, DC.

Painting and Decorating Craftsman's Manual and Textbook, Painting and Decorating Contractors of America, Fairfax, Virginia.

Preventing the Fall: a Compliance Guide for OSHA's Fall Protection Standard, American Subcontractors Association, Inc./Foundation of the American Subcontractors Association, Inc., Alexandria, Virginia.

Supervisor's Safety Manual, Painting and Decorating Contractors of America, Fairfax, Virginia.

REVIEW/PRACTICE QUESTIONS

1. OSHA requires that a ladder or stairway be provided where levels of a work area are separated by more than:
   a. 12 inches (30.48 cm).
   b. 24 inches (60.96 cm).
   c. 19 inches (48 cm).
   d. 36 inches (91.44 cm).

2. Portable ladders must support at least _____ times their maximum intended load.
   a. three
   b. four
   c. five
   d. six

3. A ladder marked with a Type III duty rating is designed to carry a maximum weight of:
   a. 200 pounds (90.72 kilograms).
   b. 300 pounds (136.07 kilograms).
   c. 250 pounds (113.39 kilograms).
   d. 225 pounds (102.06 kilograms).

4. Which ladder accessory may be used in pairs and attached to the side rails of straight or extension ladders to form a light-duty scaffold?
   a. Ladder hooks
   b. Ladder shoes
   c. Ladder stabilizers
   d. Ladder jacks

5. An extension ladder is resting against a wall with its rails on a support point 20 feet (6.09 meters) above the ground. The feet of the ladder should be positioned _____ from the wall.
   a. 3 feet (0.91 meter)
   b. 4 feet (1.22 meters)
   c. 5 feet (1.52 meters)
   d. 6 feet (1.82 meters)

6. A scaffold designed for medium-duty use can safely support loads up to and including _____ pounds per square foot.
   a. 50
   b. 75
   c. 100
   d. 25

7. Built-up scaffolds must be able to support at least _____ times their maximum intended load.
   a. three
   b. four
   c. five
   d. six

8. A scaffold that is 10 feet above the ground or working surface must be equipped with:
   a. toeboards.
   b. guardrails.
   c. gates.
   d. both a and b.

9. Rigging devices used to support a swing scaffold that extends beyond the edge of a roof are called:
   a. cornice hooks.
   b. roof hooks.
   c. outriggers.
   d. parapet clamps.

10. A swing scaffold designed for a working load of 750 pounds (337.5 kilograms) can safely hold up to _____ workers.
    a. four
    b. three
    c. two
    d. one

11. The kind of scaffolding most probably used when painting the underside of a bridge is:
    a. beam-suspended.
    b. tubular welded frame.
    c. tube and coupler.
    d. swing.

12. A light-duty or medium-duty scaffold that is suspended from outrigger beams anchored inside a building is called:
    a. a two-point suspension scaffold.
    b. an outrigger scaffold.
    c. a window jack scaffold.
    d. a multiple-point suspension scaffold.

13. A Bosun's chair is an example of a:
    a. swing scaffold.
    b. multiple-point suspension scaffold.
    c. two-point suspension scaffold.
    d. single-point suspension scaffold.

14. A scaffold you want to use is marked with a yellow tag. This means that:
    a. it is safe for use.
    b. it is in the process of being erected or taken down and should not be used.
    c. you must wear fall protection equipment when working on the scaffold.
    d. only authorized persons can work on the scaffold.

15. A cherry picker is a type of:
    a. aerial lift.
    b. manufactured built-up lift.
    c. boom lift.
    d. both a and c.

16. Use of guardrails and personal fall arrest systems are necessary to prevent workers from free falling more than:
    a. $3\frac{1}{2}$ feet (1.07 meters).
    b. 6 feet (1.83 meters).
    c. 10 feet (3.05 meters).
    d. 8 feet (2.44 meters).

17. OSHA requires that the deceleration distance for a fall arrest system be no more than:
    a. $3\frac{1}{2}$ feet (1.07 meters).
    b. 2 feet (0.61 meters).
    c. 4 feet (1.22 meters).
    d. 6 feet (1.83 meters).

18. Full-body harnesses equipped with front chest D-rings are typically designed for use in:
    a. personal fall arrest systems.
    b. ladder climbing systems.
    c. personal positioning systems.
    d. both b and c.

19. Body belts should not be used in personal fall arrest systems because:
    a. the fall arresting forces are concentrated in the victim's abdominal area.
    b. they are banned from use as of January 1, 1998.
    c. both a and b.
    d. they have a weight limit of 3,000 pounds (13.3 kilonewtons).

20. The entire weight of a personal fall arrest system is supported by the:
    a. harness.
    b. anchoring device.
    c. deceleration device.
    d. lanyard.

ACKNOWLEDGMENTS

Figures 1, 16, and 20 courtesy of Painting and Decorating Contractors of America.

NCCER CRAFT TRAINING USER UPDATES

The NCCER makes every effort to keep these manuals up-to-date and free of technical errors. We appreciate your help in this process. If you have an idea for improving this manual, or if you find an error, a typographical mistake, or an inaccuracy in the NCCER's Craft Training Manuals, please write us, using this form or a photocopy. Be sure to include the exact module number, page number, a description of the problem, and the correction, if possible. Your input will be brought to the attention of the Technical Review Committee. Thank you for your assistance.

Instructors – If you found that additional materials were necessary in order to teach this module effectively, please let us know so that we may include them in the Equipment/Materials list in the Instructor's Guide.

Write: Curriculum and Revision Department
National Center for Construction Education and Research
P.O. Box 141104
Gainesville, FL 32614-1104

Fax: 352-334-0932

Craft ______________________ Module Name ______________________

Module Number ______________________ Page Number(s) ______________________

Description of Problem

____

____

____

(Optional) Correction of Problem

____

____

(Optional) Your Name and Address

____

notes

Identifying Surface/Substrate Materials and Conditions

Module 07104

Painting Trainee Task Module 07104

IDENTIFYING SURFACE/ SUBSTRATE MATERIALS AND CONDITIONS

PAINTING AND DECORATING CONTRACTORS OF AMERICA

NATIONAL CENTER FOR CONSTRUCTION EDUCATION AND RESEARCH

Objectives

Upon completion of this module, the trainee will be able to:

1. Identify various substrates used in construction:
   - Wood
   - Masonry, concrete, and stucco
   - Plaster/drywall
   - Synthetic
   - Metal
2. Identify the surface condition of substrates and coatings:
   - New
   - Aged
   - Previously coated
3. Be aware of the basic surface preparation methods and coatings required for various substrates.

Prerequisites

Successful completion of the following Task Modules is required before beginning study of this Task Module: Core Curricula, Painting Level 1, Modules 07101 through 07103.

Required Trainee Material

1. Trainee Module
2. Appropriate Personal Protective Equipment

Course Map

This course map shows all of the *Wheels of Learning* task modules in the first level of the Painting curricula. The suggested training order begins at the bottom and proceeds up. Skill levels increase as a trainee advances on the course map. The training order may be adjusted by the local Training Program Sponsor.

Course Map: Painting, Level 1

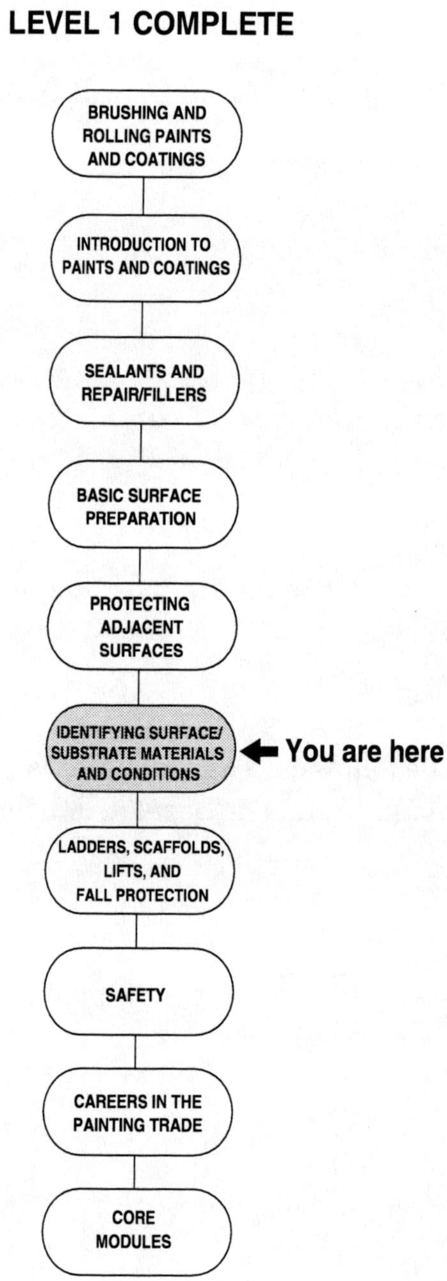

CMAP104.EPS

TABLE OF CONTENTS

Trade Terms Introduced In This Module

Alkaline cleaner: A cleaner that saponifies (converts oil, grease, or organic compounds to a water-soluble soap) and washes away other types of contaminants.

Alligatoring: A paint surface defect that forms cracks resembling alligator hide. Also known as *crocodiling*.

Blasting: Short term for abrasive blasting using compressed air or pressurized water.

Block filler: A heavily pigmented coating used to fill the pores and voids of cinder or concrete block.

Calcimine: A finish composed of water, calcium carbonate, and glue or size that was used in older construction. It was used primarily on plaster ceilings, but was also used on walls or interior masonry and sometimes tinted for decorative purposes.

Chalking: The decomposition of a paint film into a loose powder on the film surface.

Concrete: A mixture of portland cement, aggregates, water, and, at times, special additives.

Consolidant: A liquid epoxy consisting of a resin and hardener that, when mixed properly, can be used to repair rotted, dried out, or spongy wood. When dry, it may be sanded, sawed, drilled, and finished like wood.

Detergent: A synthetic organic cleaning agent that is liquid or water soluble and has wetting and emulsifying properties.

Efflorescence: An encrustation of soluble salts, commonly white, deposited on the surface of coatings, stone, brick, plaster, or mortar. It is caused by the leaching of salts or alkalis from mortar or adjacent concrete as moisture moves through them.

Laitance: A thin, weak, brittle layer of cement on a concrete surface. It is usually caused by an overly wet or overworked mixture.

Neutralized: The process used to eliminate excess alkalinity or acidity in concrete or masonry. It is measured in units of pH, with 7 being the desired neutral value.

Parge coat: A mixture of water, portland cement, and sand that is spread onto masonry or poured concrete walls in a thin layer to provide a seal and a smooth finish.

pH: A measure of the alkalinity or acidity of a liquid or solid material. A pH reading of 7 is neutral; greater than 7 is alkaline (basic); less than 7 is acidic. The further from 7 the reading, the more acidic or alkaline the liquid or material is.

Primer: 1. The first coat in a painting operation. It is designed to promote adhesion of subsequent coats. 2. A coating applied to a substrate to improve the adhesion of wallcovering adhesive.

Profile: Contour or roughness of a surface. Also known as *surface profile*, *anchor profile*, or *anchor pattern*.

Substrate: Any material to which paint, coating, or wallcovering is applied.

Surface: 1. The portion of the substrate to which paint, coating, or wallcovering is applied. 2. The finish obtained after paint, coating, or wallcovering has been applied.

Wallcovering: Any type of paper, vinyl, fabric, or specialty material that is pasted onto a wall or ceiling for decoration and/or protection.

Wash coat: A thinned coat of paint or stain applied as a primer prior to the final finish.

White coat: A thin, fine plaster coat, also known as a putty coat, applied over a coarse plaster base to obtain the final plaster finish. Years ago, this coat was sometimes tinted to the desired room color to eliminate painting.

1.0.0 INTRODUCTION

One of the most important tasks of the painter is to properly identify the surfaces and substrates that are to be painted. Without a knowledge of **surface** (the portion of substrate to which paint, coating, or wallcovering is applied or the finish after they have been applied) and **substrate** (the material to which paint, coating, or wallcovering is applied) types, it would be difficult to choose compatible coating systems. Before identifying the surfaces that are to be painted, we must first examine the environment in which the surface exists, exterior or interior. Next, we must determine if the substrate is made of wood, metal, masonry/concrete, plaster/drywall, or synthetics. Lastly, we must determine if the surface of the substrate is new, aged, or previously coated. This module discusses all three of these important tasks. Basic substrate and surface preparation, while outlined briefly in this module, is covered in more detail in a later module.

2.0.0 NEW EXTERIOR SUBSTRATES

The primary substrates used in exterior construction are:

- Wood
- Metal
- Masonry, concrete, and stucco
- Synthetic

Outside surfaces of these substrates are subjected to a variety of extremes. Temperatures may vary from very hot in the summer to extremely cold in the winter. Wind, rain, hail, sleet, snow, and the sun's ultraviolet rays also attack the surface and affect the life and appearance of the coating. In urban areas, chemicals produced by vehicles and factories further attack and destroy the coating. Some coatings are subjected directly to very corrosive materials such as in chemical plants.

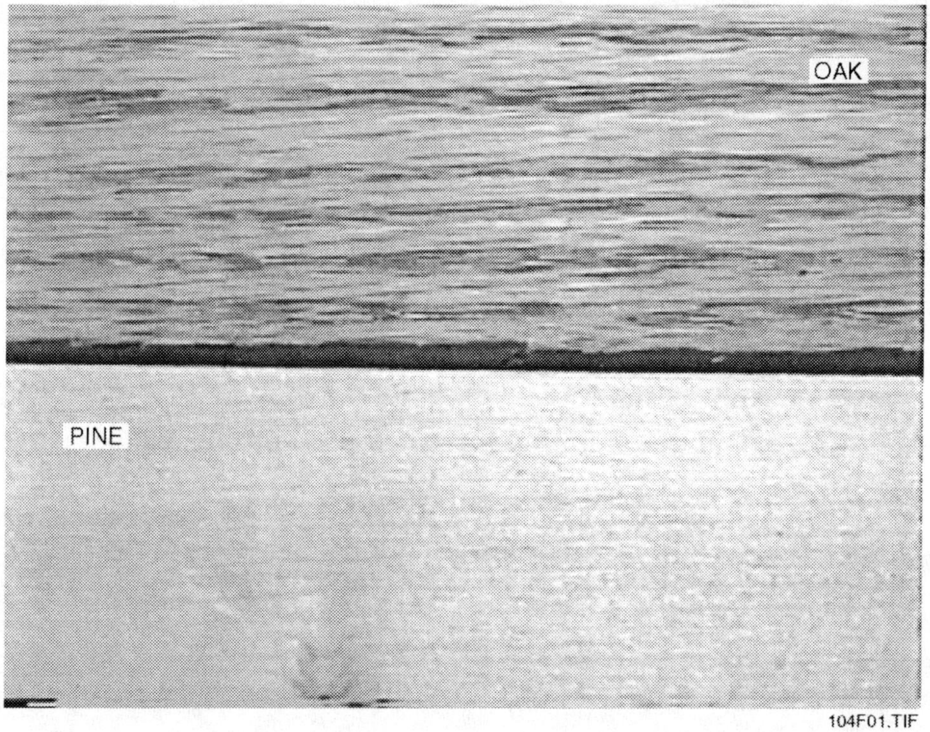

Figure 1. Close-Up View Of Open-Grain Oak And Close-Grain Pine

TYPE	COLOR	GRAIN	HARDNESS	DECAY RESISTANCE	USES	OTHER CHARACTERISTICS
Red Cedar	Dark Reddish Brown	Close Medium	Soft	Very High	Exterior	Cedar Odor
Cypress	Orange Tan	Close Medium	Soft to Medium	Very High	Exterior	
Fir	Yellow to Orange Brown	Close Coarse	Medium to Hard	Medium	Framing Millwork Plywood	
Ponderosa Pine	White with Brown Grain	Close Coarse	Medium	Low	Millwork Trim	Pine Odor
Sugar Pine	Creamy White	Close Fine	Soft	Low	Patternmaking Millwork	Large Clear Pieces
Western White Pine	Brownish White	Close Medium	Soft to Medium	Low	Millwork Trim	
Southern Yellow Pine	Yellow Brown	Close Coarse	Soft to Hard	Medium	Framing Plywood	Large Amount of Pitch
Redwood	Reddish Brown	Close Medium	Soft	Very High	Exterior	Light Sapwood
Spruce	Cream to Tan	Close Medium	Medium	Low	Siding Subflooring	Spruce Odor

Table 1. Common Types Of Softwood And Their Characteristics

2.1.0 EXTERIOR WOOD SUBSTRATES

Wood is classified as either *hardwood* or *softwood*. There are several different methods of classifying wood. The most common method is by its source. Hardwood comes from trees that shed their leaves each year. Softwood is cut from cone-bearing trees, commonly known as *evergreens*. In this method of classifying wood, some types of softwood may actually be harder than hardwood. For example, fir, a softwood, is harder and stronger than basswood, a hardwood.

Wood may also be divided into two groups according to cell structure. *Open-grain wood* (*Figure 1*) has large cells that show tiny openings or pores in the surface. All softwood is classified as *close-grain wood*. Refer to *Table 1* for types of softwood and their characteristics. *Table 2* lists common types of hardwood.

TYPE	COLOR	GRAIN	HARDNESS	DECAY RESISTANCE	USES	OTHER CHARACTERISTICS
Ash	Light Tan	Open Coarse	Hard	Low	Tool Handles Oars Baseball Bats	
Basswood	Creamy White	Close Fine	Soft	Low	Drawing Boards Veneer Core	Imparts No Taste or Odor
Beech	Light Brown	Close Medium	Hard	Low	Food Containers Furniture	
Birch	Light Brown	Close Fine	Hard	Low	Furniture Veneers	
Cherry	Light Reddish Brown	Close Fine	Medium	Medium	Furniture	
Hickory	Light Tan	Open Medium	Hard	Low	Tool Handles Diving Boards	
Luan	Light Reddish Brown	Open Medium	Soft	Low	Veneers Paneling	
Mahogany	Russet Brown	Open Fine	Medium	High	Quality Furniture	
Maple	Light Tan	Close Medium	Hard	Low	Furniture Flooring	
Oak	Light Brown	Open Coarse	Hard	Medium	Flooring Boats Millwork Trim	
Poplar	Greenish Yellow	Close Fine	Medium Soft	Low	Furniture Veneer Core	
Teak	Honey	Open Medium	Medium	Very High	Furniture Boat Trim	Heavy Oily
Walnut	Dark Brown	Open Fine	Medium	High	High Quality Furniture	

Table 2. Common Types Of Hardwood And Their Characteristics

Identifying different kinds of wood can be very difficult because some closely resemble each other. For example, ash and white oak are hard to distinguish from each other, as are some pine and spruce. Not only are they the same color, but the grain pattern and weight are about the same. Only the most experienced workers are able to tell the difference.

It is possible to get some clues to identifying wood by checking its color and grain, pressing it with your fingernail to see if it dents easily (softwood) or barely dents (hardwood), or by smelling a newly-sanded section for a characteristic odor. Most exterior wood is used in residential environments and thus is subject only to normal weathering and atmospheric conditions.

Construction wood, including plywood, particle or flake board, composite wood siding, pressure-treated wood, and fire-retardant wood is generally made of porous softwood. This usually means that it must be sealed prior to being coated. If wood has a high oil content, sealers must be used to keep it from bleeding. Many pine varieties contain a sap or *pitch* which bleeds through many coatings (particularly water-based emulsion paints). Alkyd-oil sealers are usually used to prevent the bleeding. Pressure-treated wood should be allowed to dry sufficiently before coating. Fire-retardant treated wood must be coated with a fire-retardant coating.

Moisture content is also a problem with new wood surfaces. If wood is too wet, the moisture will try to escape through the surface. If the surface is waterproof, the moisture will probably lift the coating. New wood that has been subjected to rain should be allowed to dry sufficiently before applying a **primer**, which is the first coat of a painting operation. A moisture meter should be used to make sure that the moisture content is sufficiently low before applying primer. If new wood has been exposed to the sun's rays for more than several days, it should be sanded before applying a primer.

In general, all new wood should be primed before painting. Wood that bleeds must be primed with a primer-sealer before painting.

When considering coating systems for exterior wood surfaces, you must consider the following environmental factors:

- Level of sun exposure
- Extent of heat or cold exposure
- Amount of humidity
- Other atmospheric or environmental (e.g., corrosive) conditions

2.2.0 EXTERIOR METAL SUBSTRATES

Both ferrous (steel) and non-ferrous (no iron content) metals are used in construction today. If the metal substrate is not known, it may be generally classified using a magnet. A magnet will be attracted to ferrous metals but will have little or no attraction to non-ferrous metals, such as brass. Carefully sanding a hidden portion of a metal to the bare surface may reveal zinc or tin plate on steel or the color and surface characteristics of non-ferrous metals. Non-ferrous metals include:

- *Aluminum* – Whitish with a dull and soft surface easily scored by sanding.
- *Stainless steel* – Shiny like steel with a relatively hard surface.
- *Copper* – Characteristic bright copper color.
- *Brass* – Generally bright yellow or gold color.
- *Bronze* – Reddish yellow in color (between brass and copper).

2.2.1 Steel

Steel has three contaminants that must be removed prior to painting (*Figure 2*). The first one is rust, which is caused by oxidation of the metal. The second is mill scale or *slag*, which is produced during manufacture. The third is a combination of oil and dirt, which is caused by handling.

Contaminants can be removed in a variety of ways, including hand sanding, power tools, chemical cleaning, or abrasive blasting. These techniques will be discussed in detail in other modules.

Once the surface has been cleaned, a primer is selected and the surface is primed. The selection of the primer is dependent on several items, such as the surface corrosion pretreatment required, the top coating to be used, and the environment in which the metal surface exists.

Figure 2. Steel Substrate

Generally, alkyd-oil paints and latex-emulsion paints are used in mild environments. Where heavy corrosion is a concern, corrosion treatment (using phosphate conversion and/or zinc chromate coatings) is applied before the primer. The primer is followed by top coats (where legally permitted) of epoxy, urethane, silicone, chlorinated rubber, vinyl, or phenolics.

2.2.2 Aluminum

Aluminum is a whitish, soft metal that oxidizes much more slowly than steel. Its oxidation actually provides a protective coating against further corrosion. These oxides sometimes provide a good surface for paint so little preparation is needed after surface cleaning. Aluminum is cleaned primarily by hand or mechanical means. This can be done using power tools, **blasting** (compressed air abrasive blasting or pressurized water), or chemical cleaning. New aluminum may need to be etched by chemical means to provide the proper **profile** (contour or roughness of a surface) for paint. Zinc chromates are excellent primers for aluminum; however, there are many others. Be sure to follow the coating manufacturer's recommendations.

2.2.3 Galvanized Steel

Galvanized (zinc coated) steel is often used for siding, fencing, and roofing (*Figure 3*). The zinc coating itself is rust resistant. However, galvanized steel coats better if it is weathered. Washing it with an etching solution helps to "weather" or give tooth to the surface. Special primers are also available.

104F03.TIF

Figure 3. Close-Up View Of Galvanized Steel Substrate

2.2.4 Copper, Brass, Bronze, And Stainless Steel

Most other metals, such as copper, brass, bronze, and stainless steel do not need to be coated for protection. However, a clear finish may be applied to a clean surface to provide sheen and abrasion protection.

2.3.0 MASONRY, CONCRETE, AND STUCCO SUBSTRATES

These substrates are found in residential, commercial, and industrial construction. **Concrete**, which is a mixture of water, portland cement, aggregates, and at times, special additives, is used for foundation walls, ceilings, and floors. It may be in either slab or block form (*Figure 4*). Brick, stone, stucco, gunnite, and fiber-cement panels are used for walls and veneers. Brick and stone are also used for floors.

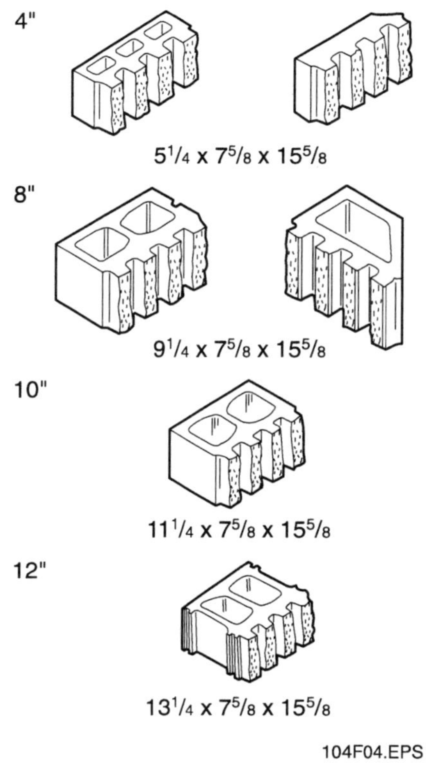

Figure 4. Decorative Concrete Block

Masonry products all produce alkali and require the use of alkali-resistant coatings. Where concrete or gunnite is used for water immersion such as swimming pools, waterproof coatings like chlorinated rubber and epoxies are used where permitted.

New masonry, stucco, gunnite, or concrete must cure for at least 28 days to eliminate moisture. The surfaces must then be cleaned of any **efflorescence** (salts deposited on the surface that are leached from mortar and concrete by water), curing compounds, and form release agents. Concrete, gunnite, stucco, fiber-cement panels, or masonry walls should be mechanically or hand cleaned and should not be acid etched. All cracks and voids must be filled (*Figure 5*).

This can be done using a **parge coat** (a mixture of water, portland cement, and sand) or a commercial **block filler** (a heavily pigmented coating).

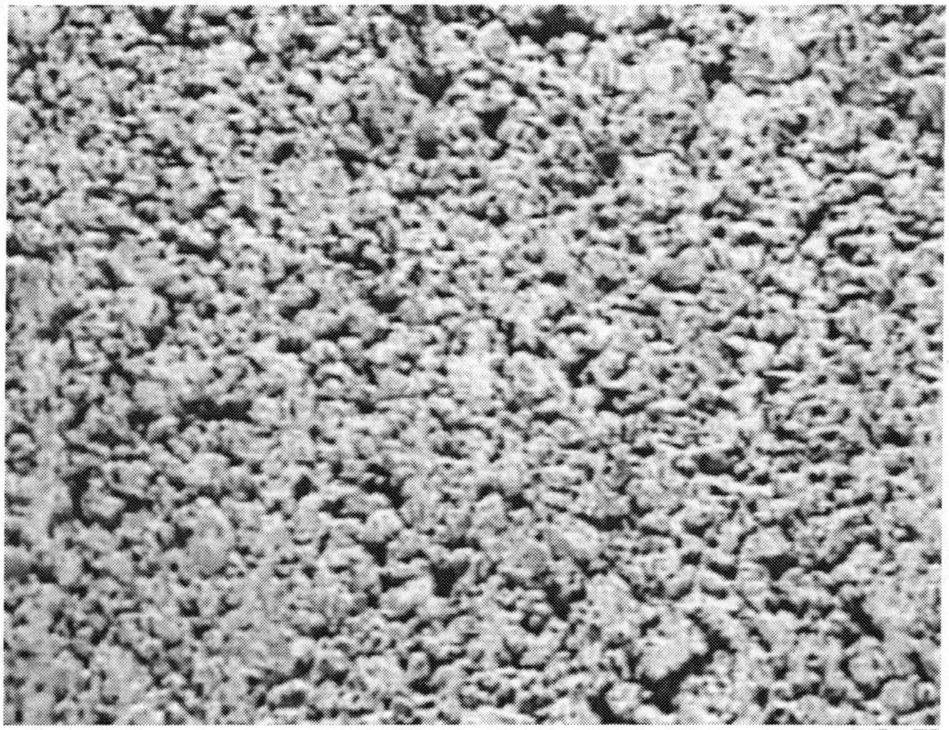

104F05.TIF

Figure 5. Close-Up View Of Voids In Concrete Block Substrate

The surface must be rinsed and checked for the proper **pH** range, which is the measure of acidity or alkalinity of the surface. After the surface is dry, it must be checked for moisture content before being coated.

New concrete floors should be blasted or cleaned with hand or power tools. Also, they may be acid etched to remove **laitance** (a thin, brittle layer of cement on a concrete surface) and loose surface material and provide the proper profile (*Figure 6*). A good surface is obtained when water will soak into it. Concrete is usually etched with muriatic acid solution and then neutralized.

After the surface has been rinsed and checked for an acceptable pH range and has thoroughly dried, it must be checked for moisture content before being coated.

New brick should be allowed to age for one year before coating. Brick surfaces can be cleaned with a **detergent** or by blasting. Acid etching is not recommended for vertical surfaces. Coatings used for brick are (where legally permitted) acrylics, epoxy, urethane, chlorinated rubber, silicone, and phenolics.

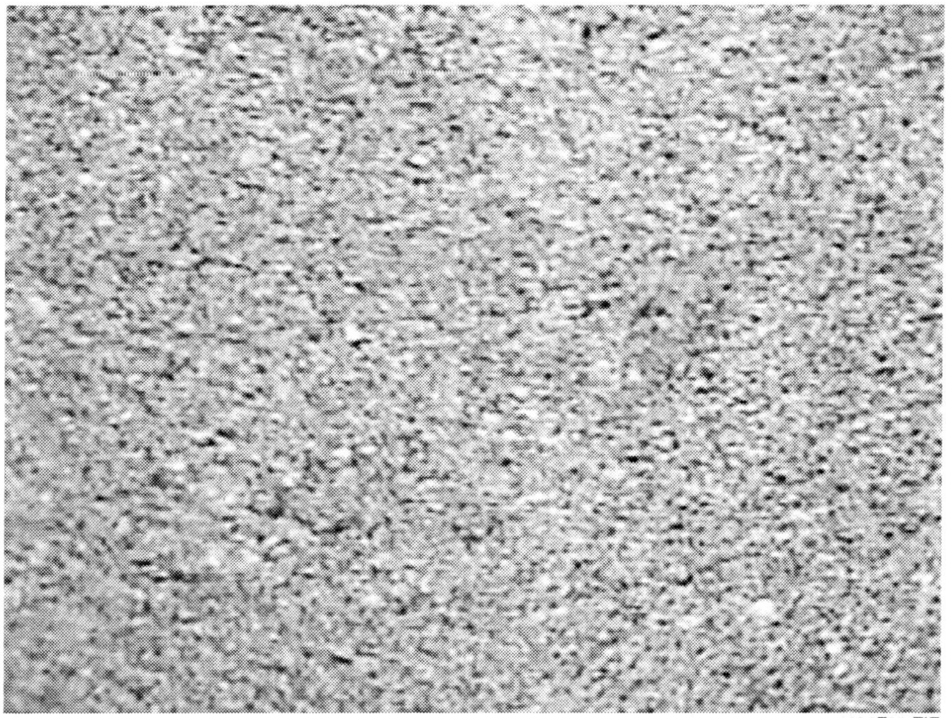

104F06.TIF

Figure 6. Close-Up View Of Etched Concrete Substrate Showing Fine Sandpaper Effect (Profile)

2.4.0 SYNTHETIC SUBSTRATES

A painter is sometimes required to work on a synthetic (man-made) surface. It may be made of fiberglass or various types of plastic. Vinyl siding is an example of a synthetic surface.

For proper preparation and finishing, the type of plastic should first be identified. Next, reference books or the manufacturer should be consulted to obtain instructions on surface repair, preparation, priming, and painting. Some plastics, such as teflon, do not coat well; others, such as flexible vinyl, require special primers or paint additives. Vinyl siding and plastic (ABS, PVC, etc.) pipe can usually be coated with a high-quality latex primer and paint after the surface is scuff-sanded with sandpaper or deglossed with a chemical deglosser.

For new construction, the general contractor or architect should be able to furnish the information necessary to identify the plastic.

3.0.0 NEW INTERIOR SUBSTRATES

The primary substrates of interior construction are:

- Drywall/plaster
- Wood
- Masonry, concrete, or stucco
- Metal
- Synthetic

Generally, interior substrates are not as troublesome as exterior substrates. They are not subjected to extreme weather and atmospheric conditions, and usually experience less chemical interference. When preparing or painting interior substrates, you must take precautions to prevent dust or vapor contamination of adjacent areas. Occupants in the work area or in adjacent areas are often relocated while work is in progress.

3.1.0 PLASTER SUBSTRATES

Two-coat plaster (*Figure 7*) is becoming less common in new construction. New, wet plaster has a high alkalinity which deteriorates oil-based coatings. Wet plaster should be allowed to cure thoroughly before coating. This may take anywhere from 4 to 5 weeks. Plaster should be checked for an acceptable pH range. It must also be checked for moisture before being coated. If the plaster is dry and has an acceptable pH, water-based emulsion paints should be used or acrylic primer sealers should be applied.

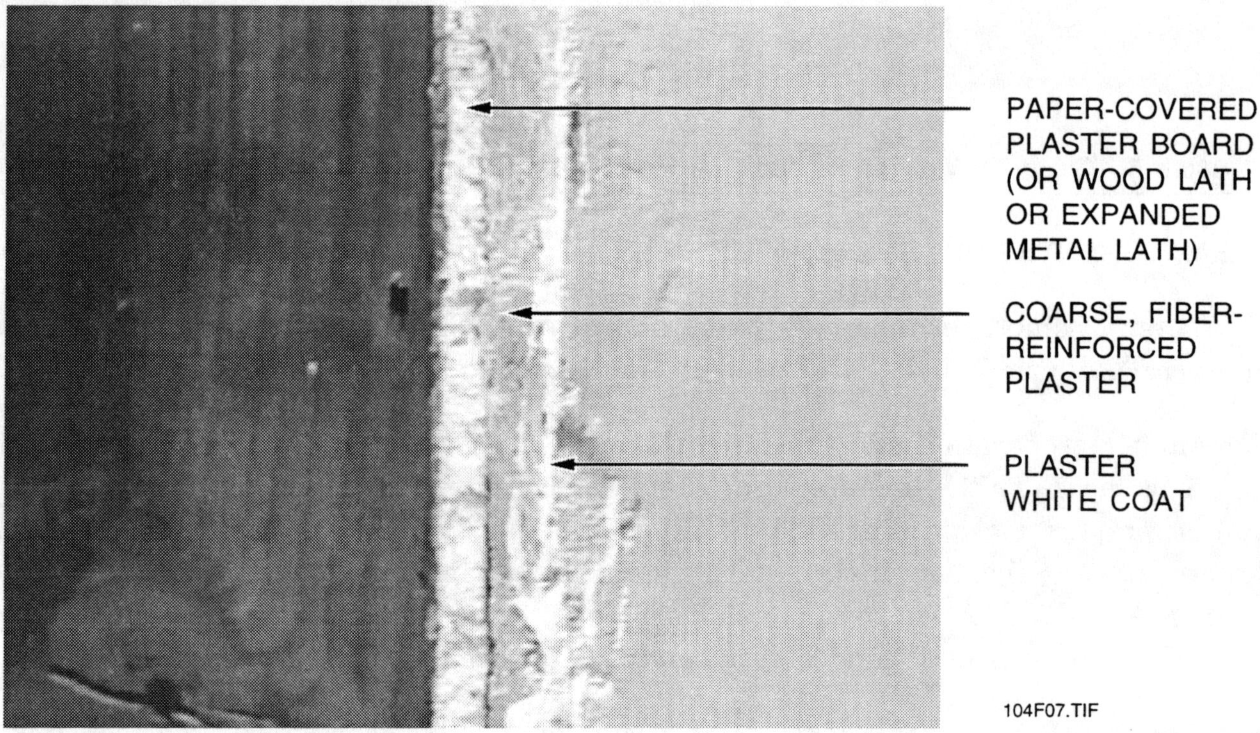

Figure 7. Typical Two-Coat Plaster System

3.2.0 DRYWALL (WALLBOARD) SUBSTRATES

Today, most interior walls are constructed of drywall. Drywall is a sandwich of gypsum in between coverings of paper. These materials are produced in sheets of various sizes and thicknesses. They are screwed, nailed, or glued to wall studs and finished with paper or fiber-mesh tape and/or joint compound to eliminate joints and fastener heads (*Figures 8* and *9*).

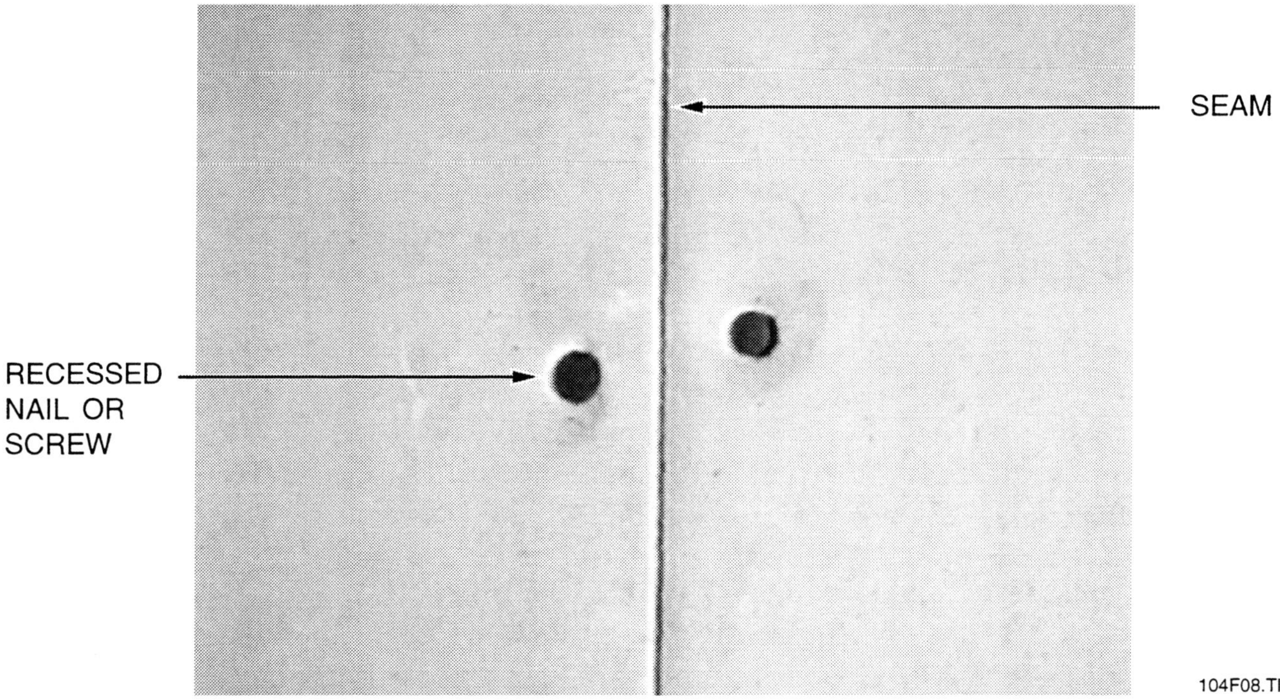

Figure 8. Unfinished Drywall Seam

Figure 9. Recessed Drywall Nail Or Screw Covered By Joint Compound And Sanded Flush With Surface

When painting drywall, some water-based paints may be applied directly to the surface without primer. Oil-based paints should be used with the correct primers and sealers. Under critical lighting conditions or when a high-luster finish or a thin wallcovering will be applied, drywall is covered with a thin skim (veneer) coat of joint compound and then lightly touched up by sanding to provide a very smooth surface.

3.3.0 WOOD SUBSTRATES

Interior wood substrates are generally not exposed to as harsh an environment as exterior wood substrates. However, surface preparation and paint selection may be just as demanding. *Tables 1* and *2* described the characteristics of common types of wood. Oak, for example, is an open-grain hardwood. Because of this, oak will not produce a smooth surface unless it is filled. Pine, on the other hand, does not need to be filled.

An exception to this is where a stain and/or a clear finish is used. Clear finishes will not hide minor imperfections in the wood. In this case, surface preparation becomes critical. With stain finishes, a **wash coat** of clear stain base or clear sealer may have to be applied to softwoods first to prevent rapid or uneven stain absorption. This is generally not true of hardwoods.

3.4.0 MASONRY OR CONCRETE SUBSTRATES

New interior masonry and concrete are prepared and finished like exterior masonry or concrete.

3.5.0 METAL SUBSTRATES

New interior metal substrates are cleaned like exterior metal substrates. However, corrosion pretreatment is generally not required unless the metal will be subjected to a harsh interior environment. Generally, alkyd-oil or latex-emulsion primers and paints are used on interior metal.

3.6.0 SYNTHETIC SUBSTRATES

New interior substrates are identified and handled in the same way as new exterior synthetic substrates.

4.0.0 UNCOATED AGED SUBSTRATES

Aged substrates are harder to paint than new substrates. They may involve materials that have not been painted or coated since they were installed. An aged surface is painted like a new surface, except that the condition of the surface must be carefully analyzed.

For example, old wood has a tendency to dry out, decompose, and crack. Paint will not fill these cracks, so spackle, putty, or caulking must be used. In addition, decomposed wood must either be replaced or it must be repaired using **consolidants** (epoxy hardners) and fillers. Concrete and masonry tend to form a whitish powder on their surfaces that must be removed prior to painting. Ferrous and non-ferrous metals corrode as time passes, and deep pits develop on their surfaces. The pits and corrosion must be cleaned, filled, and smoothed or the paint may not cover them adequately (due to possible shrinkage) or may not adhere at all. *When painting aged surfaces, good surface preparation is critical.*

5.0.0 PREVIOUSLY COATED SUBSTRATES

This type of surface can either be one of the easiest or hardest surfaces to paint. If the old coating is in good condition, it is easy to repaint with only a minimum of preparation. If the old coating is in poor condition, it may be very difficult to repaint.

WARNING! Caution must be used when removing or preparing single or multi-layered coated surfaces applied prior to 1978. These coatings may contain lead or other toxic heavy metals. Test kits should be used to determine if any layer of the coating contains lead or other toxic metals. If so, the required environmental and personal protection must be used if pressure washing, abrasive blasting, power or hand dry sanding, or scraping is used in surface preparation.

WARNING! Asbestos was used in older residential, commercial, and industrial buildings as a flocked ceiling finish, plaster fiber, acoustical tile, fire-stop board, insulation on pipes or heating ducts, asbestos-cement siding or shingles, and other applications. Do not sand or abrade these surfaces when repainting unless samples have been laboratory tested and confirmed as asbestos free. If they contain asbestos and must be abraded, special permits and special environmental and personal protective equipment must be obtained and used.

5.1.0 PAINTED WOOD

When recoating exterior wood surfaces, many things must be considered. The type and condition of the existing coating is very important. Most paints tend to chalk (form a loose powdery film) with exposure to outside conditions. **Chalking** is a condition where the surface of the paint deteriorates to the point where it can be washed away by rain (*Figure 10*). Years ago, chalking was a property built into almost all exterior paints because it kept the surface clean and fresh looking. Today, paints are designed to be chalk resistant.

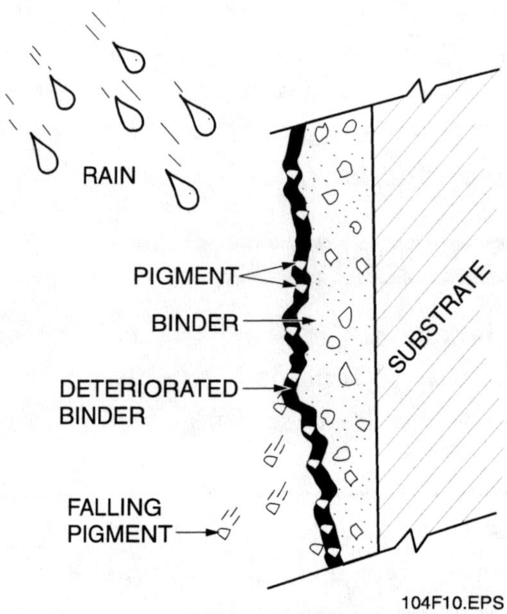

Figure 10. Chalking

Chalking must be removed completely or, if this is not possible, a penetrating primer/sealer must be applied when water-based paints are used. This is because water-based paints adhere mechanically to the surface. Chalk acts as a loose coating between the original coat and the new coat, which greatly hampers this adhesion.

Chalking is washed off with a strong detergent or alkaline cleaner either manually or by pressure washing. Chalking is a major problem and only one form of coating failure. Most other coating failures require removal of the old paint prior to repainting. Some of these conditions are checking, cracking, peeling, flaking, crawling, blistering, **alligatoring**, and wrinkling. Some of these are shown in *Figures 11* through *14*. These failures usually relate to age or improper surface preparation, mixing, and/or application. These conditions and their remedies are discussed in detail in a separate module.

104F11.TIF

Figure 11. Blistering And Later Stage Peeling (Inset)

104F12.TIF

Figure 12. Alligatoring And Early Stages Of Alligatoring (Inset)

104F13.TIF

Figure 13. Checking

104F14.TIF

Figure 14. Cracking And Later-Stage Flaking (Inset)

Interior painted wood surfaces are not usually subject to these failures except around windows. Generally, interior wood surfaces must be de-glossed using sandpaper or a chemical de-glosser and then thoroughly cleaned with detergent and water and allowed to dry before painting.

5.2.0 PAINTED METAL

Recoating metal surfaces involves many considerations. The surface may have a buildup of chemical salts which require chemical cleaning or removal. The surface may be rusting through and the coating may have to be partially removed. Generally, where a lot of rust is showing through on a steel surface, the entire surface must be cleaned. Other coating failures can result from improper preparation, coating selection, and/or application (*Figure 15*). If the substrate metal is unknown, the checks described previously should be used to determine the type of metal and hence the proper preparation and finish. Where coating failures are present, the cause of the failure must be found before recoating takes place.

5.3.0 PAINTED DRYWALL OR PLASTER

Recoating drywall surfaces requires proper cleaning of the old surface. If the surface is a semigloss or gloss surface, it should be "roughed up" with abrasive paper or chemical treatment before painting. All dirt and grease should be washed off with detergent and water. If repair work is done with joint compound, it should be sealed before finished coats are applied. On rare occasions, ceilings or walls may be encountered that have a **calcimine** finish (a finish made of water, calcium carbonate, and glue or size). To test this finish, wipe the surface with a wet cloth. If the finish comes off on the cloth, all traces of the calcimine finish must be removed before recoating the surface with any paint.

Recoating plaster surfaces is the same as for drywall unless the paint is peeling from the putty coat or **white coat** (a thin, fine plaster coat over a coarse plaster base) substrate (*Figure 16*). In this case, most of the original paint and primer may have to be removed and the surface cleaned. If the surface is soft, it may have to be hardened with a white-vinegar wash. After the plaster has hardened and dried, it should be refinished like new plaster.

5.4.0 PAINTED MASONRY, CONCRETE, OR STUCCO SUBSTRATES

Recoating concrete, masonry, gunnite, stone, fiber-cement, or stucco walls requires removing efflorescence, scale, loose alkali, and flaking coatings (*Figure 17*). Blasting, power tool cleaning, scraping, and wire brushing are all used on previously painted concrete, masonry, gunnite, stone, fiber-cement, and stucco walls. All voids and cracks should be filled before recoating the surface. In addition, acid etching may be used on concrete floors.

104F15.TIF

Figure 15. Peeling From Galvanized Metal Caused By Improper Substrate Preparation

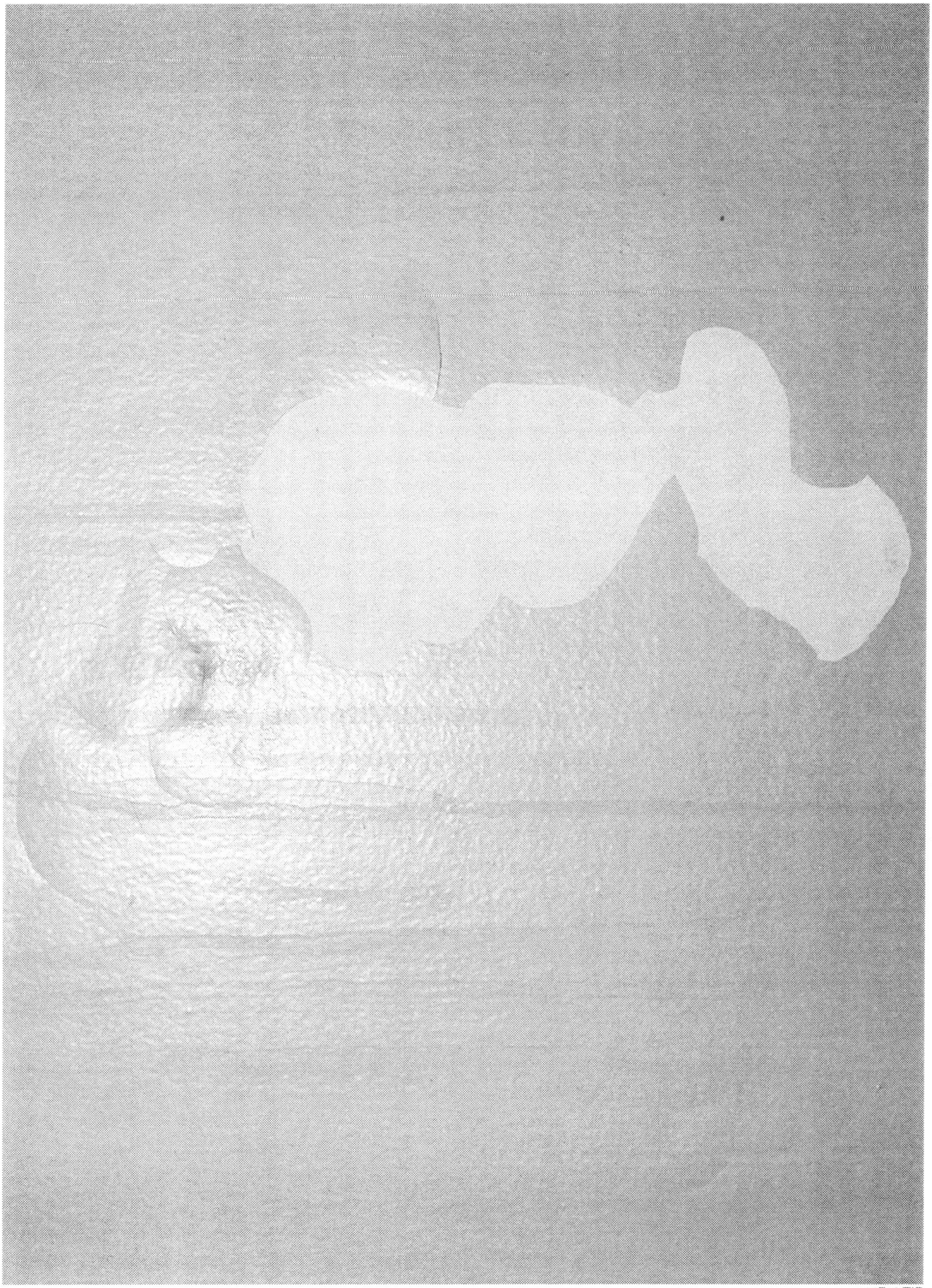

104F16.TIF

Figure 16. Peeling From Plaster Caused By Moisture Or Chalking Or Softening Of The Plaster White Coat

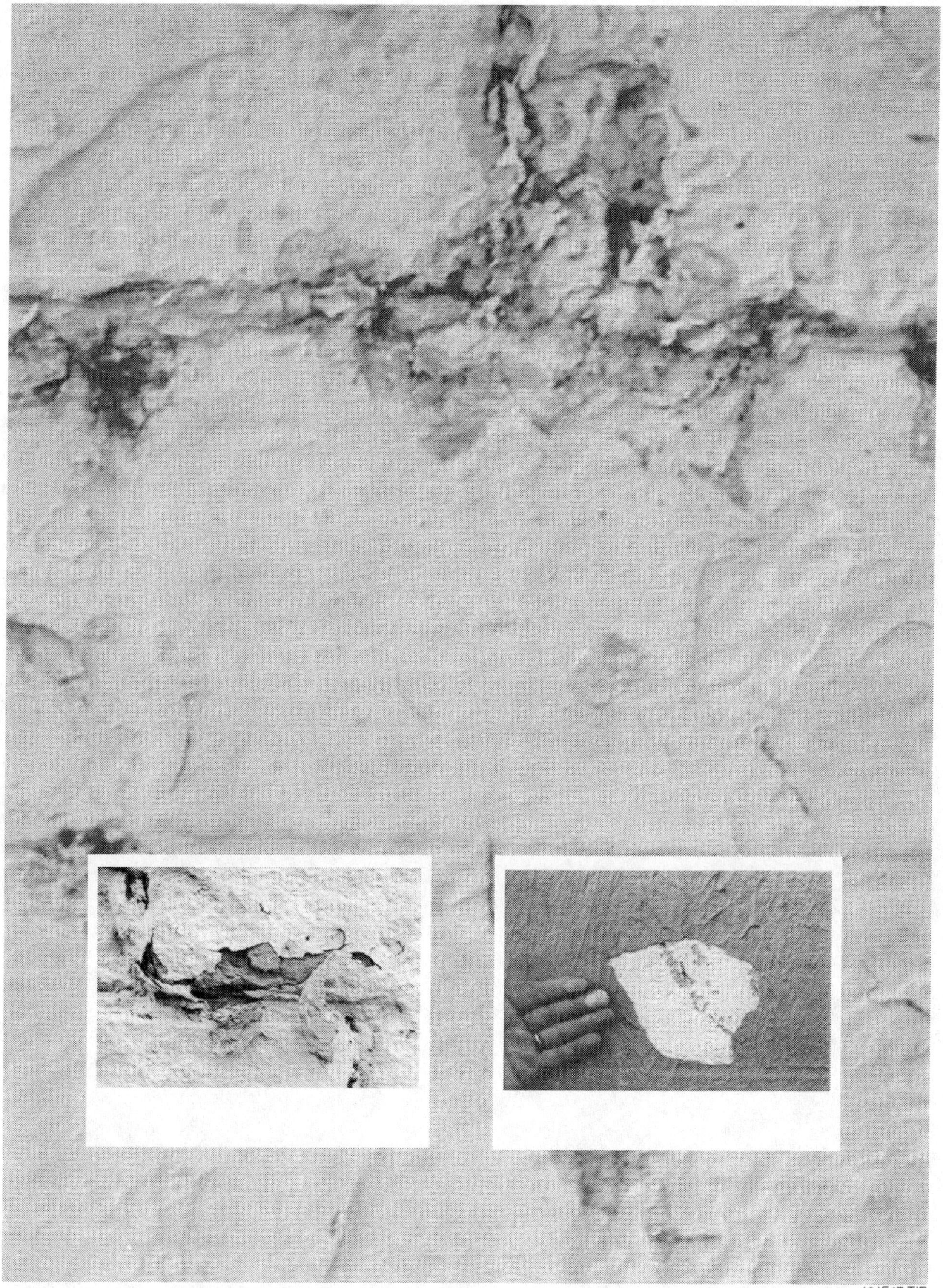

104F17.TIF

Figure 17. Peeling From Masonry Concrete Or Brick Wall Caused By Efflorescence Or Crumbling Substrate (Left Inset) Or Chalky Surface (Right Inset)

5.5.0 WALLCOVERED SUBSTRATES

Wallcoverings include any paper, vinyl, fabric, or other material pasted onto a wall or ceiling. Wallcovering should not be painted—it should be removed. If it has to be painted (often because it has been applied directly to drywall) extreme care must be used. Most wallcoverings are applied with a water-based or water-soluble glue or paste. Most latex paints contain water. Applying water-based paint to wallcovering usually results in the paste loosening and bubbles developing in the paint. If wallcovering has to be painted, it should first be cleaned with a mild detergent and painted with an appropriate primer/sealer. The primer/sealer will adhere to the wallcovering, seal the surface, and provide a tooth for the paint to adhere to. Textured or flocked wallcoverings cannot be painted without the texture or flock showing through.

5.6.0 PAINTED SYNTHETIC SUBSTRATES

If the synthetic substrate is rigid and the existing coating is in good condition, clean the surface and dull it before painting. If the substrate is exposed and appears to be damaged, the coating must be removed and the substrate repaired.

SUMMARY

When identifying surfaces/substrates, keep the following in mind:

- Study the environment in which the surface/substrate exists.
- Look at the surface/substrate itself and understand its special characteristics (i.e., painted, hardwood, softwood, alkaline, non-alkaline, etc.)
- Select the appropriate preparation method(s).
- Select the primer and topcoat suitable for the surface/substrate and its environment.
- Select an appropriate method of applying the coating(s).

Never underestimate the importance of good surface/substrate preparation. Paint will only cover (not hide) imperfections. Paint is only as good as the surface to which it is applied. A poorly-prepared surface or improper paint selection and/or application will provide a less than professional job.

References

For advanced study of topics covered in this task module, the following books are suggested:

Painting and Decorating Craftsman's Manual and Textbook, Painting and Decorating Contractors of America, Fairfax, Virginia.

Painting and Coating Systems, The Sherwin-Williams Company, Cleveland, Ohio.

REVIEW/PRACTICE QUESTIONS

1. Which of the following is not generally used as an exterior substrate?
   a. Masonry
   b. Concrete
   c. Wood
   d. Drywall

2. Which of the following is generally *not* used as an exterior wood substrate?
   a. Spruce
   b. Poplar
   c. Redwood
   d. Cedar

3. Which factor is *least* important when considering a wood coating for exterior use?
   a. Level of sun exposure
   b. Extent of heat or cold exposure
   c. Amount of humidity
   d. None of the above

4. Which contaminant is *not* considered when preparing raw steel for a coating?
   a. Rust
   b. Chalking
   c. Mill scale
   d. Oil and dirt

5. Which metal substrate should be weathered before a coating is applied?
   a. Steel
   b. Copper
   c. Galvanized steel
   d. Bronze

6. Concrete floors are sometimes acid etched, primarily to remove:
   a. oil and grease.
   b. paint.
   c. chalking.
   d. laitance and to provide profile.

7. When painting drywall, which of the following is true?
   a. Some water-based paints may be applied without primer.
   b. A paper surface should be abraded before priming.
   c. Oil-based paints may be applied without primer.
   d. A paper surface must always be primed.

8. Open-grain new interior hardwoods need not be _____ before a stain is applied.
   a. lightly sealed
   b. filled
   c. sanded
   d. bleached

9. Chalking paint that is in otherwise good condition is usually cleaned by:
   a. abrasive blasting.
   b. wire brushing.
   c. sanding.
   d. washing with a strong detergent cleaner.

10. Wallcoverings should be _____ before painting the wall.
    a. washed
    b. primed
    c. removed
    d. abraded

ACKNOWLEDGMENTS

Figures 11 through 17 courtesy of Painting and Decorating Contractors of America.

NCCER CRAFT TRAINING USER UPDATES

The NCCER makes every effort to keep these manuals up-to-date and free of technical errors. We appreciate your help in this process. If you have an idea for improving this manual, or if you find an error, a typographical mistake, or an inaccuracy in the NCCER's Craft Training Manuals, please write us, using this form or a photocopy. Be sure to include the exact module number, page number, a description of the problem, and the correction, if possible. Your input will be brought to the attention of the Technical Review Committee. Thank you for your assistance.

Instructors – If you found that additional materials were necessary in order to teach this module effectively, please let us know so that we may include them in the Equipment/Materials list in the Instructor's Guide.

Write: Curriculum and Revision Department
National Center for Construction Education and Research
P.O. Box 141104
Gainesville, FL 32614-1104

Fax: 352-334-0932

Craft ______________________ Module Name ______________________

Module Number ______________________ Page Number(s) ______________________

Description of Problem

____

____

____

____

(Optional) Correction of Problem

____

____

____

(Optional) Your Name and Address

____

____

notes

Protecting Adjacent Surfaces

Module 07105

Painting Trainee Task Module 07105

PROTECTING ADJACENT SURFACES

PAINTING AND DECORATING CONTRACTORS OF AMERICA

NATIONAL CENTER FOR CONSTRUCTION EDUCATION AND RESEARCH

Objectives

Upon completion of this module, the trainee will be able to:

1. Describe the tools and materials required for protecting surfaces:
   - Tape dispensers
   - Types of tape
   - Types of masking material, such as paper, film, light-duty plastic sheeting, and liquid or gel masking
   - Paint shields
   - Covering materials, such as dropcloths, netting, and heavy-duty plastic sheeting
2. Describe the methods of applying interior and exterior masking and coverings to various surfaces.
3. Understand the importance of proper cleanup.

Prerequisites

Successful completion of the following Task Modules is required before beginning study of this Task Module: Core Curricula, Painting Level 1, Modules 07101 through 07104.

Required Trainee Material

1. Trainee Module
2. Appropriate Personal Protective Equipment

Course Map

This course map shows all of the *Wheels of Learning* task modules in the first level of the Painting curricula. The suggested training order begins at the bottom and proceeds up. Skill levels increase as a trainee advances on the course map. The training order may be adjusted by the local Training Program Sponsor.

Course Map: Painting, Level 1

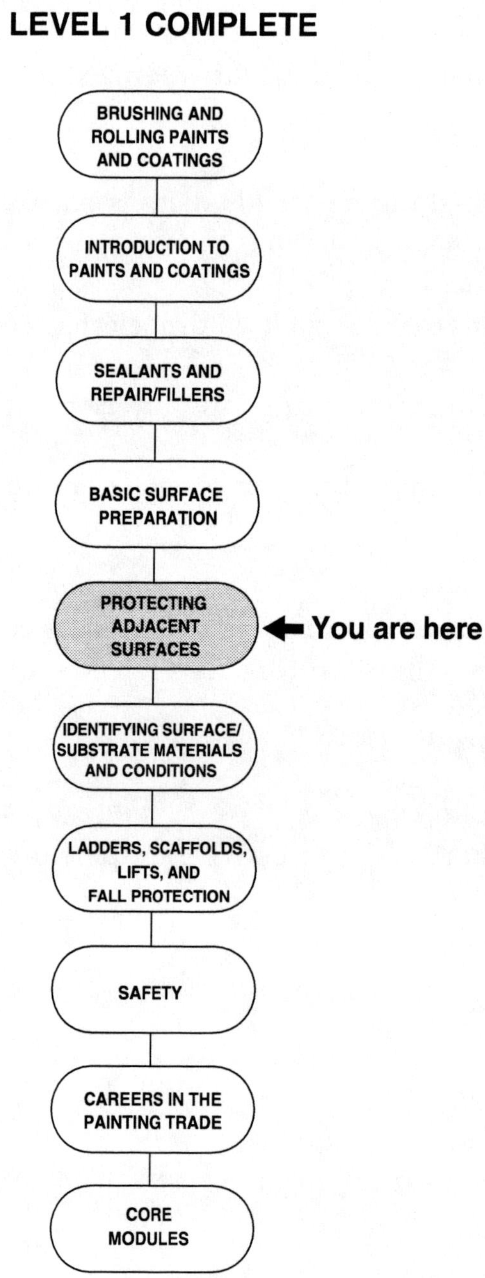

CMAP105.EPS

TABLE OF CONTENTS

Trade Terms Introduced In This Module

Containment: A term representing the various methods of sealing a work area to prevent the escape of dust, hazardous materials, or vapor to other areas or into the environment.

Dropcloths: Protective, reusable coverings made from cotton canvas or other absorbent material.

Masking: The protection of a surface or object by the application of paper or plastic film that is secured by masking tape or another adhesive. Also, the protection of a surface by the application of liquid masking or masking gel.

Tarps: An abbreviated term for tarpaulins (waterproof sheet material made from treated canvas or fiber-reinforced plastic). Tarps usually have rope-reinforced edges that are fitted with grommets for tie-down purposes.

1.0.0 INTRODUCTION

Painters must be concerned with more than just the surfaces they are painting. Proper care must be taken to protect all surrounding surfaces and areas. If a Persian rug gets painted along with a wall, profits and popularity will diminish. A little effort in preparing and protecting surfaces can save time and money in the long run. It will also help to give the painting crew and the contractor a good name. This means they will be in demand.

The painter is said to be in the care, custody, and control of the property being painted/coated. Because of this, a painter can be held legally responsible for damages that are caused. Common sense is the best guide to protecting surfaces and areas. *A good rule to follow is: if it doesn't get painted, move it or cover it.*

Timing is also very important to proper and economical protection of surfaces and areas. In new construction, it is much easier, safer, and more economical to prepare and paint a room before the fixtures are in, the rug is down, and the ceiling grid is up. Obviously, it will not always be possible to get to the job at exactly the right moment. The object is to strive toward good timing.

2.0.0 TOOLS AND MATERIALS

2.1.0 TAPE AND MEDIA DISPENSERS

Tape and media dispensers are the most common and fastest way of applying paper or film **masking** media to a surface. They are three to five times faster than hand application. Both hand-held and bench-mounted dispensers are available. *Figure 1* shows a typical hand-held tape and media dispenser with a manual tape dispenser attached to it along with a cutoff blade. The manual tape dispenser on this model allows the painter to obtain strips of tape to hold the masking media in place before securing the untaped edges to the surface.

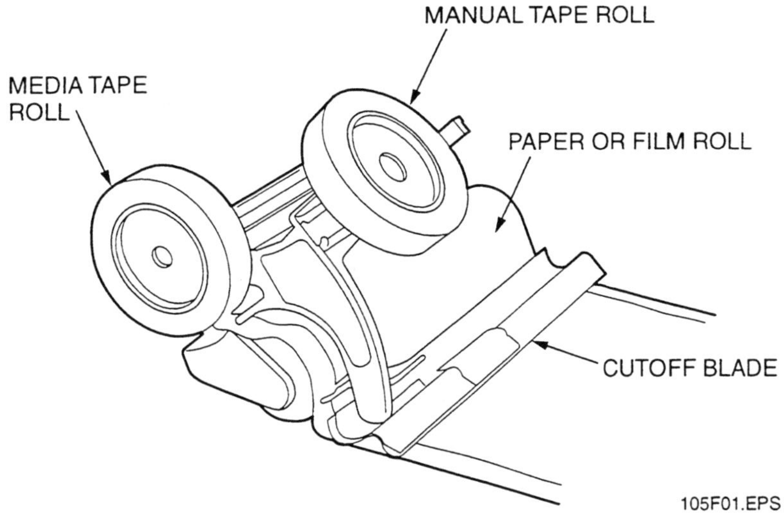

Figure 1. Typical Hand-Held Tape And Media Dispenser

Cutoff blades are available in different lengths for varying widths of masking media. *Figure 2* shows a typical bench-mounted tape and media dispenser. Both devices function to apply various widths of masking tape to one edge of the masking media at an adjustable distance from the edge of the media. In the case of the hand-held model, this can be done simultaneously with the application of media directly to the surface to be protected.

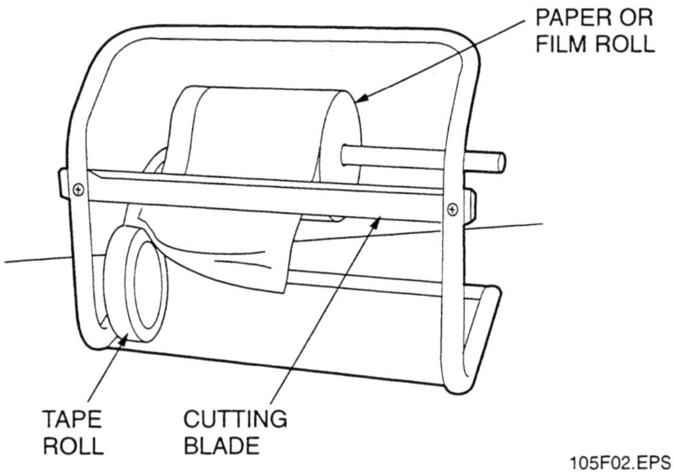

Figure 2. Typical Bench-Mounted Tape And Media Dispenser

Figure 3 shows a typical hand-held molding masking tape dispenser. It may be used to apply tape to the loose edges of masking media previously applied with a hand-held or bench-mounted tape or media dispenser. It is primarily used to apply tape to the edges of moldings, floors, windows, etc. As shown in *Figure 4*, the rollers of some of these devices have one raised edge that presses the edge of the tape down tightly against the surface, thus eliminating the use of a putty knife or fingernail to seal the edge of the tape. In addition, tape tracking is adjustable on the rollers so that tape may be applied above or below the molding edge as shown in *Figures 5* and *6*.

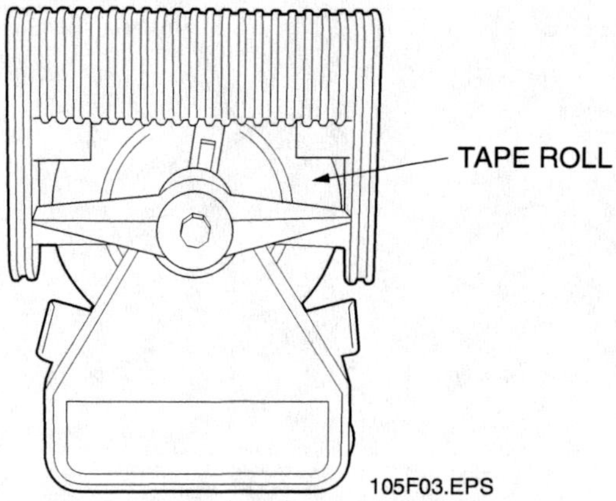

Figure 3. Typical Hand-Held Molding Masking Tape Dispenser

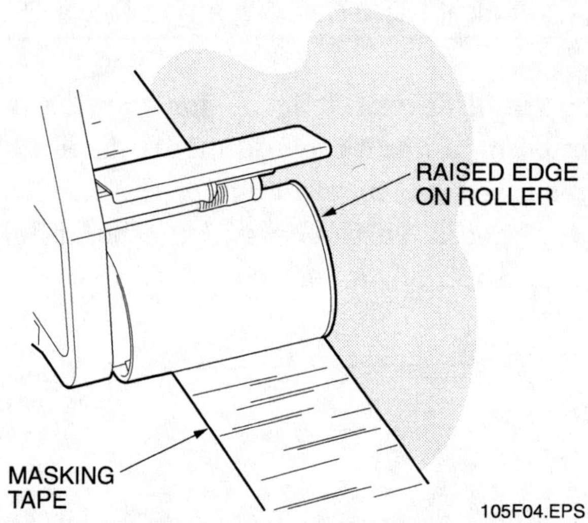

Figure 4. Raised-Edge Roller On Hand-Held Molding Masking Tape Dispenser

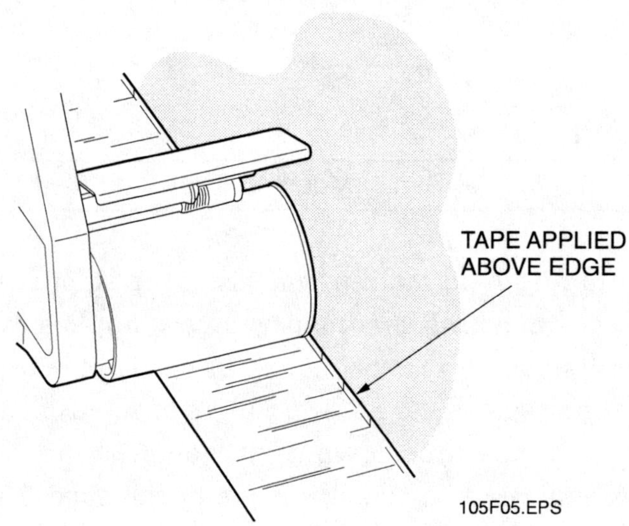

Figure 5. Masking Tape Applied Above Molding Edge

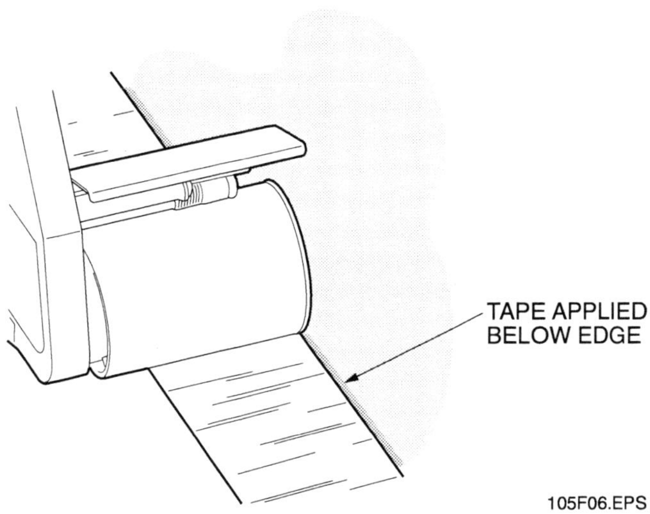

Figure 6. Masking Tape Applied Below Molding Edge

Figure 7 shows a special hand-held dual tape dispenser. This device simplifies securing the loose edges of masking media applied with a masking tape and media dispenser or when hanging large sheets of untaped or non-folded media. It creates a double-sided masking tape by laminating a regular masking tape, which affixes to the surface, to the edge of high-tack tape which is reversed so that the adhesive surface faces out. The masking media is then applied over and pressed into the high-tack adhesive as shown in *Figures 8* and *9*.

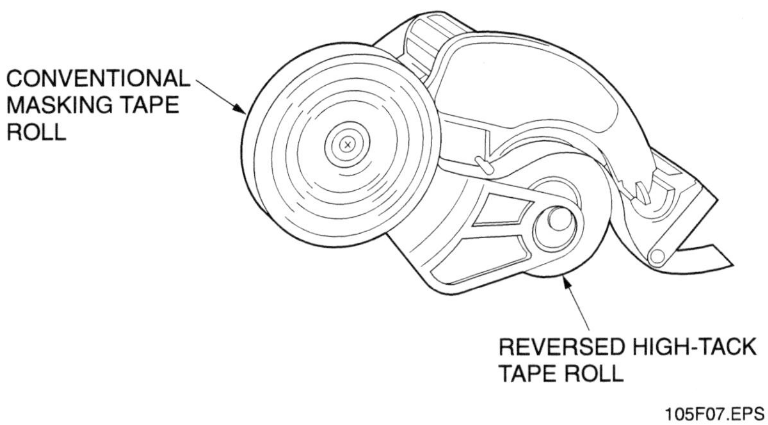

Figure 7. Hand-Held Dual Tape Dispenser

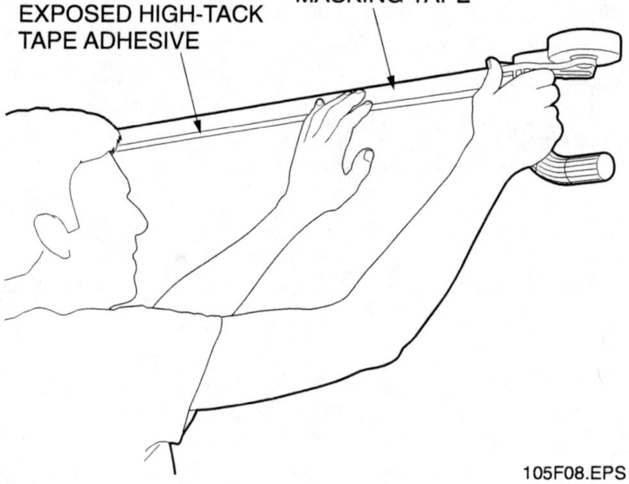

Figure 8. Applying Dual Tape To A Surface

Figure 9. Applying Masking Media To Exposed High-Tack Tape Adhesive

2.2.0 MASKING TAPE

Masking tape (*Figure 10*) used in the painting trade generally ranges from 3/4 inch (19 mm) to 3 inches (76.2 mm) wide and is usually supplied in 60 yard (54.9 m) rolls for use in tape dispensers. It is also available with various adhesives for use in specific applications. Common types of masking tape include:

- *High-adhesion/24-hour tapes* – These tapes provide instant adhesion and excellent holding power. They also provide superior paint lines, but are designed to be removed within 24 hours.

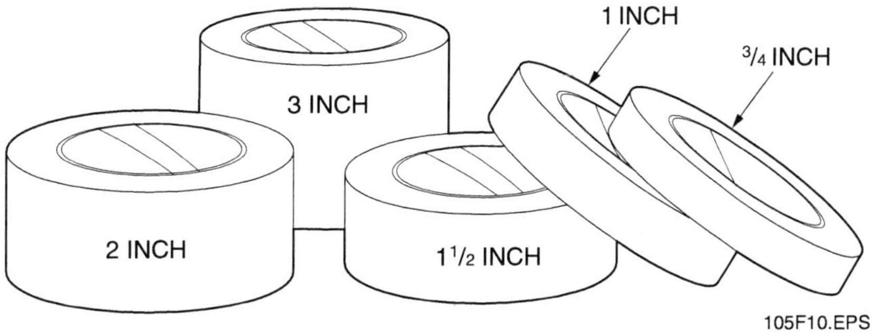

Figure 10. Typical Masking Tape Roll Sizes

- *Medium-adhesion / 24-hour tapes* – These tapes provide easy removal and dispensing characteristics. They also provide sharp paint lines with minimal paint-edge buildup. They must be removed within 24 hours.
- *Safe-release / 7-day tapes* – These tapes provide a low-tack adhesive that is ideal for delicate surfaces such as wallpaper, unpainted wallboard, and freshly cured paint. They also provide extra sharp paint lines that are useful for graphic and striping applications. They are designed for removal within 7 days.
- *Medium-adhesion, sunlight-stable / 7-day tapes* – These tapes are designed for indoor/outdoor applications where the tape is exposed to sunlight. They should be removed within 7 days.
- *Dual-tack tape sets / 7-day medium-adhesion sunlight-stable or 24-hour high-adhesion tapes* – These tape sets combine a high-tack tape and a conventional 7-day or 24-hour masking tape and are used with the dual tape dispenser discussed earlier.

2.3.0 NON-REUSABLE MASKING MEDIA

In the following paragraphs, various types of non-reusable masking media are discussed. They are generally provided in lengths that are usable in hand-held dispensers.

2.3.1 Masking Paper

Figure 11 shows common paper roll widths. The rolls are generally 60 yards (54.9 m) long. They are available in a variety of grades, some formulated to be pinhole free to minimize bleedthrough. Others are specially made for use with latex, oil, or urethane paints.

2.3.2 Masking Film

Figure 12 shows common masking film roll widths. The rolls are generally 180 yards (164.6 m) long for use in hand-held dispensers. The films are thin and lightweight, but are high density and non-porous to paints, stains, urethanes, and chemical strippers. They are also treated to prevent liquids from running off the surface.

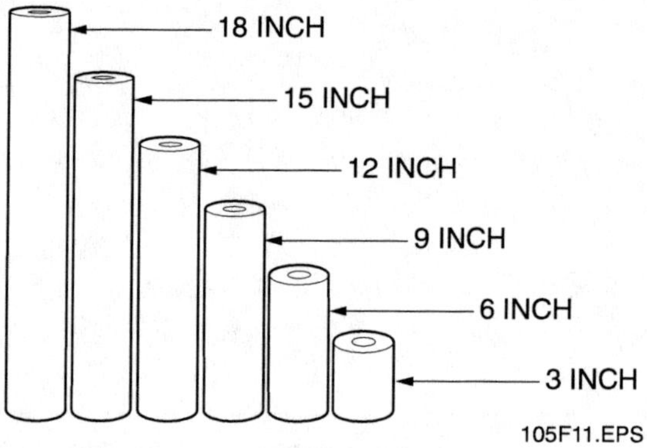

Figure 11. Typical Masking Paper Roll Width Sizes

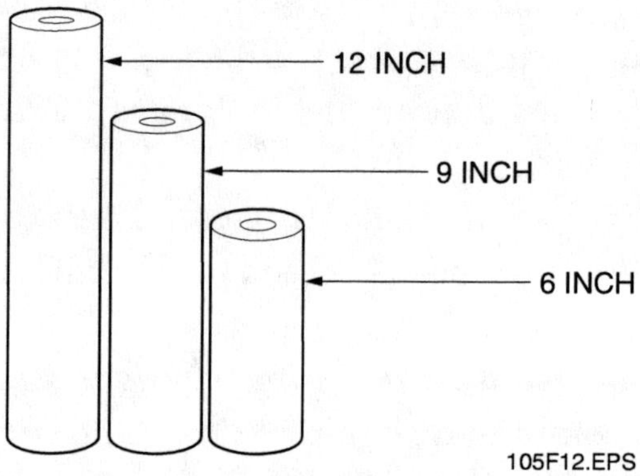

Figure 12. Typical Masking Film Roll Width Sizes

2.3.3 Pre-Folded Masking Film

Figure 13 shows pre-folded masking film roll widths. They are similar to the masking film discussed earlier, with the additional advantage that they can be unfolded to the various widths shown in *Figure 13* once the taping edge has been secured to a surface. They are usually taped horizontally across the top of the surface and then unfolded downward. They are ideal for covering large areas and are much easier and faster to apply than large unfolded sheets of polyethylene plastic. The rolls are usually supplied in the following lengths for the indicated widths:

- 60 yards (54.9 m) for 24-inch (610 mm), 36-inch (912 mm), and 48-inch (1.22 m) wide rolls
- 100 feet (30.5 m) for 72-inch (1.83 m) and 99-inch (2.5 m) wide rolls

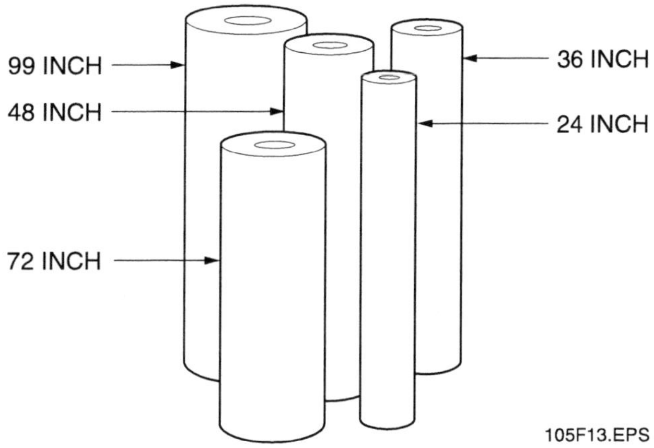

Figure 13. Typical Pre-Folded Masking Film Roll Width Sizes

2.3.4 Bulk Polyethylene Sheeting (Light Duty)

This plastic sheeting is available in 1-mil to 2-mil thicknesses and in rolls with widths up to 30 feet (9 m) folded. This light-duty material can be used to mask large areas or as interior cover material for objects in a room. It is not recommended as a floor cover because it is slippery and does not provide stable footing. Some polyethylene may contain excessive amounts of *slip agent*, which doesn't allow masking tape to adhere to the film surface properly.

2.3.5 Liquid Masking Or Masking Gel

A number of brushable or sprayable liquid masking or masking gel products are available. These are primarily used on hard, non-porous surfaces such as glass or metal, or for objects that may be difficult to mask with tape because of their shape or location. Some of these products dry to a strippable rubbery film and others do not dry at all and must be wiped or washed off. The drying type may be painted over with a brush. The non-drying types are suitable only for spray painting use. Some painters find that these products are too expensive and more trouble than they are worth except for limited use on irregularly shaped objects. Liquid dishwashing soap or petroleum jelly will perform the same function for considerably less money. *Figure 14* illustrates a dryable type liquid masking with a built-in applicator for masking glass or similar surfaces. However, a razor blade scraper used on freshly dried paint is just as effective.

Figure 14. Liquid Masking With Built-In Applicator

2.3.6 Self-Adhesive Masking Material

Several self-adhesive masking paper or film products are available in rolls of various widths. These products are relatively expensive and are usually coated with a medium-tack or low-tack adhesive along one edge which causes some paint weeping at the edge. If hand-held dispensers are not available, these products can be used in situations where there are no irregular edges or over delicate surfaces such as wallpaper.

2.4.0 REUSABLE MASKING/COVERING MEDIA

2.4.1 Dropcloths

Dropcloths are generally made from cotton canvas or other absorbent material and are available in light-duty and heavy-duty weights. Pliable plastic-coated paper is also available for use as a dropcloth. Dropcloths are supplied as large rectangular area coverings or as runner cloths for use along walls. The advantage of dropcloths is that they absorb paint to prevent tracking and provide stable footing when used as a floor covering over plastic sheeting. The lightweight cloths can also be used as a covering over shrubbery or plants for outdoor painting or over room objects for indoor painting. Since they are porous, they allow dust, air, and water to pass through. If a dust-free, watertight covering is required, plastic sheeting or plastic-coated paper should be used. *Figure 15* is an example of a runner-type dropcloth. Interior and exterior dropcloths should not be used interchangeably nor should they be turned over and used when wet paint is still present on one side.

105F15.EPS

Figure 15. Runner-Type Dropcloth

WARNING!	Dropcloths, especially those not treated with a fire retardant, can be a fire hazard. Careless disposal of smoking materials can cause a fire long after painters have left for the day.

2.4.2 Plastic Sheeting (Heavy Duty)

An alternative to the dropcloth is heavy-duty plastic sheeting (6 mils or thicker). It is available in bulk rolls and a wide range of widths. The sheeting is relatively inexpensive and, for exterior use, the heavier grades can be reused several times. Since it is non-porous, dust, paint, water, and chemical sprays will not penetrate it. Because the material is slippery, it should be covered with a canvas dropcloth when used as a floor covering to provide stable footing and to prevent tracking of paint. It can also be used to cover large objects such as machinery and furniture that must be kept dust free.

For exteriors, plastic sheeting may be used as masking for adjacent structures or on large areas of buildings when spray painting. It may also be used on shrubbery, foliage, or lawn areas for short periods of time (up to three hours) in the immediate work area when spray painting, or when working with a chemical cleaner that may cause damage, or when removing hazardous materials from a substrate.

Heavier grades of sheeting, when adequately reinforced, can be used for interior or exterior dust, hazardous material, or vapor **containment** enclosures with or without negative pressure. Negative pressure enclosures are evacuated by vacuum fans coupled with appropriate filters. This prevents dust, hazardous materials, or vapor from escaping the enclosure (*Figure 16*).

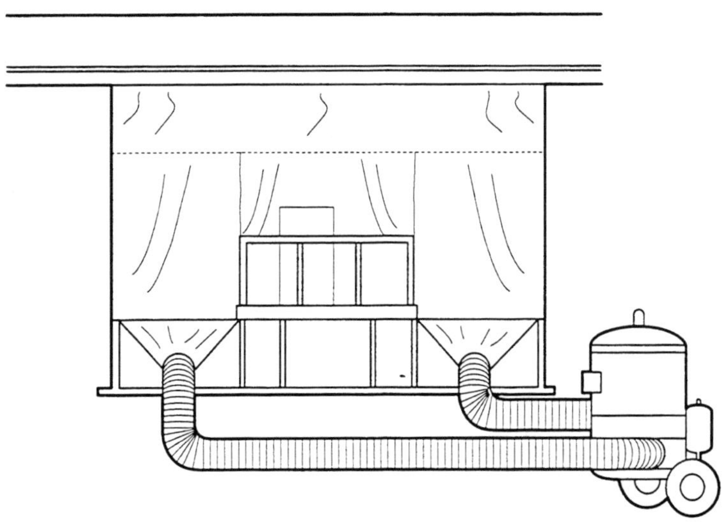

105F16.EPS

Figure 16. Example Of Total Enclosure With Negative Pressure

2.4.3 Fine-Mesh Netting

This material is available in various sizes and is used in low-pressure water washing (without chemicals) to cover shrubbery, lawns, and foliage for an extended period. It prevents debris from covering the shrubbery, lawn, or foliage, but allows fine dust and water to drain through and air to reach the covered areas.

2.4.4 Spray Shields

Spray shields (*Figure 17*) are hand-held tools used to protect adjacent surfaces from overspray during spray painting, texturing, or flocking operations. They vary in size from approximately 9 inches by 24 inches to 9 inches by 48 inches, with the most popular size being the 36-inch long version. They are available with handles that vary in length from about 8 inches to 18 inches. The shield is usually made of aluminum or in some cases, plastic, both of which are available in stiff or flexible form. Some shields have a 30° angle for easier handling. Others are furnished with a paint-run guard at the handle edge to alleviate paint dripping. Some aluminum shields are available with a coating that simplifies clean up and eliminates the black marks sometimes caused by untreated aluminum. Shields are a quick and easy way of protecting as you go. However, the edges must be wiped after each use or the paint will weep under the edge. In addition, shields are usable on only flat, true surfaces. Most professional painters prefer other methods.

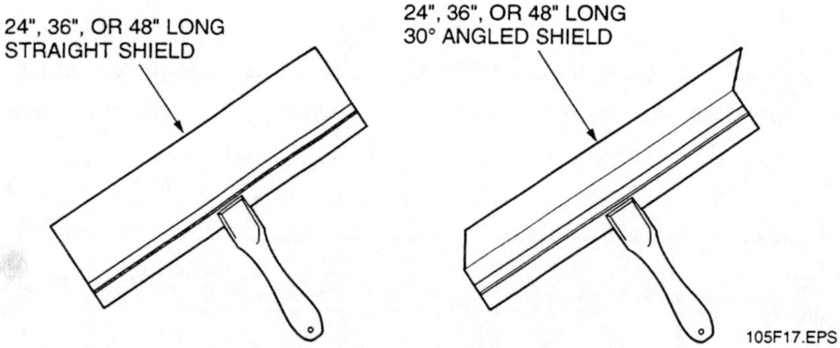

Figure 17. Spray Shields

3.0.0 MASKING AND COVERING METHODS AND GUIDELINES

3.1.0 INTERIOR MASKING AND COVERING

WARNING! Many paints applied prior to 1978 contain lead or other toxic heavy metals. In addition, some older flocked finishes, acoustic tile, duct wrapping, and plaster contain asbestos fiber. If laboratory tests or MSDSs confirm the presence of any of these, total space containment and dust or vapor collection equipment, along with approved personal protection, are required.

Before masking and covering an interior space for preparation or painting, all movable furniture, appliances, lamps, and decorations should be removed if possible. Otherwise, they should be grouped in narrow rows near the center of the space to allow access to the ceilings if the ceilings are also to be finished. If possible, small fixed lighting fixtures should be loosened and dropped down, then covered if not used for illumination. If desired and if possible, switch and receptacle plates should be removed along with door and window hardware (knobs, door bolt plates, striker plates, locks, etc.) and any other removable hardware. Exposed receptacles, switches, door hinges (if not removed), and any other hardware should be masked.

Heavy or non-removable lighting fixtures, if not used for illumination, or plumbing fixtures must be bagged or draped and taped at the ceiling or wall mounting as detailed below. Sprinkler heads should be bagged or protected as shown below. All other door trim, cabinets, floor trim, window trim, and any other surface not to be refinished should be masked by one of the methods shown below. Interior surfaces to be masked must be clean and dry; otherwise, the masking tape will not adhere to the surfaces. Finally, furniture, rugs, floors, and other objects should be covered as outlined below. If hazardous materials are present, or will be present during the refinishing process, drape and seal all external vents or openings to the space and employ proper dust or vapor collection methods as well as appropriate personal protection.

3.1.1 Paper Masking

Figure 18 shows typical methods of applying paper masking to various surfaces using a hand-held dispenser. Once the border applications are completed, dropcloths or additional paper must be applied to complete the masking of surfaces, if required.

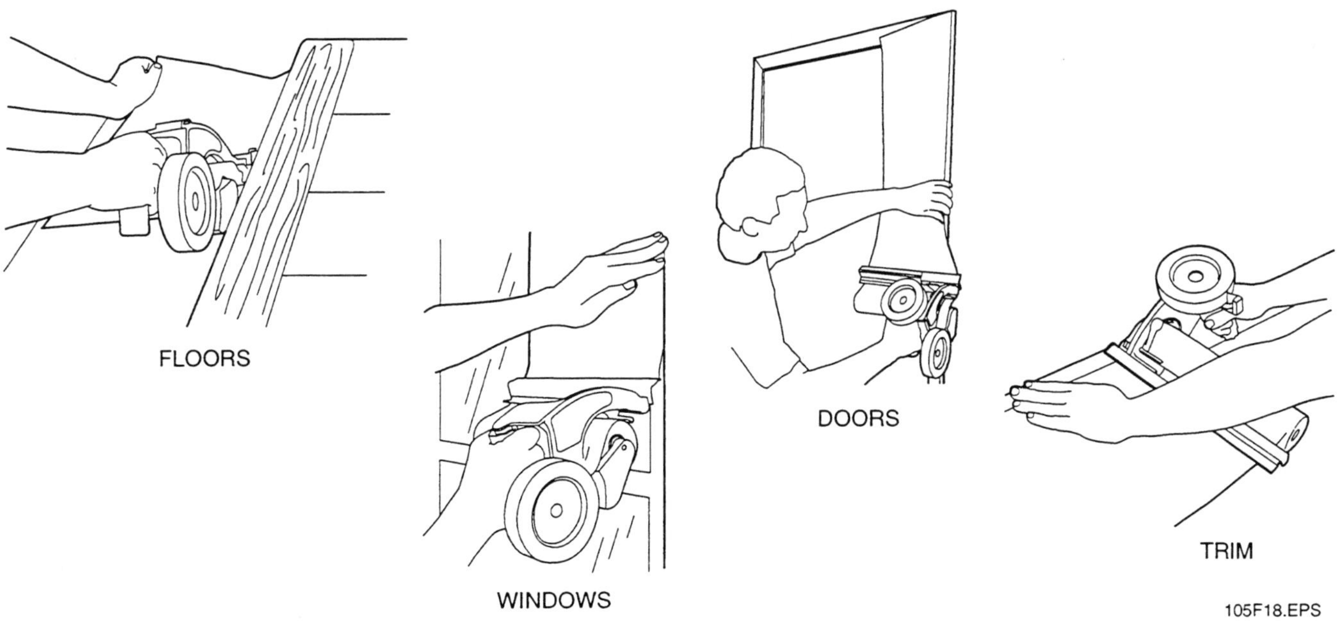

Figure 18. Applying Border Paper Masking With A Hand-Held Dispenser

3.1.2 Film Masking

Figures 19 through *21* show methods of applying pre-folded film to various surfaces using hand-held dispensers. This masking method is the fastest and most economical.

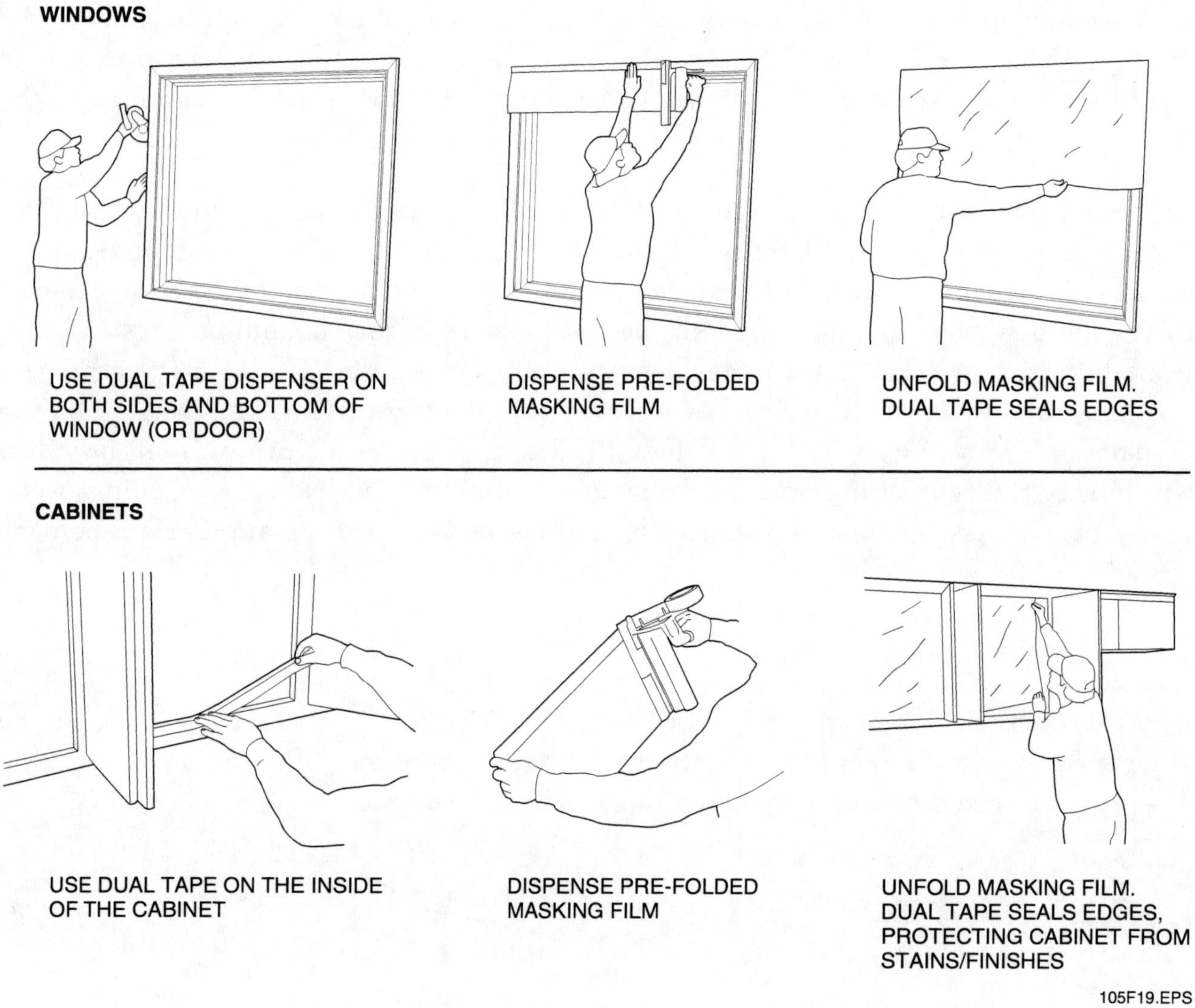

Figure 19. Applying Pre-Folded Film To Various Surfaces

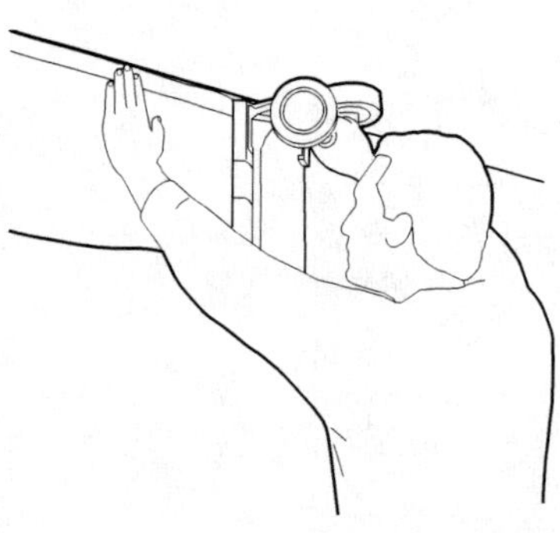

Figure 20. Applying Pre-Folded Film To Walls

105F21.EPS

Figure 21. Film Taped And Draped Over Kitchen Cabinets

3.1.3 Molding Masking

Figures 22 through *24* show methods of masking moldings with or without the use of pre-folded or unfolded film.

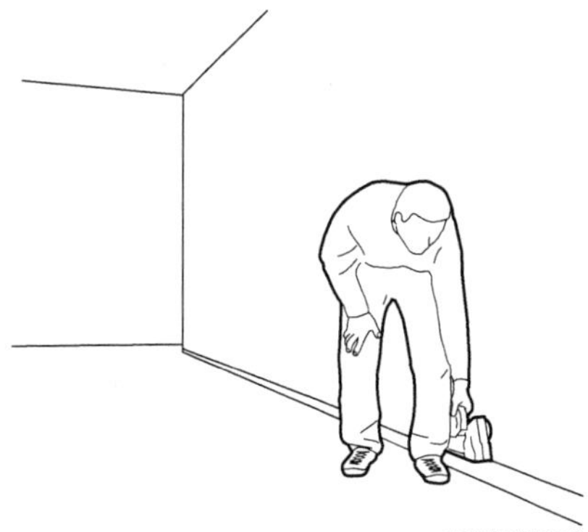

105F22.EPS

Figure 22. Applying Tape To Vinyl Molding

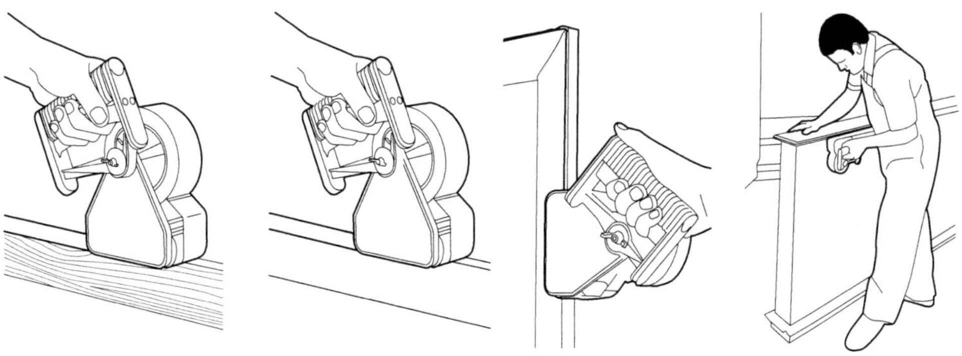

105F23.EPS

Figure 23. Taping Molding, Doors And Windows, And Counters Or Wall Caps

TOP, SIDES, AND BOTTOM TAPED WITH 1 1/2 OR 2 INCH WIDE TAPE

PRE-FOLDED FILM APPLIED ACROSS TOP FRONT

FILM UNFOLDED AND SECURED BY FOLDING 1 1/2 OR 2 INCH TAPE OVER FILM ALL AROUND

105F24.EPS

Figure 24. Taping With A Molding Tape Applicator And Pre-Folded Masking Film

3.1.4 Bagging Or Draping Non-Removable Lighting Or Plumbing Fixtures

WARNING! Make sure that any bagged or draped fixtures cannot be turned on while covered. Bagged lighting fixtures must not be used for illumination since the trapped heat will shorten the life of the lamps and may cause a fire, especially when used on high-intensity lamp fixtures such as quartz, sodium, mercury vapor, or metal halide.

Many non-removable fixtures may be bagged using different sized plastic bags, including large and small garbage bags (*Figure 25*). The bags should be taped and sealed at the base of the fixture. Chandeliers, other hanging light fixtures, and ceiling fans may be protected by draping plastic sheeting around them and securing it with tape at the ceiling mount. If lighting fixtures must be used for illumination or cannot be disabled at night by removing the bulbs or locking off the power, the masking or bags must be removed.

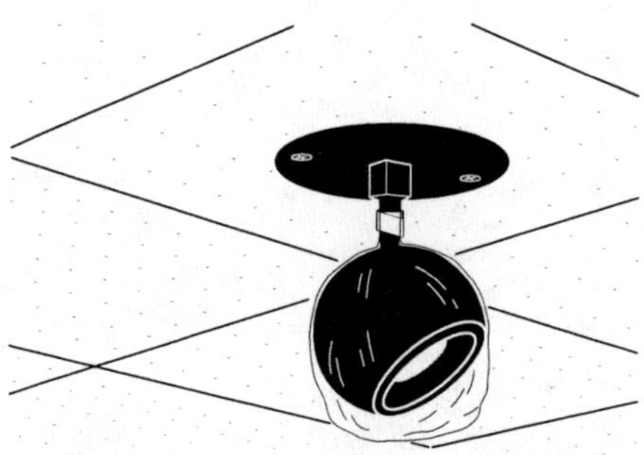

105F25.EPS

Figure 25. Bagged Lighting Fixture

3.1.5 Bagging Sprinkler Heads, Nozzles, and Fire Alarm Sensors

WARNING! If fire alarm sensors, sprinkler heads, or nozzles on active fire alarm systems, sprinkler systems, or dry chemical/halon dump systems must be bagged or covered, the proper authorities must be notified and permission granted prior to covering the sensors, sprinkler heads, or nozzles. Dry chemical/halon dump system or fire alarm system zones in the work area should be disabled if the zone sensors are being covered. Plastic bags or masking film should not be used to cover sprinkler heads or nozzles.

CAUTION Do not disturb the trigger mechanism and meltable link on sprinkler heads when covering a fixture.

Paper bags or paper cups that are not plastic coated may be taped over exposed sprinkler heads or nozzles. Secure the tape around the base of the head or nozzle. Make sure that the trigger mechanism and meltable link of sprinkler heads are not disturbed when covering the fixture. Do not use plastic materials since they may interfere with the operation of the sprinkler head or nozzle if a fire occurs. If the detection sensors for fire alarm systems or dry chemical/halon dump systems are also going to be covered, the zones of the system in the work area should be disabled. Photometric or ionization-type zone sensors may be covered with plastic bags or material to prevent accidental contamination or triggering of the sensor. At night, the coverings over the sensors should be removed and the fire system should be reactivated.

If pressure washing, spray painting, sweeping, or other cleaning activities are conducted that result in mist or dust clouds in areas where photometric or ionization-type fire detection zone sensors are located, the fire alarm system or dry chemical/halon dump system could accidently be activated. In these cases, the zone sensors should be covered and deactivated after notification of the proper authorities. They should be uncovered and reactived after the cleaning or spray painting activities are completed.

3.1.6 Covering, Draping, and Sealing

WARNING! Unguarded floor openings should never be covered by plastic sheeting and/or dropcloths. Do not climb or step on any covered items. Damage to the item or injury to the painter from falls caused by slippage of the covering may occur. If dust, hazardous material, or vapor infiltration will present a problem, the owners/occupants of the building must be notified, all interior ventilation equipment for the affected area must be turned off and disabled, and all openings to other areas must be sealed off with plastic and masking. Proper environmental containment equipment and personal protection equipment must also be used.

Once all items not to be finished or refinished are masked, dropcloths or plastic sheeting or plastic-coated paper should be used to cover floors, furniture, appliances, and other items left in the space. Floors and rugs should be covered using heavy grade plastic that is then covered with heavyweight area or runner type dropcloths to provide proper footing. If scaffolding will be erected or rolled about in the work area, plywood or similar sheeting should be positioned over the floor covering to prevent any possible damage to the floors or floor covering. Use lightweight plastic, plastic-coated paper, or dropcloths to cover objects left in the room. Make sure that the covering will not crush or damage the objects. If dust prevention is desired, use only plastic sheeting or plastic-coated paper over the objects. Make sure that the sheeting is secured with tape. *Do not climb or stand on furniture or other covered objects.* If dust, hazardous material, or vapor infiltration in areas outside the space to be refinished will present a problem, all external openings should be draped with plastic and sealed with tape. Building owners and occupants should be notified and interior ventilation systems for the work area should be shut off and disabled. Proper environmental containment equipment and personal protection equipment must also be used.

3.2.0 EXTERIOR MASKING AND COVERING

WARNING! Many paints applied prior to 1978 contain lead or other toxic heavy metals. In addition, some older flocked finishes, acoustic tile, duct wrapping, and plaster contain asbestos fiber. If laboratory tests or MSDSs confirm the presence of any of these, total space containment and dust or vapor collection equipment, along with approved personal protection, are required.

When preparing or painting exteriors, the most common problem a painter must deal with is the weather. Wind can carry dust, debris, and spray mist chemicals from the preparation activities, or paint mist or spatter from the painting operation a considerable distance. If the winds are strong, it may be best to postpone the job until it is calmer. If the job cannot be delayed, extensive masking and covering of adjacent surfaces and structures must be done or containment structures must be erected to prevent damage to other property. In addition, dry fall or dry fog paint usage may have to be considered. These paints are formulated so that mists from spraying dry before settling on a surface. If hazardous materials are involved or the weather is a problem in the preparation or painting operation, containment structures will almost always have to be fabricated and used. *Figure 26* is an example of an elaborate total containment enclosure erected around a water tower. The enclosure illustrated can be raised or lowered as needed to prevent wind-loading stress on the structure.

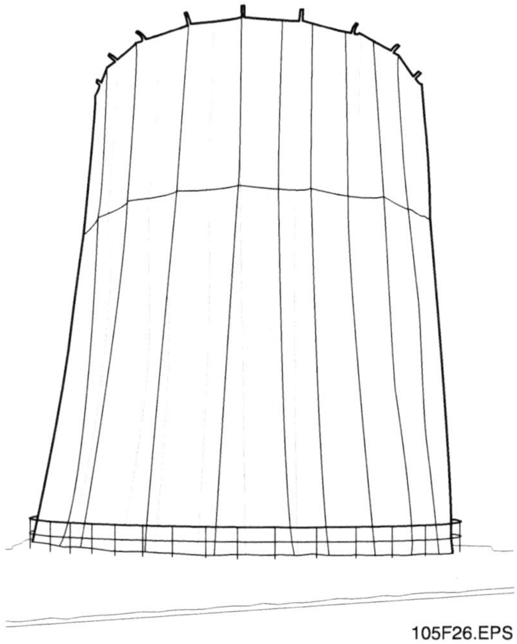

105F26.EPS

Figure 26. Total Structure Enclosure

3.2.1 Exterior Masking

CAUTION Exterior air conditioning/heat pump heat exchangers and ventilation fans must be disabled before covering and the owners/occupants of the building must be notified that the systems are disabled.

All surfaces in or near the work area that are not to be finished or refinished should be masked with sunlight-stable tape and film or plastic sheeting using the same methods as described for interior surfaces. The surfaces to be masked must be clean and dry or the tape will not adhere to the surface.

3.2.2 Exterior Covering

If containment enclosures are not used or required, all movable objects should be moved a safe distance from the work area. All other adjacent structures, shrubbery, foliage, lawn areas, and driveways should be draped with the appropriate material. Lawns, shrubbery, and other foliage should be covered with porous, lightweight dropcloths or fine-mesh netting. Temporary use of plastic sheeting over the netting (less than two or three hours) may be required in some work areas when cleaning surfaces with chemical low-pressure washers or when painting is in progress in that area. The plastic sheeting should be moved as each work area is completed. It should be completely removed if it has been in place longer than three hours. Exterior air conditioning/heat pump heat exchangers or ventilation system fans must be covered to prevent damage to the units or the entrance of contaminants into the structure. These units must be disabled to prevent their operation while covered. In addition, the owners/occupants of the building must be notified that the systems are disabled.

3.2.3 Containment Enclosures

Containment is a way to limit dust, debris, paint chips, spent abrasives, and overspray from contaminating the environment. This is more than just a good idea; it is often required by local, state, or federal agencies. The type, concentration, and toxicity of the contamination will determine the extent of containment required. Extensive containment enclosures will require engineered approaches and will generally be erected by others.

A common containment method is to build a temporary enclosure around the work area or around the entire structure with materials such as **tarps**, drapes, or screens. Some enclosures are designed so that they can be raised or lowered quickly to reduce the effect of wind stress on the structure. Here are four general levels of containment:

- Free-hanging enclosures are loosely attached to the structure to reduce the escape of airborne contamination from the work site. Free-hanging means that the enclosure is attached to the top and sides of the structure, but not to the bottom.
- Partial structure enclosures surround the entire work area with materials supported by rope, cables, staging, or outrigging. They are not tightly attached to the structure. An example is shown in *Figure 27*.

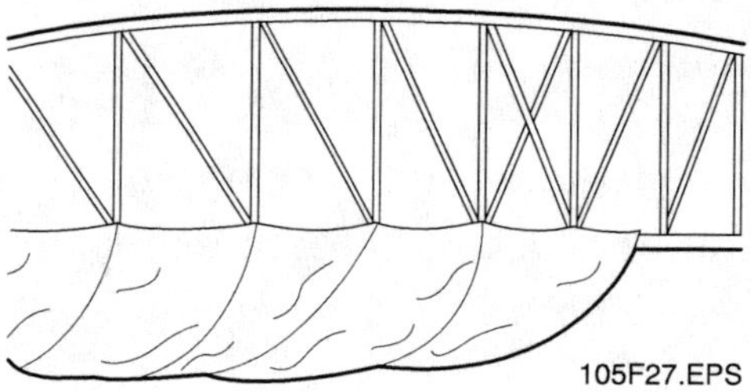

Figure 27. Partial Structure Enclosure

- Total structure enclosures form a sound containment around the entire structure with the support of scaffolding or framing made of wood or steel. They attach to the structure to minimize the escape of contaminated debris. *Figure 26* showed containment of a water tower. The enclosure is designed to be raised or lowered as necessary.
- Total enclosures with negative pressure are the same as total structure enclosures with the addition of vacuum fans or dust collectors. Air movement must be sufficient to prevent airborne particles from escaping the enclosure. Dust collectors with high efficiency particulate air (HEPA) filters may be required to treat discharged air.

Containment of dust, debris, and spent abrasives is not enough. They must also be collected and removed from the work site. One method is to let the material collect on ground tarps, decking, or the support system, and then collect it from there. Another method that works on elevated enclosures is to use vertical shields or funnels to channel the material into waste containers or storage hoppers below. Working over waterways can require creative ways to prevent contamination of the environment, such as anchoring a barge beneath a bridge enclosure to collect debris. Sometimes, booms are used on streams or rivers to catch stray contaminants.

If the waste involves lead-based paint or asbestos, it must be dealt with in accordance with strict health and environmental standards. The EPA has rules for handling, storing, labeling, shipping, and disposing of hazardous waste. In addition, state and local environmental laws may apply. Failure to follow these rules is not only hazardous but can also result in penalties.

In addition, OSHA has established regulations to protect the health and safety of workers in these environments, especially when removing lead-based paint or asbestos products. Besides wearing the appropriate protective gear and respirator, you must follow all prescribed safety procedures, which may include using on-site wash stands, decontamination facilities for changing clothes and showering, and undergoing physical examinations, including blood lead level readings.

4.0.0 CLEANUP

Cleaning up the work site after the job is done is extremely important, especially when dealing with hazardous material residues. All masking must be removed and disposed of properly. All coverings must be carefully folded from the outside edges inward to trap residue and debris. The residue and debris must also be disposed of properly. Vacuum up all remaining dust and debris. Dispose of used paint buckets, brushes, and other materials properly. Replace any fixtures, switch and receptacle plates, door and window hardware (knobs, door bolt plates, striker plates, and locks), and any other hardware that was removed prior to masking. Replace all objects (furniture and appliances) in their original position. The basic rule is to leave the work site cleaner and better looking than you found it. The extra effort will assure repeat business and referrals from the customer.

SUMMARY

Remember that common sense is the best guide to protecting adjacent surfaces and areas. If you stop to consider what you are doing and what may occur because of what you are doing, most problems can be avoided. It should be noted that repairing usually takes longer than preparing. A little effort before the job will assure minimum effort after the job.

When you are finished, clean the work site. Nothing lessens the value of a good paint job like poor cleanup. Always leave the area cleaner than you found it.

References

For advanced study of topics covered in this task module, the following books are suggested:

Painting and Decorating Craftsman's Manual and Textbook, Painting and Decorating Contractors of America, Fairfax, Virginia.

Painting and Decorating Encyclopedia, The Goodheart-Willcox Company, Inc., Tinley Park, Illinois.

REVIEW/PRACTICE QUESTIONS

1. Painters are said to be responsible for:
   a. care of the property.
   b. custody of the property.
   c. control of the property.
   d. all of the above.

2. The best guide for protecting adjacent surfaces and areas is:
   a. the use of the most expensive methods.
   b. common sense.
   c. spending extra time.
   d. the use of only plastic masking.

3. The device used to directly apply paper or film masking is a:
   a. bench-mounted dispenser.
   b. dual tape dispenser.
   c. tape and media dispenser.
   d. molding masking tape dispenser.

4. Tape widths normally used by painters are:
   a. $^{3}/_{4}$, 1, 1$^{1}/_{2}$, 2, and 3 inches.
   b. $^{3}/_{4}$, 1$^{1}/_{2}$, 2, 3, and 4 inches.
   c. 1, 1$^{1}/_{2}$, 2$^{1}/_{2}$, 3, and 3$^{1}/_{2}$ inches.
   d. $^{3}/_{4}$, 1, 1$^{1}/_{2}$, 2, and 3$^{1}/_{2}$ inches.

5. Exterior sunlight-stable masking tape is usually:
   a. high-adhesion/24-hour tape.
   b. medium-adhesion/24-hour tape.
   c. low-adhesion/7-day tape.
   d. medium-adhesion/7-day tape.

6. Masking film is usually supplied in widths of:
   a. 3, 6, 9, or 12 inches.
   b. 6, 9, 12, or 15 inches.
   c. 6, 9, or 12 inches.
   d. 6, 12, or 24 inches.

7. Liquid masking is:
   a. inexpensive.
   b. preferred by most painters.
   c. preferable for flat surfaces.
   d. none of the above.

8. Fine-mesh netting is used to protect:
   a. furniture.
   b. foliage.
   c. structures.
   d. floors.

9. The most economical method of masking is:
   a. paper masking.
   b. film masking.
   c. pre-folded film masking.
   d. plastic sheet masking.

10. Bagged lighting fixtures may:
    a. not be turned on.
    b. be used intermittently.
    c. be used continuously.
    d. none of the above.

11. The primary hazard associated with floor coverings is:
    a. slippery surface.
    b. hidden floor openings.
    c. a tripping hazard.
    d. all of the above.

12. When performing exterior cleaning or painting activities where wind-blown dust or paint residue is a problem and the wind speeds are high, the easiest solution is to:
    a. erect total containment enclosures.
    b. use extra covering.
    c. postpone operations, if possible.
    d. none of the above.

13. Plastic sheeting may be used over shrubbery:
    a. in emergencies.
    b. for extended periods of time.
    c. for up to 3 hours.
    d. none of the above.

14. The preferred containment enclosure is a:
    a. total structure.
    b. free-hanging enclosure.
    c. partial structure enclosure.
    d. none of the above; the level of containment is determined by the type, concentration, and toxicity of the contaminants.

15. One of the most important aspects in completing a job is:
    a. custody of the job.
    b. clean painting practices.
    c. proper cleanup of the site and disposal of waste materials.
    d. none of the above.

NCCER CRAFT TRAINING USER UPDATES

The NCCER makes every effort to keep these manuals up-to-date and free of technical errors. We appreciate your help in this process. If you have an idea for improving this manual, or if you find an error, a typographical mistake, or an inaccuracy in the NCCER's Craft Training Manuals, please write us, using this form or a photocopy. Be sure to include the exact module number, page number, a description of the problem, and the correction, if possible. Your input will be brought to the attention of the Technical Review Committee. Thank you for your assistance.

Instructors – If you found that additional materials were necessary in order to teach this module effectively, please let us know so that we may include them in the Equipment/Materials list in the Instructor's Guide.

Write: Curriculum and Revision Department
National Center for Construction Education and Research
P.O. Box 141104
Gainesville, FL 32614-1104

Fax: 352-334-0932

Craft ________________ Module Name ________________

Module Number ________________ Page Number(s) ________________

Description of Problem

____

____

____

(Optional) Correction of Problem

____

____

(Optional) Your Name and Address

____

Basic Surface Preparation

Module 07106

Painting Trainee Task Module 07106

BASIC SURFACE PREPARATION

Painting and Decorating Contractors of America

National Center for Construction Education and Research

Objectives

Upon completion of this module, the trainee will be able to:

1. Describe preparation tools and materials:
   - Cleaning agents
   - Surface conditioning agents
   - Repair agents
   - Hand tools
   - Power tools
2. Describe or demonstrate preparation methods:
   - Washing and cleaning
   - Hand tool cleaning
   - Power tool cleaning
   - Etching and neutralization
   - Vacuuming
   - Checking for moisture in concrete, stucco, masonry, wood, or plaster substrates
   - Repair/replacement of substrates
3. Describe or demonstrate general preparation procedures for various types of surfaces/substrates:
   - Wood
   - Concrete and masonry
   - Plaster and drywall
   - Metal
   - Synthetic

Prerequisites

Successful completion of the following Task Modules is required before beginning study of this Task Module: Core Curricula, Painting Level 1, Modules 07101 through 07105.

Required Trainee Material

1. Trainee Module
2. Appropriate Personal Protective Equipment

Course Map

This course map shows all of the *Wheels of Learning* task modules in the first level of the Painting curricula. The suggested training order begins at the bottom and proceeds up. Skill levels increase as a trainee advances on the course map. The training order may be adjusted by the local Training Program Sponsor.

Course Map: Painting, Level 1

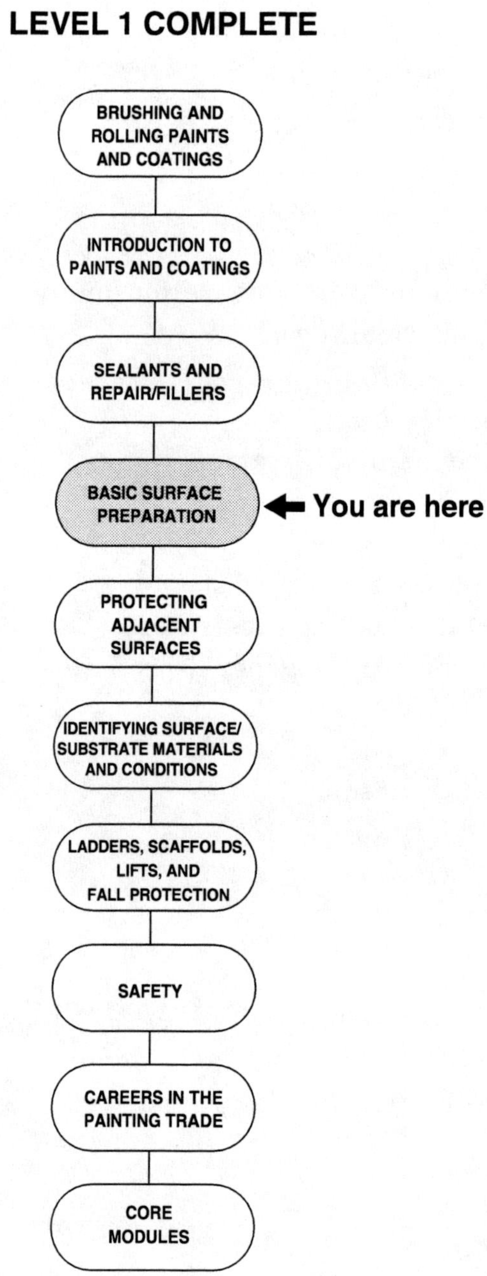

CMAP106.EPS

TABLE OF CONTENTS

TABLE OF CONTENTS

TABLE OF CONTENTS

Trade Terms Introduced In This Module

Abrading: The roughening of a surface by use of hand or power tools, or by abrasive blasting.

Abrasive: A material used for wearing away a surface.

Alkaline cleaner: A cleaner that emulsifies or saponifies a surface.

American Society for Testing and Materials (ASTM): A not-for-profit organization that develops voluntary consensus standards for materials, products, systems, and services.

Consolidant: A liquid, two-part epoxy that penetrates and restores damaged or disintegrated wood.

Emulsify: To break up and hold a contaminant as a tiny droplet so it can be washed away.

Flash rusting: The almost immediate reoxidation of bare steel cleaned with water. Also known as *rust-back*.

High-efficiency particulate air (HEPA) filter: An air filter designed to remove 99.97% of particles above 0.3 microns.

Mildewcide: An additive for a paint or primer, or a separate coating that retards the formation of mildew on painted or treated surfaces.

NACE: An association that deals with corrosion prevention and control for all materials.

Saponification: The process of converting an organic substance into a water-soluble soap by chemical reaction with an alkali.

Steel Structures Painting Council (SSPC): An organization which publishes surface preparation (SP) standards related to the cleaning and coating of steel and concrete in industrial environments.

1.0.0 INTRODUCTION

Applying the finish coat and seeing the final, new, clean surface emerge may be the most rewarding and perhaps the least time-consuming part of a painting job. The most important, and perhaps the most time-consuming, is the surface/substrate preparation. Paint cannot, and will not, make a bad surface or substrate into a good surface. A finish applied over a poor surface or substrate will not provide a lasting coating. It may temporarily make the surface or substrate look good, but in some cases, it will not even do that.

Surfaces or substrates should be inspected before painting. Once painted, you and your employer automatically accept responsibility for the final surface.

Since one primary purpose of painting is protection, it is foolish to invest a large amount of money in painting a surface/substrate that is not prepared properly when new, or has deteriorated to the point where it is not worth protecting. Paint is only as good as the surface or substrate to which it is attached.

This module will cover three general actions necessary to prepare a surface or substrate for coating:

- Cleaning the surface/substrate
- Repairing the surface/substrate (if necessary)
- Pretreating the surface/substrate (if necessary)

All three steps may be performed, or only one or two may need to be done. They may or may not be done in any specific order. They may or may not need to be repeated. The surface and its condition determine which of the three actions should be used, the sequence in which they should be done, and whether they should be repeated. The end result is to obtain a surface or substrate that is clean, dry, and dull.

2.0.0 TYPES OF SURFACES/SUBSTRATES

There are many types of surfaces/substrates that may be painted. To list them all would be almost impossible. Therefore, surfaces have been classified into the following five categories:

- *Cementitious* – Includes all surfaces made from masonry/concrete. Concrete walls and floors, concrete blocks, bricks, stucco, gunnite, stone, and fiber-cement board surfaces are examples.
- *Synthetic* – Includes all man-made materials. Fiberglass impregnated with plastic is one example.
- *Metal* – Includes all surfaces made from metal. Steel, aluminum, and copper are examples.
- *Wood* – Includes all surfaces made from wood. Wood siding and floors and plywood are examples.
- *Plaster/drywall* – Includes all surfaces with a gypsum base material and reinforced with paper or fibers.

Each category will be treated separately and the general rules for preparing them will be outlined.

3.0.0 BASIC PREPARATION TOOLS, MATERIALS, AND METHODS

Previously, you were introduced to three basic actions for preparing a surface/substrate; cleaning, repairing, and pretreating. Before we provide the general preparation procedures for various types of surfaces/substrates, we will examine what materials and tools to use and some of the methods involved in surface/substrate preparation.

1. Surface/substrate preparation tools and materials:
    - Cleaning, surface preparation, and repair agents
    - Hand tools
    - Power tools
2. Surface/substrate preparation methods:
    - Washing and cleaning
    - Hand tool cleaning
    - Power tool cleaning
    - Etching and neutralizing
    - Vacuuming
    - Checking for moisture in concrete, stucco, masonry, or plaster substrates
    - Repair/replacement of substrates

4.0.0 SURFACE/SUBSTRATE PREPARATION TOOLS AND MATERIALS

4.1.0 CLEANING, SURFACE PREPARATION, AND REPAIR AGENTS

Surface/substrate preparation, the foundation of a good painting job, should always begin with washing away dust, dirt, water-soluble contaminants, oil, grease, and any other residue that may interfere with paint adhesion. Make every effort to clean the surface/substrate as well as possible before painting or repainting, or before using other surface/substrate preparation methods. Cleaning agents include detergents, solvents, and chemical treatments such as alkaline cleaners and acids. In addition, surfaces/substrates can be treated with disinfectants and rust and mildew retardants or repaired with patching agents. This section looks at some of these agents, how they work, and when to use them.

4.1.1 Detergents

A detergent is a cleaner used on all types of surfaces to **emulsify** and disperse dirt and oil. To *emulsify* means to break up and hold the contaminant as a tiny droplet so it can be washed away. Detergents work best in water heated to 150°F (66°C) or more. They can be applied by hand (using a clean rag or sponge) or by machine (using steam or high-pressure water) to remove debris loosened by the chemical action. Hand scrubbing with a fiber, wire, or plastic brush is an effective way to remove stubborn spots. Remember to rinse well with water to avoid leaving a film that may interfere with good paint adhesion.

4.1.2 Solvents

Solvent-soaked rags are good for removing dust from wood surfaces, such as floors that have been sanded for refinishing. Solvents, including naphtha and mineral spirits, also dissolve and remove oil, dirt, grease, soil, and waxes from metal surfaces. Solvents are also used for spray cleaning, soaking, and cleaning surfaces by hand. After cleaning with a solvent, always rinse the surface with fresh solvent and a clean rag to remove any residue. Do not use alphatic (wax-based) hydrocarbon solvents under certain acrylic coatings unless the surface is subsequently cleaned with a detergent or alkaline cleaner and then thoroughly rinsed with water before the coating is applied.

WARNING! Most solvents are hazardous and flammable. Using them properly requires strictly following safety precautions, including the use of eye, respiratory, and skin protection. Also, be sure to store used rags in a covered, fire-resistant container until they are removed from the job site.

4.1.3 Surface Preparation Agents

Alkaline cleaners, such as trisodium phosphate (TSP), remove oil and grease and also work on old paint by emulsifying or saponifying it. **Saponification** is a process of converting a coating into a soap by chemical reaction with an alkali. Since the soap formed is water soluble, it rinses away.

Note Do not use strong alkaline cleaners on stainless steel or aluminum.

Since TSP is banned for use in certain areas of the country, sodium metasilicate-based cleaners or other environmentally-safe cleaners may be used (*Figures 1* and *2*). These cleaners are very powerful and effective in their undiluted form and may remove or dull existing painted surfaces. Most are low-foaming cleaners and may be used in low-pressure water washers.

Always rinse a cleaner thoroughly; residue from the cleaner may soften or loosen existing paint or damage a new coating. A litmus test can be used to determine the alkalinity of the surface. Continue rinsing until the pH of the surface is the same as fresh rinse water.

Figure 1. Examples Of Industrial Heavy-Duty All-Purpose Cleaners/Degreasers

106F02.EPS

Figure 2. Examples Of Industrial Environmentally-Safe Cleaners/Degreasers

Acid, another surface preparation agent, is used on masonry and metal. The acid chemically attacks and dissolves certain contaminants. Muriatic acid, a diluted form of hydrochloric acid, will clean spattered mortar from masonry, but is not currently recommended for this purpose. Phosphoric acid will convert rust on steel to an inert iron phosphate. Muriatic acid is used to etch concrete floors to remove efflorescence and laitance and to provide the proper profile for paint. As with alkaline cleaners, always rinse the surface well after an acid wash to remove any residue and to minimize the effect the acid may have on a new coating. Use litmus paper to check the acidity of the surface.

WARNING! Working with any chemical cleaners calls for extra care to protect against harmful vapors, skin burns or irritation, and eye damage. Always wear appropriate protective clothing and follow prescribed safety procedures.

4.1.4 Disinfectants

Household bleach is a good disinfectant for treating surface mildew or bacteria before painting. Mildew is the common term for mold fungus and the damage it causes to paint and surfaces. Mildew is usually black or gray, although it may have different colors. It grows best in warm, humid conditions. Painting over mildew without removing it may result in coating failure as mildew will grow through the new paint.

For milder cases of mildew, clean the surface with a mixture of one part household (5%) bleach to three or four parts water. The bleach will kill the mildew and turn it white. For severe cases, use a stronger solution. Sometimes, an aggregate-type cleaner is added for extra strength.

WARNING! Never mix bleach with ammonia or detergents containing ammonia; doing so will create a poisonous gas.

Some sodium metasilicate-based cleaners, such as Knight's Spray Nine®, are coupled with an Environmental Protection Agency (EPA) registered disinfection capability and are extremely powerful spray-and-wipe germicidal cleaners when used in accordance with the manufacturer's instructions.

After cleaning, always rinse the surface well to prevent residue from interfering with the coating. Avoid splashing bleach or chemical cleaner onto plants or any nearby surface that could be damaged. For personal safety when working with bleach or chemical cleaners, wear rubber gloves and safety glasses or goggles.

4.1.5 Rust And Mildew Retardants

Paint additives can help retard mildew. A **mildewcide** attacks fungal spores—the seed-like particles that cause mildew—when they land on a treated surface, and helps prevent them from growing. Many paints contain a mildewcide. If a paint does not contain a mildewcide, it can be stirred in before application. Always check with the coating manufacturer before adding a mildewcide to a paint. Mildewcides can be added to the primer as well as to the finish coat. There are also mildewcides that can be applied to a surface before painting. In all cases, follow the recommendations of the coating manufacturer.

Note Mildewcides will not kill existing mildew. They are only effective on surfaces where mildew already has been eliminated by washing with bleach or where mildew has not occurred, but could be a problem in the future.

For heavy-duty industrial applications, there are additives for the water used in water washing systems to inhibit **flash rusting** (*rust-back*) of a steel surface until it can be painted. The inhibitor must be compatible with the new coating because traces of the chemical may remain on the surface after cleaning. If the inhibitor is not compatible, or if too much is used, it may cause blistering or adhesion problems. Too little inhibitor, on the other hand, may not prevent flash rusting. Follow the manufacturer's instructions for the proper proportion and use of the inhibitor. Also, make sure the chemicals in the inhibitor will not corrode the water washing equipment and that they are compatible with the coating used on the job.

4.1.6 Low-Pressure Washer Abrasives

Abrasives are the materials used in low-pressure water washer systems. They come in a range of types, shapes, sizes, and degrees of hardness. Given the multitude of options, selecting the right abrasive for the job becomes a matter of determining which material will give a surface the desired finish, which means the proper degree of cleanliness and roughness as well as the type of texture needed for the coating. Other factors to consider are the type of surface to be blasted and its present condition, the consumption rate of the abrasive, its cost, and any environmental regulations that may affect the blasting operation.

There are four types of abrasives: metallic, mineral, byproduct, and special purpose. Of these, mineral abrasives and, to a much lesser extent, byproduct abrasives are used with low-pressure washer, wet-abrasive injectors.

- *Mineral abrasives* – Sand has long been used for abrasive blasting. Silica sand, in particular, is readily available, which makes it economical. Its range of angular and spherical shapes gives it good cleaning action. However, it breaks down quickly and produces a large amount of silica dust, which causes a dangerous lung disease known as *silicosis* when the dust is small enough to be respirable (less than 10 microns). Wet blasting greatly reduces the dust problem, but dust-laden mist is still a hazard for the operator. Nonsilica or heavy mineral sands used in blasting include magnetite, staurolite, and olivine rutile, which are tough, dense, and smaller than silica sand but contain no dangerous free silica. These abrasives are generally not recycled because of how quickly they break down in blasting. Garnet, a tough abrasive, can be recycled a number of times. Other mineral abrasives, such as zircon and novaculite, are generally used for specialty blasting because of their cost and the finish they leave. Zircon, for example, leaves a smooth, matte finish, and novaculite leaves a satin luster.
- *Byproduct abrasives* – Slag, the refuse from smelting or other processes, makes a good abrasive because of its availability, relatively low cost, and low silica content. Its sharp, angular shape gives it good cutting power, and it comes in a full range of sizes. Blasting slags include coal slag (*Figure 3*), copper slag, and nickel slag.

106F03.TIF

Figure 3. Coal Slag Abrasives

4.1.7 Fiberglass And Plastic Repair Kits

Fiberglass repair kits are available for repair of fiberglass-reinforced plastic (FRP) panels or deteriorated steel or concrete structures. These kits consist of glass or fiber mats or cloth or loose fibers along with epoxy resins and hardeners for use on steel, concrete, or fiberglass; or with polyester resins and hardeners for use on fiberglass only. Each of these components may be obtained separately in bulk form from automotive paint supply wholesalers, plastics wholesalers, or building supply wholesalers.

If a damaged plastic substrate can be identified, repair kits may be available from a plastics manufacturer or distributor or an automotive paint supply wholesaler. For example, polypropylene (PP) plastic may be heat welded using a polypropylene filler rod and a hot smoothing tool. The same is true for vinyl (V) or polyvinyl chloride (PVC) plastics using filler rods or liquid resins that are bonded with a hot tool.

To repair minor surface imperfections (scratches, small dents, nicks, etc.) on fiberglass or opaque rigid plastic, automotive polyester body filler or spot putty may be used.

4.1.8 Wood Consolidants

A **consolidant** is a liquid wood epoxy that reinforces or restores damaged or disintegrated wood. It consists of resin and a hardener that, when mixed together properly, can repair rotted, dried out, or spongy wood. It can also be used to reinforce other porous building materials such as plaster.

A consolidant can be brushed, rolled, sprayed, or poured onto a surface. Once it has cured, the restored material can be sawed, nailed, carved, sanded, or planed like any other piece of wood. It can be painted, or stain can be added while it is being mixed.

Also available are two-part epoxy fillers that, when properly mixed, form a putty-like, adhesive material which can be tooled into just about any shape. For that reason, these materials are often used in wood restoration work.

4.1.9 Etching Acids

A metal or concrete surface can be prepared for painting by mechanical methods such as blast cleaning or power tool cleaning, and by chemical methods that use either an acid or alkali to attach and alter the surface of the substrate so it is more suitable for painting. Removing dirt, oil, and grease from a surface is a prerequisite for proper wetting and cleaning with a chemical pretreatment.

- *Acid cleaning of steel* – There are two types of acid surface treatments for steel—phosphate and chromate. Phosphate cleaners are usually based on phosphoric acid attacking the steel and forming an iron phosphate conversion coating that is highly receptive to subsequent coats of paint. There are a number of different phosphate formulas. All involve the use of a phosphoric acid solution that remains in contact with the steel surface until sufficient phosphate passivation occurs (the steel darkens due to the formation of iron phosphate). The surface is then rinsed and dried. Paint application can then proceed. Chromate coatings, based on phosphoric acid, zinc chromate pigments, and vinyl butyral resins, are used in the preparation of steel, aluminum, tin, lead, and other metals. Zinc chromate vinyl butyral wash primers provide very good results for galvanized metal. This wash primer should be applied to a clean surface. After it dries, the surface is ready for coating.

- *Acid etching of concrete* – Concrete floors may be prepared for painting by acid etching to provide the proper surface texture and to clean and remove dirt and other impurities. The concrete surface to be etched should be thoroughly cleaned with water and detergent. If water beads on the concrete, it is likely that oil, grease, or a curing compound is present. These materials will prevent wetting by the acid etching solution, making it ineffective, so they must be removed. Muriatic acid or a 5% to 10% solution of concentrated hydrochloric acid in water can be used to etch the concrete surface.

WARNING! When mixing acid solutions, always add the acid to the water; never reverse the process or a violent, explosive reaction may occur.

4.1.10 Plaster Patching Compounds

Some holes and defects in plaster and other surfaces may be too big or deep to be filled easily with spackling compound. In those cases, patching material such as Plaster of Paris or quick-setting compounds with a glue base can be used.

Plaster of Paris is calcium sulfate or gypsum. It is mixed with water and applied with a putty or broad knife of appropriate width. It sets very rapidly, but adding a small amount of slaked lime or mixing with glue size may prevent premature hardening. It should be smoothed out immediately.

Quick-setting compounds with a glue base also come in powder form and mix with water. As the name implies, they start to dry quickly, so mix only as much material as can be used within the setting time specified on the label. This material is good at repairing deep holes because it does not sag, and it sets fast with little or no shrinkage.

4.1.11 Spackling Compound

Spackling compounds are materials used to fill cracks, nail holes, and dents in order to provide a smooth surface for painting or wallcovering. A desirable quality of these patching materials is that they be nonshrinking. Spackle comes in a premixed, ready-to-use form and in powder form that must be mixed with water at the time of application. Most spackle is made solely for interior use, but some exterior grades are available. For exterior use, epoxy, acrylic latex, and other exterior-type filler materials can also be used.

4.1.12 Joint Compound

Joint compound comes in both dry powder and premixed paste. Dry powder compounds should be mixed with clean, drinkable water to the proper consistency following the manufacturer's instructions. Mix only as much as is needed for the job. Premixed compounds can be applied straight from the container. They eliminate variable results that job-mixed powders may produce, but the choice of which type to use is up to the user.

Drying time varies with the type of compound, and such factors as ambient temperature, humidity, and ventilation in the work area. Check the manufacturer's instructions for information on drying time and how soon the compound may be sanded, recoated, or painted.

Some dry powder compounds are rapid-setting materials. They can be used for filling joints, embedding tape, or both. They generally have a gypsum base that sets in a fixed position and holds it when dry. These compounds have a short *wet life*, which means they must be used within a specified time limit.

Both dry powder and premixed joint compound come in three types:

- Taping compound
- Topping compound
- All-purpose compound

For repair work and use as a spackle, all-purpose compounds serve as both taping and topping materials. They combine the features of both types, and can be used for all coats to embed and top drywall tape, finish metal corner beads, and conceal fasteners. They can also be used to texture a surface.

Other types of joint compound are designed for special uses, such as with water-resistant drywall panels.

WARNING! When mixing dry powder joint compound, or when sanding compound after it has dried, wear the appropriate type of dust mask or respirator to protect against inhalation of powder or dust particles.

4.1.13 Joint Tape

Tape used to join drywall seams may be made of paper, woven glass, fiber, or cloth, but the most widely used is a strong fiber paper with feathered edges. It offers good strength both with and across the grain. Joint tape measures about 2 inches wide and comes in various lengths, including rolls of 75, 250, and 500 feet. For repair work or when covering a small number of seams, an adhesive-backed woven glass or fiber mesh tape is the quickest and easiest to apply.

4.1.14 Caulks, Sealants, And Putties

There are a large variety of caulks, sealants, and putties available. They are generally applied after surface preparation and priming have been completed. These products are covered in detail in another module.

4.2.0 HAND TOOLS

Hand tools are most effective for cleaning small areas. In addition, they are a good choice when power tools are not available, or in confined areas that are inaccessible with power tools. The use of hand tools is cost-effective in maintenance painting programs when only part of the surface needs attention. However, since the user of the tool is the power source, hand-tool cleaning can be slow and tiring work.

Common hand tools include cloth rags, sponges, abrasive pads, sandpaper, scrapers, broad knives and putty knives, wire brushes, and chipping hammers. The best tool to use depends on the job. Before using scrapers, brushes, or hammers, wash away any oil or grease using detergents or solvents. Afterwards, brush, sweep, or vacuum away any debris created by the tools.

WARNING! Working with hand tools is relatively safe, but still requires attention to basic safety guidelines such as:

- Wear protective glasses or goggles to protect eyes from flying debris or irritating detergents and solvents.
- Keep tools in good condition. Replace or repair any defective equipment.
- Do not leave unused tools on scaffolds, ladders, or overhead work areas where they could fall on someone working below.
- Never throw or drop tools from one place to another.
- Avoid using tools that may cause sparks near volatile vapors or combustible material. In some cases, brass tools may be needed to avoid sparks.

Remember that hand tools require you to work with your hands, using tools that are harder, rougher, and sharper than your hands, and working on surfaces that can cause cuts, punctures, and scrapes. Even a splinter of wood or metal or a chip of paint can cause a painful injury. Always wear gloves and goggles when using hand tools and keep alert to the danger of potential injury.

4.2.1 Cloth Rags And Wipers

The simple rag or cloth wiper is perhaps the most widely used surface preparation device. When used with water, detergent, alkaline cleaner, or solvent, it can clean dirt, grease, oil, mud, markings, chalking paint, and other washable contaminants from all types of surfaces. This simple cleaning method is recommended before any other surface preparation work.

Rags may be purchased in bulk or recycled from old household items. Cotton fabrics, such as sheets, towels, or diapers, work better than less absorbent synthetic materials. Always start with a clean rag. A dirty rag may spread contamination.

When using solvents, rags become contaminated quickly, so replace them as necessary. Never leave rags soaked in solvent or other flammable material indoors after quitting time or near any possible source of ignition.

WARNING! According to the Occupational Safety and Health Administration (OSHA) regulations on disposal of waste materials (29 CFR 1926.252), all solvent waste, oily rags, and flammable liquids must be kept in fire-resistant, covered containers until removed from the work site. If rags have become contaminated with a hazardous material such as lead, they must be tested and, if necessary, disposed of as a hazardous waste.

4.2.2 Sponges

Sponges come in natural and man-made form. They are a useful alternative to rags for cleaning washable surfaces. However, rough surfaces may cause a sponge to tear, leaving small pieces on the surface. Another use for a damp sponge is to smooth freshly-dried plaster or joint compound. After each use, wash the sponge thoroughly, rinse, and squeeze it dry.

4.2.3 Tack Cloths

A tack cloth is treated to make it tacky to attract and hold dust. It is very useful for cleaning abrasive dust from woodwork before or between coatings. To make a tack cloth, sprinkle a clean, soft, lint-free piece of cotton with varnish diluted with turpentine. Wring out the cloth until almost dry. Ready-made tack cloths are also available. Dispose of the cloth safely after use.

4.2.4 Abrasive Pads: Wet And Dry

Abrasive pads are webs of nylon fibers imbedded with abrasive material. They can be used wet or dry for a variety of jobs, such as removing dried soil, debris, and loose paint from all types of surfaces, as well as loose rust and mill scale from steel. They can also be used to sand wood surfaces and remove fuzz or nibs between coats. They can also be used to remove dry spray or to feather the edges of a repair area to disguise the repair.

Nonwoven abrasive pads can be used alone or attached to a holder with a handle. The pads alone are flexible and can be used for contour work. They can be cut to size, as needed, and rinsed for reuse.

Sanding sponges are flexible and coated with abrasive material—some on one side, some on all sides—for use on curved or contoured surfaces. Some are available for wet or dry sanding. Like nonwoven abrasive pads, they can be rinsed clean for reuse.

4.2.5 Sandpaper

Sanding smooths rough wood, metal, and plaster, as well as putty, spackle, or other patching material.

Two main considerations in choosing a sandpaper are its abrasive coating and grit. Abrasive coatings include flint, emery, garnet, aluminum oxide, and silicon carbide. Each type has different uses. The grit indicates how coarse or fine the sandpaper is.

The grit of some sandpaper is described in words: very coarse, coarse, medium fine, very fine, extra fine, and super fine. Other types use numbers, which indicate the grams of abrasive on a given amount of backing. The higher the number, the finer the grit. For example, 220 is very fine sandpaper; 36 is very coarse. The full range of grades goes from 12 to 600.

Some sandpapers are made with water-resistant adhesives, so they can be used wet or dry.

In addition, there are two densities of grit: closed coat and open coat. *Closed coat* means the abrasive completely covers the paper. *Open coat* means the abrasive covers about 50%. Open-coat abrasives are recommended for removing paint or varnish when loading or filling with the removed material may be a problem.

Aluminum oxide and silicon carbide sandpapers can handle all sanding jobs. They tend to be more durable and work faster than sand or garnet. Aluminum oxide, which is brown, is an ideal all-around sandpaper for use on bare wood or metal. It is easy to use and produces a high-quality finish. For finishing varnish or lacquer coats, sandpaper coated with silicon carbide, a shiny black mineral, offers good results. Silicon carbide sandpaper comes in two forms—one for dry sanding and one for wet sanding with lubricants. When used with light oil as a lubricant, fine-grit silicon carbide gives lacquer a lustrous satin finish.

Unlike aluminum oxide and silicon carbide, which are both manufactured abrasives, emery, flint, and garnet are natural minerals mined from the ground. Because of its rounded grains, emery is relatively slow cutting. This dull, black sandpaper is used primarily to clean or polish metal. Flint, which is white, is a relatively dull mineral compared to the synthetics. Its softness gives it a short cutting life. Garnet sandpaper, which is reddish-brown, is commonly used for woodworking. It has medium-hard cutting edges, and it produces good results with a moderate amount of labor.

A specialty sandpaper is available from the 3M company for sanding latex paints. It is known in the trade as *green paper*. 3M furnishes it in grits from 40 through 80 (Type 230U) and 100 through 220 (Type 235U). It is a nonloading type with a relatively long life.

Always sand with the grain of the wood; sanding against the grain will cause scratches that are difficult to remove. On flat surfaces, use a sanding block to prevent unevenness and rounded edges. For curved or irregular surfaces, tear the paper into strips and use it without a block.

Start with the grade of sandpaper that will do the best to smooth the surface. Use progressively finer grades, finishing with fine or very fine sandpaper to achieve the smoothness finish necessary. Dampening the wood will raise its grain slightly; sanding the wood after it dries will give a smoother finish. After sanding any surface, always remove all dust with a tack cloth or vacuum before coating.

4.2.6 Steel Wool

Steel wool is another abrasive material used for surface preparation. It can be used wet or dry for cleaning and polishing. However, wet steel wool strands can rust. Be sure to remove them from the surface. Steel wool is graded by coarseness, ranging from grade 4 (the coarsest), to 0000 (the finest). Non-rusting stainless steel, aluminum, or brass wool is also available. Use protective gloves when working with steel wool, and dispose of it properly after use.

WARNING! According to OSHA regulations on disposal of waste materials, all solvent waste, including steel wool or other wool saturated with a flammable liquid, must be kept in fire-resistant, covered containers until removed from the work site. If the wool has been contaminated with a hazardous material such as lead, it must be tested and, if necessary, disposed of as hazardous waste.

CAUTION Do not use steel wool on wood that will be finished with water-based paint or stain. The residual steel wool particles embedded in the wood will rust.

4.2.7 Scrapers

Scrapers are one of a painter's all-purpose tools. They come in a variety of types, shapes, and sizes for different jobs. They are used to remove peeling, flaking, or blistering paint, rust, or debris from any type of surface. They are also used to strip old wallcovering that is dry or that has been treated with chemical remover, and to remove paint and varnish that has been softened for stripping. Different types and uses of scrapers are shown in *Figures 4* and *5*.

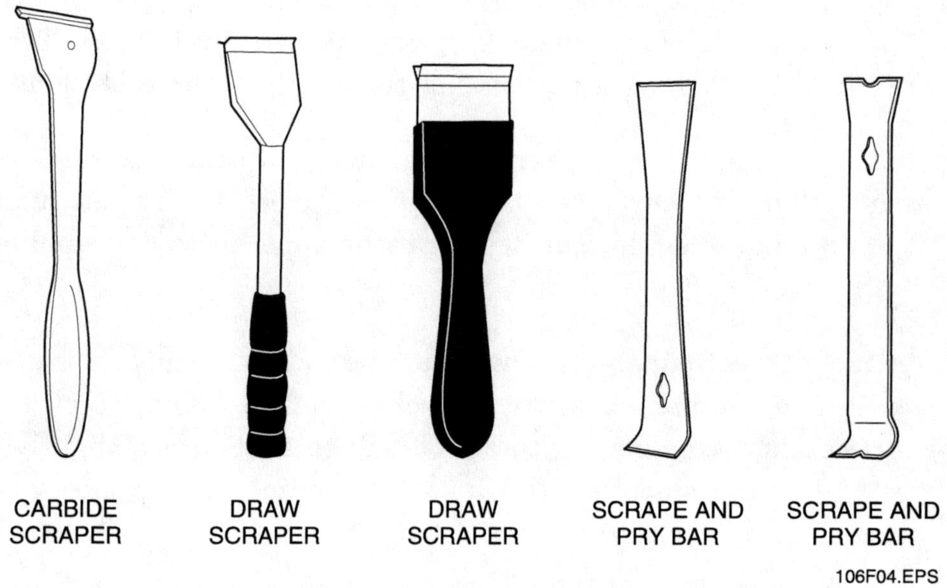

Figure 4. Typical Hand Scrapers

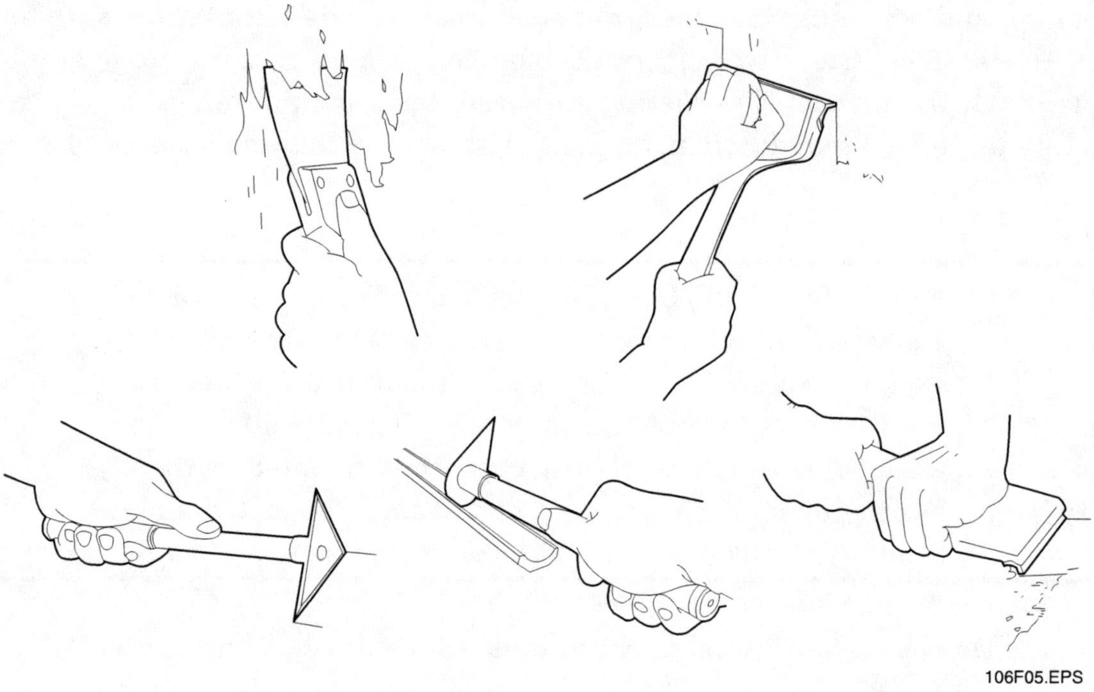

Figure 5. Examples Of Different Types And Uses Of Scrapers

One type of scraper has a permanent blade that is between 1 and 4 inches wide. The blade may be stiff or flexible, straight or bent, and the edge of the blade may be straight or angled. It is used to loosen and push material away.

Another tool, called a *draw scraper*, is made to be pulled across the surface to be cleaned. It usually has a removable blade up to about 5 inches (127 mm) wide. The handle generally is from 6 to 12 inches (152 to 305 mm) long. Long handles sometimes have a grip knob over the blade for better pressure control.

Molding scrapers have triangular or teardrop-shaped blades for getting into the grooves of woodwork. Window scrapers have a retractable single-edge razor blade in a handle. They are good for removing dried paint or caulk from glass and other smooth surfaces such as marble.

When scraping wood, plaster, or drywall with any kind of tool, be careful not to scratch or gouge the surface. Do not use a steel scraper on a masonry surface when flammable liquids are present as it may cause a spark and ignite the liquid. As with all hand tools, use scrapers with care and be alert to potential danger and injury.

4.2.8 Putty Knives And Broad Knives

In working with wood, plaster, and drywall, a variety of flat-bladed knives in different sizes and widths are available. A putty knife, which has a blade 1 to 2 inches (25 to 50 mm) wide, is used to apply putty or other patching material and to scrape off loose paint. Sometimes it may be necessary to fabricate a hand tool to clean unusual or irregular surfaces by grinding and shaping putty knives of different styles and sizes to fit a surface profile. For this reason, broken putty knives and other damaged cleaning tools are often saved.

Blades of broad knives range in size up to 24 inches (610 mm) wide. They are used to fill holes and cracks with spackling compound or patching material. In finishing drywall, they are used to apply joint compound and joint tape to the seams of drywall panels. Typically, a narrow blade is used to apply the first coat of joint compound, then wider blades are used for each successive coat. A hammer-end broad knife (*Figure 6*) has a metal tip on the butt of the handle, which is used to pound in popped nail heads in drywall panels. Broad knives are also used as a guide for trimming wallcovering and as a shield to protect adjacent areas from spatter when painting.

A versatile device called a *combination tool* or a *5-in-1 tool* (*Figure 7*) has a special blade that may be used as a scraper, putty remover, patching spreader, and paint can opener. It has a sharp point for gouging old, loose paint from crevices or open cracks for patching. Its half-round, cut-out blade is good for cleaning paint roller covers. Some uses of a combination tool are illustrated in *Figure 8*.

106F06.EPS

Figure 6. Broad Knife With Hammer End

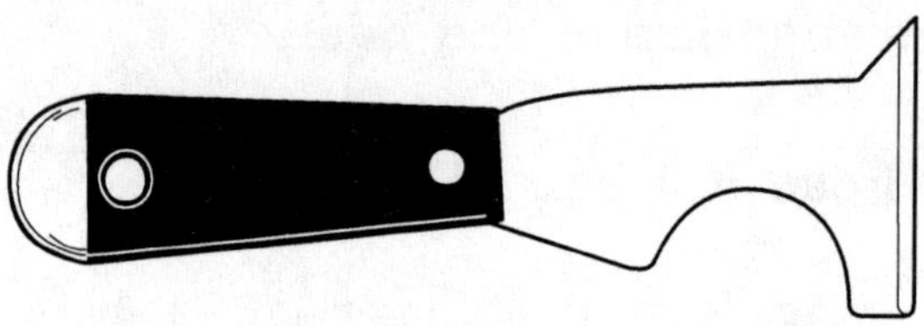

106F07.EPS

Figure 7. Combination Paint Tool

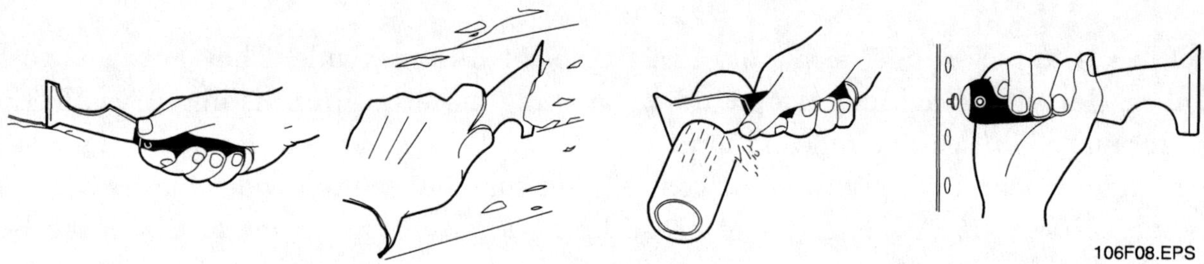

106F08.EPS

Figure 8. Various Uses Of A Combination Tool

4.2.9 Wire Brushes

Wire brushes are useful for cleaning loose rust or paint, varnish, hardened plaster spatter, and other debris from all types of surfaces. There are different types of wire brushes. Within each type, there are variations in size, number and rows of wire, as well as wire stiffness.

One type of brush is the flat block wire brush (*Figure 9*), which has bristles mounted in a flat block of wood like a scrub brush. It is used to clean away paint and rust from large, flat areas. Other types of brushes have bristles at one end of a long handle. They include an all-purpose shoe handle brush, a combination wire brush and scraper (*Figure 10*) which provides two ways to remove stubborn contaminants, and a curved handle brush (*Figure 11*) which gives better brushing control because fewer wires touch the surface at a time. The needs of each job should determine the type of brush to use.

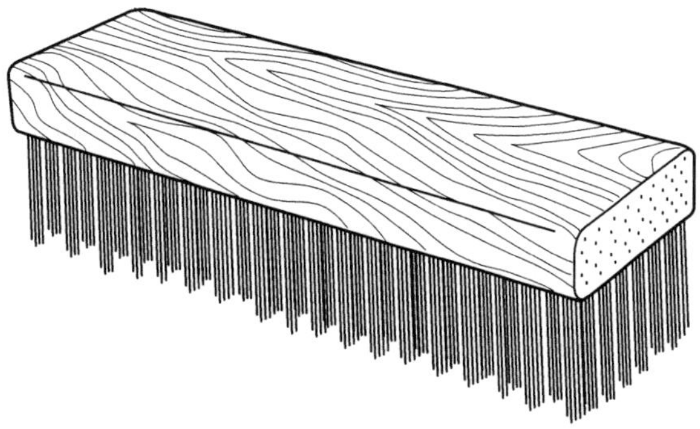

106F09.EPS

Figure 9. Flat Block Wire Brush

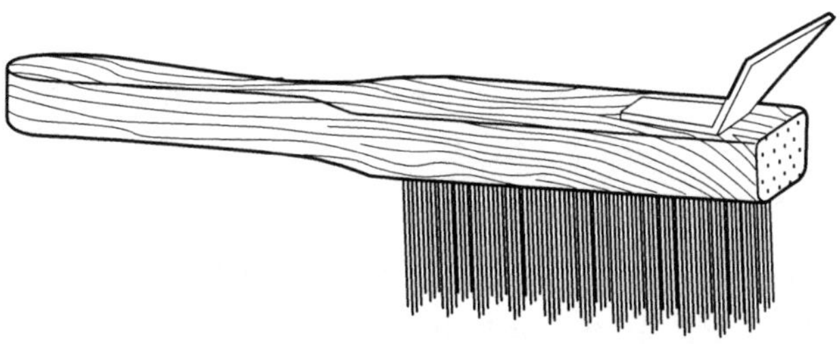

106F10.EPS

Figure 10. Combination Wire Brush And Scraper

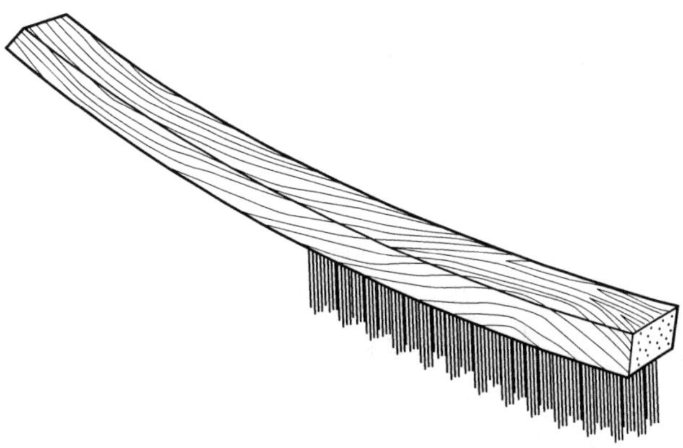

106F11.EPS

Figure 11. Curved Handle Wire Brush

4.2.10 Chipping Hammers

A chipping hammer is used to chip away layers of loose rust, paint, and mill scale from steel surfaces. It may be the first tool used to remove such contaminants, followed by other tools, such as scrapers, nonwoven abrasives, and wire brushes. The heads of chipping hammers come in different configurations. Some have two wedge-shaped faces, one pointed, the other flattened, which make the tool adaptable to different uses. When using a chipping hammer, avoid marking or raising burrs on the surface that may interfere with proper coating.

4.3.0 POWER TOOLS

Power tools are used when hand tools are too slow or ineffective in cleaning a surface adequately. For industrial painters, power tools are often selected when abrasive blasting is impractical, too expensive, or cannot be used because of environmental factors.

Generally, power tools will remove old paint, dirt, and hard, brittle material from all surfaces; spattered mortar, protrusions, fins, and bulges from concrete and masonry surfaces; roughness from wood; and loose mill scale or rust, weld spatter, sharp fins and edges, and weld flux from steel.

In addition to removing contaminants, power tools are useful for preparing welds on steel surfaces before blasting or coating, and grinding sharp edges, which may cause paint to draw thin.

As a surface preparation method, power tool cleaning is an important alternative to abrasive blast cleaning near sensitive machinery and in populated areas where the dust and debris generated by blasting could be damaging to the equipment or dangerous to people. Power tools also can be an effective choice in removing lead-based paint, especially when used with an integrated vacuum system, because they limit the amount of hazardous waste that is created compared to conventional abrasive blasting.

In recognition of the increased role of power tool cleaning, the **Steel Structures Painting Council (SSPC)** developed a standard, SSPC-SP11, *Power Tool Cleaning to Bare Metal*, to supplement its earlier standard, SSPC-SP3, *Power Tool Cleaning*. SSPC-SP3 requires the removal of paint, rust, and mill scale loose enough to be lifted with the blade of a putty knife. SSPC-SP11 requires removal of all detrimental material except "slight residues of rust and paint in the lower portion of pits." It also requires roughening to a 1-mil surface profile. SSPC-VIS 3, *Visual Standard for Power- and Hand-Tool Cleaned Steel*, was developed to supplement these standards by providing an approximate image of the specified degrees of cleanliness. It shows seven initial conditions—four unpainted surfaces and three previously painted surfaces—and how each of them looks in compliance with SSPC standards for hand tool cleaning, power tool cleaning, and power tool cleaning to bare metal. A summary of the various SSPC-SP standards is given in Appendix A.

In a comparison of paint removal methods, the *Industrial Lead Paint Removal Handbook* looks at the cost of equipment, quality of preparation, debris created, containment required, and production rate of basic power tool cleaning and power tool cleaning to bare metal with and without a vacuum attachment.

The comparison shows equipment costs to be inexpensive or, with a vacuum attachment, moderately expensive. The quality of preparation for painting ranks very poor to marginal for basic power tool cleaning, but good to excellent for power tool cleaning to bare metal. With a vacuum attachment, both degrees of power tool cleaning are found to produce little or no dust, little debris, and require little or no containment. Without a vacuum attachment, the results are somewhat lower. Power tool cleaning ranks low in terms of production, however, and power tool cleaning to bare metal ranks low or very low.

WARNING! Power tools require safety precautions to prevent injury. Here are some safety tips:

- Wear safety glasses or goggles to protect eyes from flying debris. Also, wear gloves or other appropriate gear, including ear protection if warranted by the noise level of the tool.
- Keep tools well maintained and lubricated.
- Operate electrical tools only in dry areas. They should be properly grounded and plugged directly into a ground fault interrupter. Keep power cords in good condition.
- Be sure that pneumatic tools have the proper air supply. Use moisture separators and filters to keep the air supply clean and dry.
- Follow the manufacturer's recommendations for maximum operating speed.
- Use recommended tool guards.
- Disconnect the tool from its power source before changing or tightening attachments.
- Wear an approved respirator in high dust areas and when working with lead-based paint or other hazardous contaminants.
- Use nonsparking tools when working around volatile vapors or explosive materials.
- Disconnect tools when not in use.
- Avoid fatigue, which can cause carelessness, by using the right tool for the right job and taking breaks as needed.

The variety of available power tools and accessories provides a wide range of options for cleaning a surface. All tools, whether powered by electricity or air pressure, can be grouped into three categories:

- Impact tools, which clean by striking a surface
- Rotary or vibrating tools, which use rotating or vibrating abrasives to clean a surface
- Rotary impact tools, which clean by both rotary and impact action

The choice of tool depends on the type of substrate to be treated, its original condition, the desired degree of cleanliness, and the work environment.

All surfaces should be free of oil, grease, and dirt before power tool cleaning. Afterwards, any dust, dirt, debris, or residue should be swept, brushed, blown, or vacuumed away before painting.

Working with power tools requires care to prevent them from damaging the surface by cutting or gouging, raising burrs, or creating sharp edges. Any of these defects can result in coating failure because if the paint does not cover them well enough, they may provide a place for corrosion to begin. Edges and irregular surfaces are often the areas where corrosion begins because of inadequate preparation and/or coating coverage.

Following are descriptions of some of the common power tools used in surface preparation.

4.3.1 Grinders And Sanders

Grinding and sanding tools use rotary or vibrating action to remove contaminants from concrete and steel, and make rough wood surfaces smooth.

Two common types of attachments for grinding and sanding tools are made of nonwoven abrasive and coated abrasive material. Nonwoven abrasive is a synthetic fiber web treated with abrasive grit in different grades and densities. It comes in the form of wheels, cups, or discs, which wear evenly to provide a constant rate of cleaning. Coated abrasive is made of abrasive material bonded to a backing material. It also comes in various forms, including sheets, rolls, discs, belts, and flaps.

There are several distinctions between these types of abrasives. One is that nonwoven abrasives are less susceptible to loading or filling up with dust when removing paint. Another is that coated abrasives can remove surface contaminants as well as base metal, which is useful for weld grinding.

Nonwoven and coated abrasive attachments can be mounted on either straight, in-line rotary tools or vertical, right-angle rotary tools (*Figure 12*). Wheel-shaped attachments work well on straight rotary tools; discs and cups work well on vertical tools.

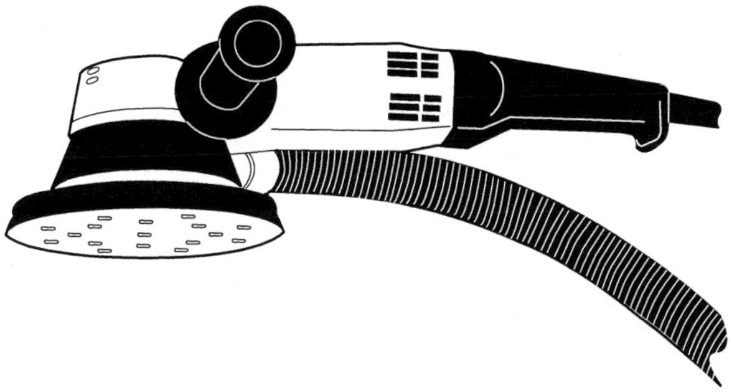

106F12.EPS

Figure 12. Rotary Sander With Vacuum System

Disc sanders and grinders have a revolving flat, circular abrasive that can be used to aggressively remove heavy coatings and other contaminants. They work well on metal or concrete, but because they can easily gouge a surface, they may be too rough for wood or plaster. A specialized, variable-speed power paint remover for wood and composition siding or wood floor decks is shown in *Figure 13*. It is equipped with a non-loading, semi-rigid, tungsten-carbide metal disc in 24, 36, or 46 grit. It also has fine adjustments for depth of cut and horizontal tracking of a siding lap. It is designed for fast removal of deteriorated surfaces. Some finer sanding may be required to produce a smooth surface after the paint or surface is removed.

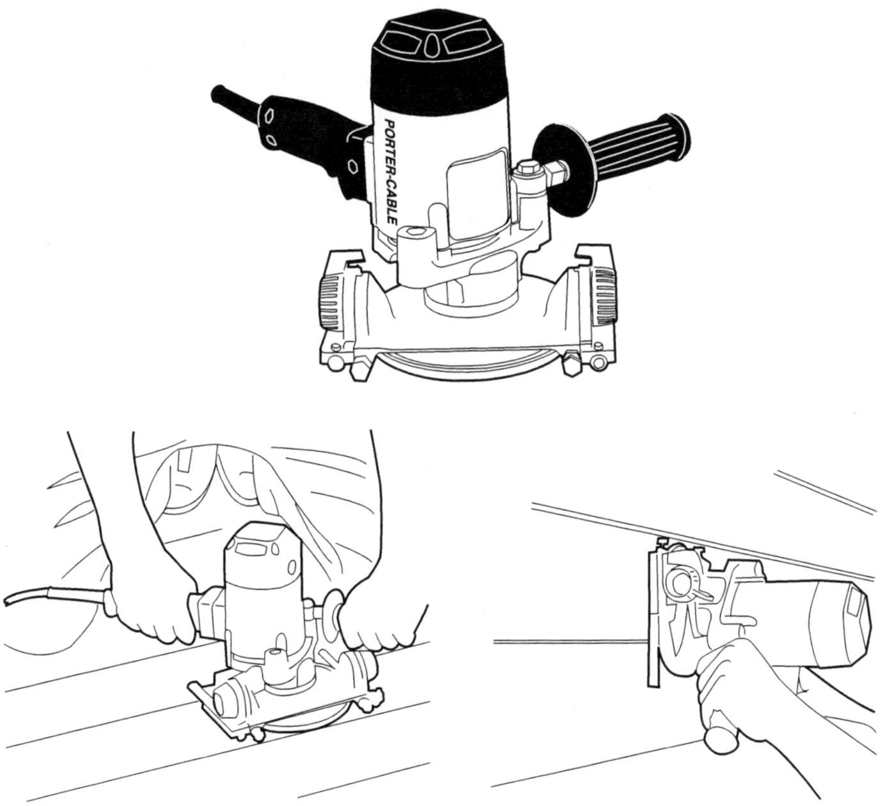

106F13.EPS

Figure 13. Power Paint Remover

Orbital sanders (*Figure 14*) can be used for coarse to very fine sanding, depending on the particular tools and the abrasive used. If used properly, they can sand without scratching or gouging a surface. The flat abrasive pad of an orbital sander rotates at high speed while moving in a small, circular motion. It can be used on wood, drywall, and other surfaces.

106F14.EPS

Figure 14. Orbital Sander

Belt sanders have a continuous belt of abrasive that moves in one direction. The width of the abrasive belt varies, depending on the size of the tool. Belt sanders can only remove limited layers of paint without gouging a surface.

Half-sheet sanders (*Figure 15*), palm sanders, and detail sanders (*Figure 16*) are all finishing tools. They are vibrating sanders with either a straight line or oscillating action. Half-sheet sanders have flat, rectangular pads to which a half sheet of sandpaper is attached. They work well on large, flat surfaces. Palm sanders, so named because they fit comfortably in the palm of the hand, are good for finishing woodwork. They work well on both flat and rounded surfaces. Detail sanders have small sanding pads attached to a handle for smoothing otherwise inaccessible corners and edges.

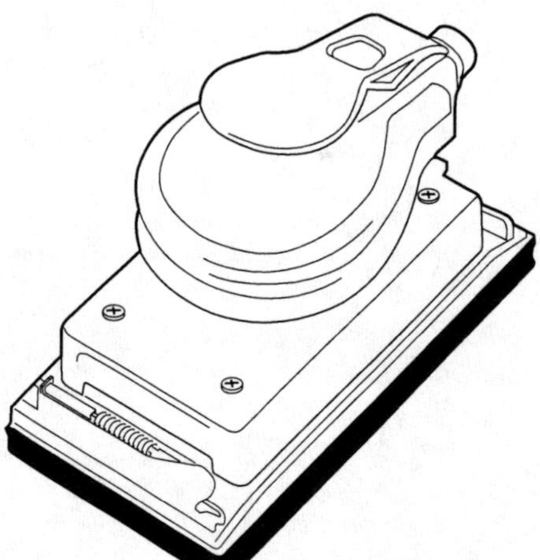

106F15.EPS

Figure 15. Half-Sheet Sander

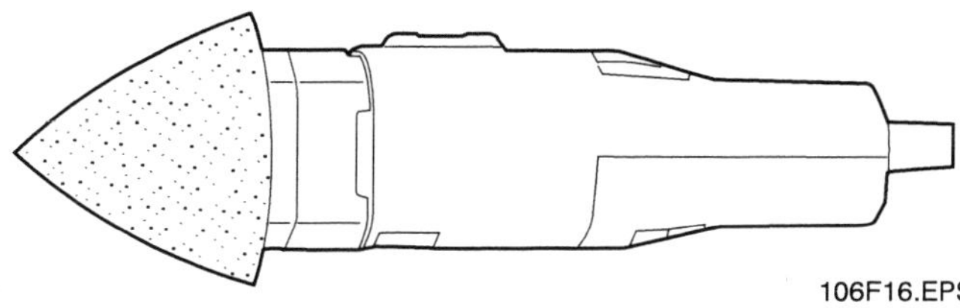

106F16.EPS

Figure 16. Detail Sander With Triangular Head

4.3.2 Power Wire Brushes

Stiff wire bristles attached to a rotary cleaning tool (*Figure 17*) will brush away old paint, dirt, mortar spatter, and other debris from concrete or masonry, as well as loose mill scale, loose rust, weld flux, slag, dirt, and old paint from steel.

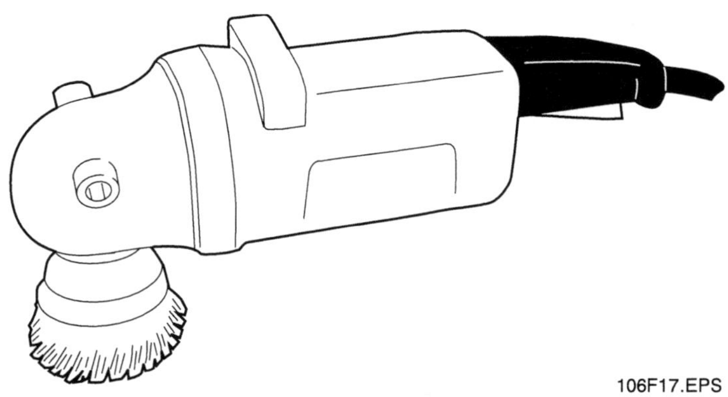

106F17.EPS

Figure 17. Power Wire Brush With Cup-Shaped Bristles

Bristles come knotted or crimped and in the shape of a wheel or a cup. They attach to either a straight rotary tool with a rotating axis in line with the handle, or a vertical rotary tool with an axis perpendicular to the handle. The tool to use depends on the job and shape of the bristles that would work best. Wheel bristles are usually used on straight rotary tools; cup bristles on vertical rotary tools.

Operating wire brushes at excessive speed or holding them in one spot for too long may burnish the surface and leave it too smooth and polished to be coated effectively.

4.3.3 Power Files

Power files are a substitute for tedious hand filing to smooth or remove burrs from metal, wood, plastic, or fiberglass surfaces. The power advantage is the regulated length of each stroke and the number of strokes per minute, which helps to achieve a uniform finish. It is also useful in doing intricate work that requires precision. The files, which are either straight or curved, can be used to reach into confined areas.

4.3.4 Needle Guns

A needle gun (*Figure 18*) is a type of compressed air-powered impact tool. It has a bundle of long, cylindrical steel rods mounted in front of a piston that strikes them against a surface. Because each needle hits the surface individually, a needle gun is good for cleaning irregular areas, such as corners (*Figure 19*) and crevices or around bolt and rivet heads. It works best on loose or brittle material.

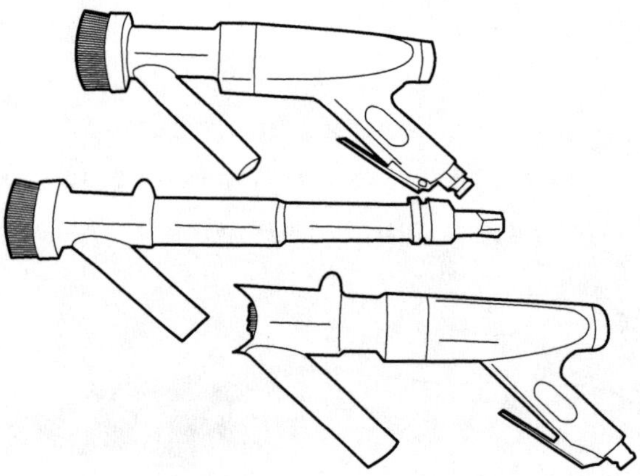

106F18.EPS

Figure 18. Needle Guns

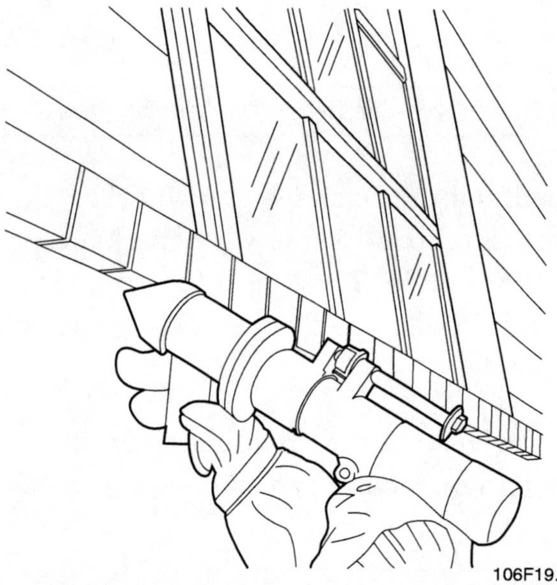

106F19.EPS

Figure 19. Needle Scaler With Attachment For Corner Cleaning

Other types of impact tools include chipping and scaling hammers, piston scalers, and power chisels (*Figure 20*). All work on the same principle as the needle gun. Some of these tools will remove heavy rust, mill scale, thick paint, weld flux, and slag.

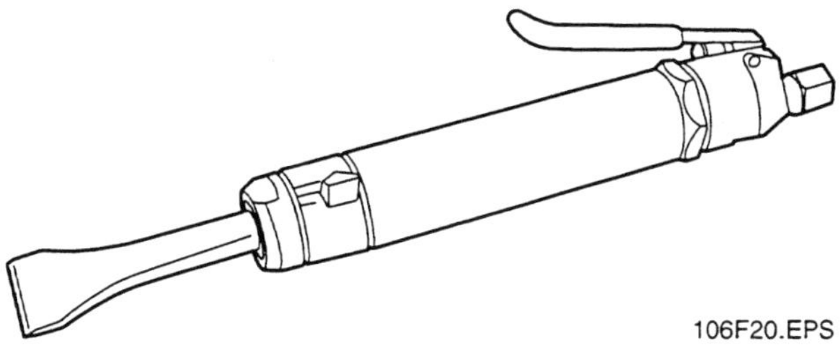

106F20.EPS

Figure 20. Power Chisel

4.3.5 Rotary Impact Power Tools

A rotary impact tool, such as the one shown in *Figure 21*, uses rotating action to throw different types of fixed abrasives against a steel or concrete surface to remove heavy coatings and contaminants. It can also be used to grind concrete. Three common types of abrasive hubs are:

- *Cutter bundles* or *stars* (*Figure 22*) which are hardened, star-shaped steel washers that rotate on spindles around the tool's axis.
- *Rotary hammers* (*Figure 23*), which are free-swinging steel plates attached to the tool's rotating hub.
- *Rotary flaps* or *peeners* (*Figure 24*), which are round carbide cleats attached to flexible flaps mounted in the rotating hub.

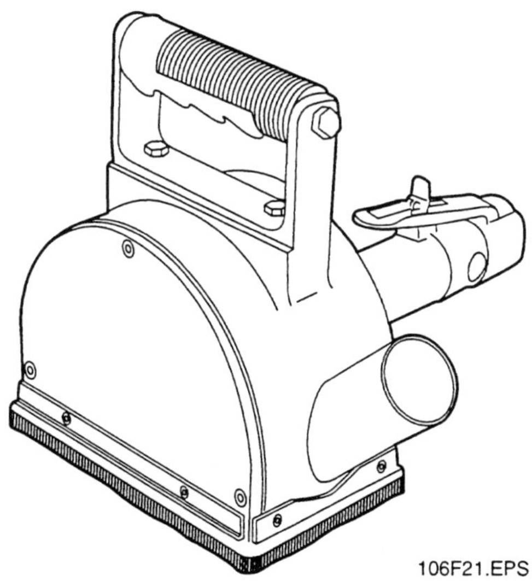

106F21.EPS

Figure 21. Rotary Impact Tool

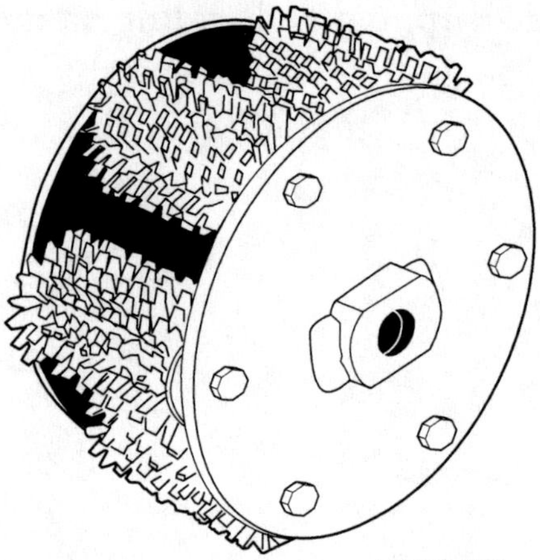

Figure 22. Cutter Bundles Or "Stars"

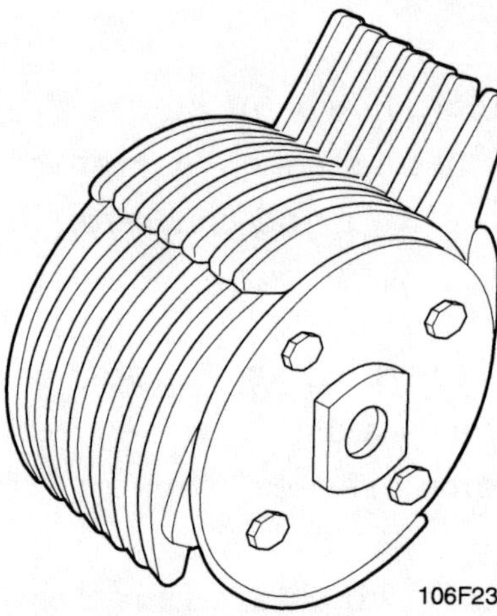

Figure 23. Rotary Hammers

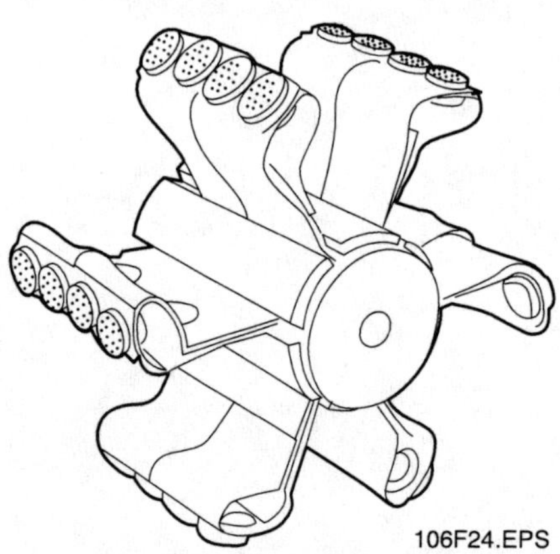

Figure 24. Rotary Flaps

In addition to hand-held models, peening equipment is also available in upright models for descaling floors.

4.3.6 Vacuum Shrouds

Some power tools are equipped with a vacuum system to capture dust, debris, and other airborne contaminants at the surface and prevent them from escaping into the environment. The head of the tool is enclosed in a shroud or casing (*Figure 25*), which is attached to a vacuum hose. A variation of this is a sander with holes in and around the sanding pad that draw dust into a vacuum hose.

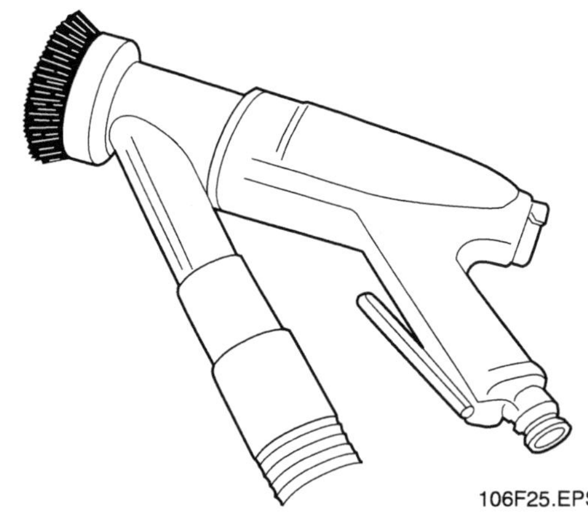

Figure 25. Power Tool With Vacuum Shroud

In some cases, the vacuum system is tool-triggered; that is, the vacuum system starts and stops with the tool. Other tools allow independent operation of the vacuum system.

Some vacuum systems are equipped with a **high-efficiency particulate air (HEPA) filter** (*Figure 26*) which is designed to remove 99.97% of all particles above 0.3 microns. To put that size into perspective, 25 microns equal 1 mil, and 1,000 mils equal 1 inch.

The benefits of using power tools with an integrated vacuum system are that by removing dust and debris, the workplace is safer and cleaner, and the worker can see better. Dust is not only a health risk, it is also an environmental hazard. Under certain conditions, it can even be combustible. Also, tools equipped with a vacuum system reduce the need for containment systems and protective gear for workers. However, the tools with which these shrouds are used are often very loud and require adequate hearing protection.

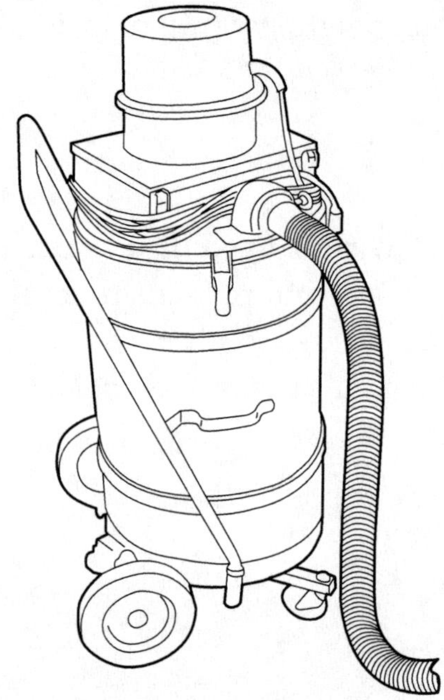

106F26.EPS

Figure 26. HEPA Vacuum

4.3.7 Blasting Systems

Blasting systems propel abrasive material at a surface with enough speed and power to clean away all contamination and leave an anchor pattern that guarantees good bonding of paint. In fact, abrasive blasting cleans and creates a profile better than any other method of surface preparation. It works effectively on the heaviest layers of rust, mill scale, paint, and other foreign material on wood, masonry, concrete, steel, and other surfaces.

Large jobs require powerful and elaborate equipment. Abrasive blasting uses air pressure, water pressure, or centrifugal force to clean and abrade a surface. This type of equipment will be covered in a later module.

4.3.8 Water Blast Cleaning Equipment

Water blasting is an alternative to conventional air abrasive blasting for cleaning wood, concrete, or metal surfaces. It is effective for removing both visible and nonvisible contaminants. Blasting with water alone will remove dirt, debris, and loose or water-soluble contaminants. Sometimes, detergents and/or alkaline cleaners are added or abrasives are injected into the water stream to make this cleaning method even more effective. Detergents and alkaline cleaners remove oil, grease, dirt, etc.; abrasives remove tightly-adhered contaminants.

As the pressure of the water increases, so does the cleaning ability of the blasting equipment, which falls into one of four categories:

- Low-pressure water cleaning, which uses water at pressures less than 5,000 psi (pounds per square inch) or in metric terms, 352 kilograms per square centimeter (352 kg/cm^2).
- High-pressure water cleaning, which uses water at pressures of 5,000 to 10,000 psi (352 to 703 kg/cm^2).
- High-pressure water jetting, which uses water at pressures from 10,000 to 25,000 psi (703 to 1,758 kg/cm^2).
- Ultra high-pressure water jetting, which is any pressure over 25,000 psi (1,758 kg/cm^2).

In this module, only low-pressure water cleaning [less than 5,000 psi (352 kg/cm^2)] will be discussed. The other types will be covered in later modules.

The type and condition of the substrate being cleaned and the size of the job will determine the kind of water blasting equipment required. *Figure 27* is a general guide showing uses for pressures ranging up to 5,000 psi (352 kg/cm^2) and water flow from 1 to 5 gallons per minute (gpm) (3.8 to 19 liters/minute) or more.

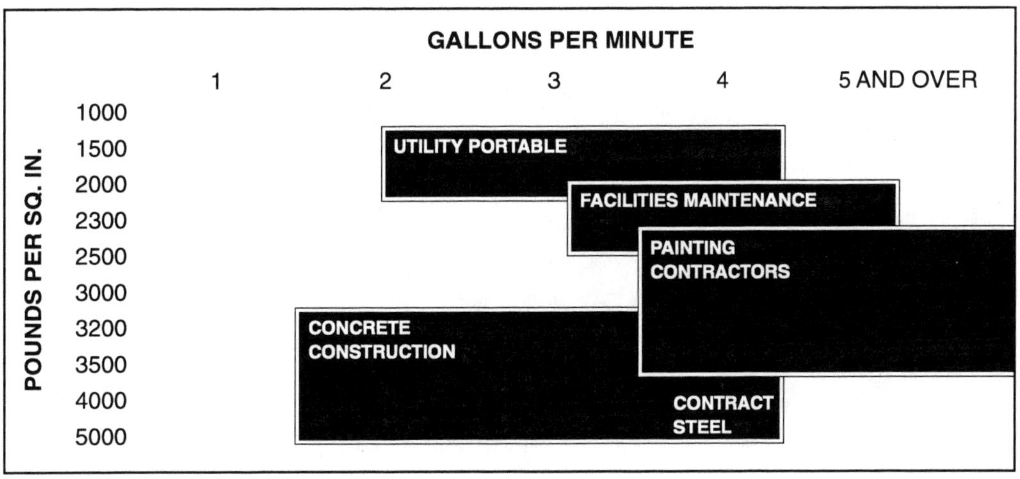

106F27.EPS

Figure 27. Guide To Low-Pressure Washers

All units include these basic components:

- A water pump to pressurize water
- A power source for the pump
- An inlet filter and pressure gauge
- A hydraulic hose and fittings to get the water from the pump to the nozzle
- A gun/nozzle equipped with a control valve, usually a trigger device to regulate blasting

Figure 28 shows a typical cold water unit of this pressure class. Externally-heated water may also be used and units with on-board heating are available.

Figure 28. Low-Pressure Washer

The nozzle of the unit may be circular, which produces a round spray, or tapered for a flat or fanned spray. The nozzle angle and stand-off distance will depend on the material being removed and the equipment being used. Various accessories can be used with the washer such as an automatic detergent and cleaning chemical in-line injector (*Figure 29*) or a rotary power scrub brush (*Figure 30*) when spray mist is objectionable or hard-to-remove contaminants are encountered.

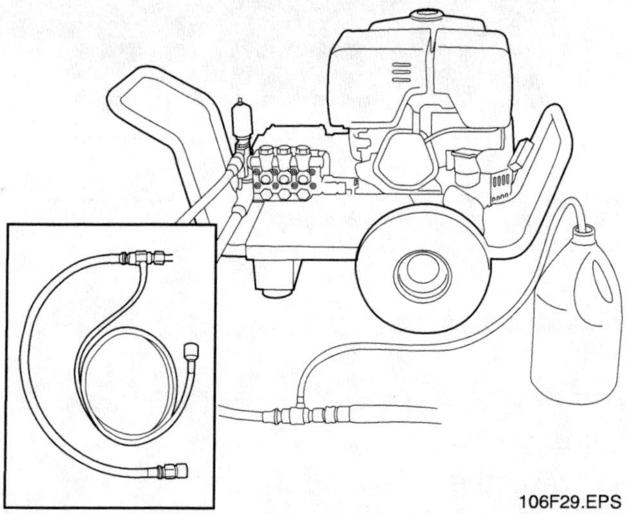

Figure 29. Automatic In-Line Injector

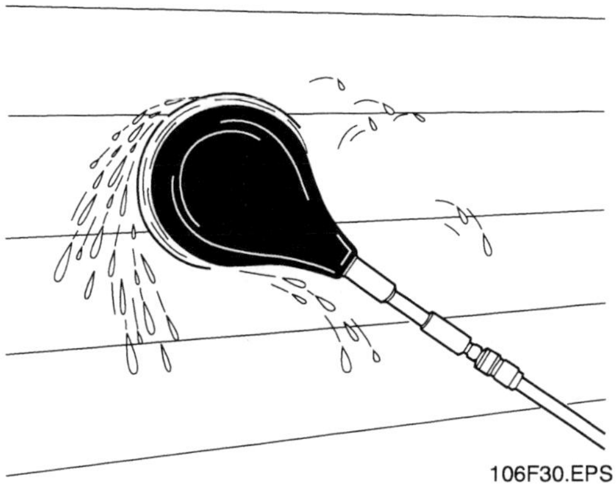

Figure 30. Rotary Power Scrub Brush

Also available are telescoping extensions (*Figure 31*), horizontal (floor) scrubbers with rotating nozzles (*Figure 32*), and water blast nozzles (*Figure 33*). Water blast nozzles use a highly concentrated steam rotating in a circular pattern that acts like a sandblaster without the sand. Wet abrasive sandblasters (*Figure 34*) are also available for dust-free sandblasting.

Figure 31. Telescoping Extension

Wet abrasive blasting with an abrasive injection is distinguished from another form of wet abrasive blasting when water is injected into an air and abrasive stream to cut down on dust. In either method, mixing water and abrasive has some drawbacks. Wet abrasive cannot be recycled easily or economically. The surface must be rinsed with clean water after blasting to remove abrasive residue. Steel surfaces may be subject to flash rusting (rust-back), although the rinse can be treated with an inhibitor to prevent this. The inhibitor must be compatible with the coating, or adhesion problems may develop. Generally, wet abrasive blasting is not the preferred method of obtaining a very high-quality surface.

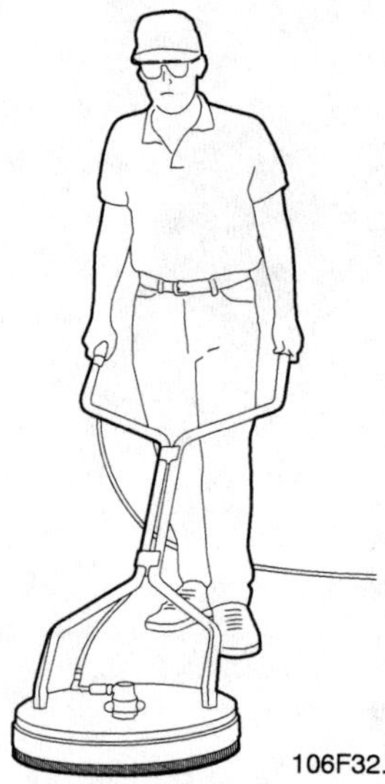

106F32.EPS

Figure 32. Horizontal Surface Scrubber

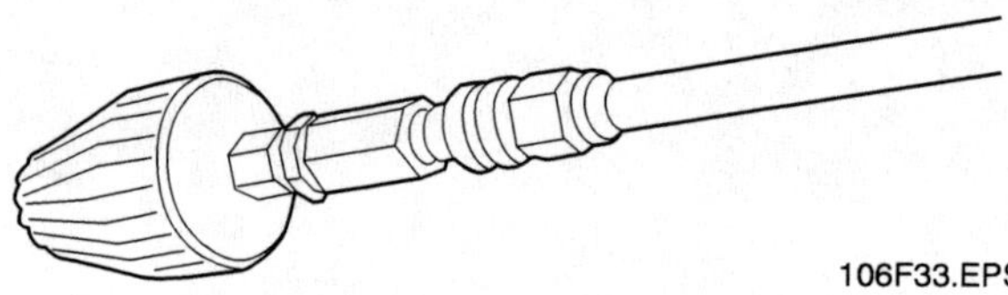

106F33.EPS

Figure 33. Water Blaster Nozzle

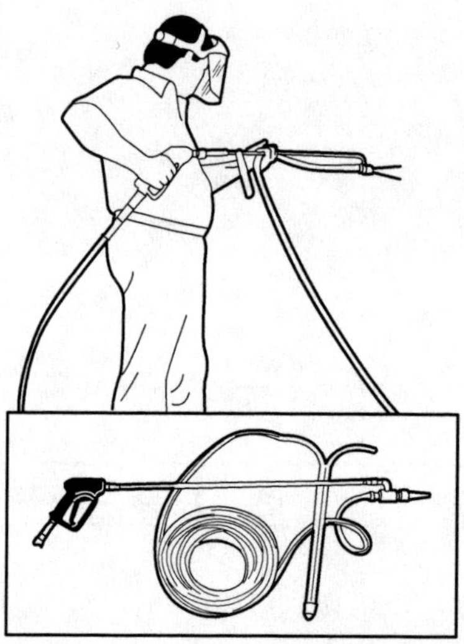

106F34.EPS

Figure 34. Wet Abrasive Sandblaster

Working with high-pressure water requires extra caution because of the potential risks to workers and the possible damage to nearby property. Operators should be well trained and equipped with appropriate protective gear.

WARNING! When using wet abrasive injectors or chemical injectors, operators must wear respirators to prevent inhalation of abrasive dust mist or chemical mist. Always follow prescribed safety rules, and watch for fatigue.

Use extra care when working from a scaffold because high-pressure equipment can cause problems with balance and footing. For the same reason, keep the work area as clean and unobstructed as possible.

Equipment should be well maintained and inspected regularly for proper reinforcements and connections. Never operate equipment beyond its recommended limits, and never use water blasting around electrical units.

5.0.0 SURFACE/SUBSTRATE PREPARATION METHODS

5.1.0 WASHING AND CLEANING (SSPC-SP1 OR ASTM D-4258)

Cleaning is an indispensable part of surface preparation. Every surface should be thoroughly cleaned so the applied coating can form a strong, firm bond. Cleaning methods range from hand washing with water to machine blasting with abrasives. A paint job may require one or more cleaning methods depending on the type of surface, the surface condition, and how clean the surface needs to be for the new coating.

Washing a surface to prepare for a new paint job is a widespread industry practice. In recent years, the number and effectiveness of special cleaning compounds have increased greatly. Cleaning compounds come in varying strengths for different classes of work—some materials for cleaning slightly soiled and greasy surfaces such as may be found in the kitchen of an average home, others for heavier duty in offices and public buildings, and still others for industrial plant cleaning.

Here is a rundown of common washing and cleaning methods used in surface preparation as defined in SSPC-SP1, *Solvent Cleaning* or **American Society for Testing and Materials (ASTM)** D-4258, *Surface Cleaning Concrete for Coating.*

5.1.1 Hydrocarbon-Based Solvent Cleaning

Solvents are liquids capable of dissolving a material. By this definition, water is a solvent. It can be used alone or with soaps or detergents to wash surfaces by hand or, more effectively, with a pressure washer. However, in the painting industry, solvent usually refers to liquids such as denatured alcohol, naphtha, mineral spirits, toluol, kerosene, and others.

After washing with a solvent, it is important to rinse the surface well with water to clean away the residue, which can prevent proper paint adhesion.

Check with the coating manufacturer before using solvents in the preparation of surfaces for painting because they may not be compatible with the coating to be applied. Solvents should not be used on porous substrates such as uncoated wood, concrete, masonry, etc. since they will cause contaminants to penetrate the surface.

Solvents generally work quickly and are easy to use. They require little equipment. A bucket and a sponge or rag are usually enough. Disadvantages of working with solvents are that many of them are hazardous; they soon become contaminated and must be replaced frequently or else they become ineffective; and they can be expensive.

With few exceptions, hydrocarbon-based solvents are flammable, explosive, and can cause serious health problems if inhaled or absorbed through the skin. Exposure can be especially dangerous to anyone with heart or liver disease. Solvents should always be used with care.

WARNING! Keep the work area well ventilated. If fresh air is not adequate, provide artificial ventilation. Headache, nausea, dizziness, and skin irritation are signs of overexposure. If they persist, get medical attention.

Wear appropriate protective clothing, including safety glasses, goggles, gloves, and the appropriate NIOSH-approved respirator.

Never use flammable solvents near heating units, motors, or machines that may spark, electrical outlets, or any possible source of ignition. Note that solvent vapors are heavier than air and can travel undetected along the floor or ground to an ignition source far away from the work area.

Clean up all solvent spills, no matter how small.

Dispose of used solvent and cleaning supplies properly.

Read the Material Safety Data Sheets for information about individual solvents.

CAUTION State and local ordinances may prohibit the use of some solvents in the cleaning and painting process.

Used solvents are considered hazardous wastes. Legal options for disposing of them properly are to pay to have them hauled away and deposited in an appropriate landfill or to recycle them using a solvent recovery system (*Figure 35*).

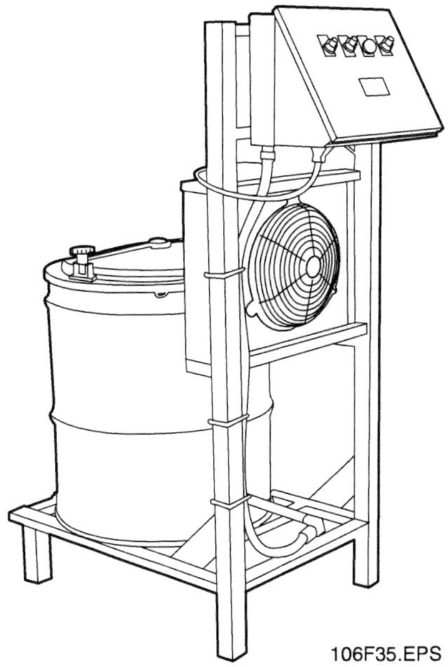

Figure 35. Solvent Recovery System

The distillation process for recycling a solvent first involves evaporating it. The vapor is then condensed back into liquid form by one of several cooling methods. This liquid is the refreshed solvent. The residue is removed and disposed of as a hazardous waste. Recycling systems come in various sizes. Some are permanent; others are portable.

In some cases, inadvertent chemical reactions must be controlled during the evaporating process. When heated, some paints emit toxic or volatile gases. In most cases, however, recycling solvents in a distillation system is simple and cost-effective.

Although some solvent is lost in the distillation process, it can be recycled repeatedly, which is economical. Also, it is much less costly to dispose of residue than an entire batch of used solvent. In addition to its other advantages, this process helps to preserve the environment.

5.1.2 Alkaline And Detergent Cleaning

As discussed previously, there are a number of powerful industrial alkaline sodium metasilicate-based cleaners with added detergent and surfactants available. These cleaners, in their concentrated spray-and-wipe forms, are strong enough to remove dried paint, oil, grease, wax, scuff marks, crayon and pencil marks, lipstick, soot, rust stains, etc. They are also available in non-foaming versions for dilution and use in pressure washing equipment.

In general, these cleaners should not be allowed to dry on the surface. They should be wiped or rinsed off thoroughly with clean water while still wet, otherwise they may leave a film that is difficult to remove. They are effective, especially when heated to 150°F (65°C) and above, for removing contaminants from all hard surfaces as well as masonry and concrete.

Note These cleaners should not be used on glass.

WARNING! Goggles and waterproof gloves should be used when handling these products. In confined areas, respirators should be worn since the fumes from the additives in some of these products are irritating. Manufacturer's instructions and warnings must be followed when using these products.

Cleaners, such as detergent, ammonia, or household stain removers, can be used with a sponge or clean rag to clean grease stains, scuff marks, dirty fingerprints, or other wall markings. For stubborn stains, use an abrasive wet pad or sand lightly and seal the surface to prevent the stain from bleeding through the coating. For rust or water stains, check for pipe leaks and surface damage, such as loose or soft plaster or drywall. Make necessary repairs. Wash and seal the surface.

The best way to eliminate mold or mildew is to treat it with household bleach in a ratio of 1 part bleach to 3 or 4 parts water. Then wash the surface with detergent and rinse it thoroughly with water. Mildewcide may be added to paint or the surface may be washed with mildewcide to prevent a recurrence, but always check with the coating manufacturer first.

WARNING! Never mix bleach with ammonia, vinegar, acids, or other household cleaners. Doing so will release hazardous gases. Also, do not use bleach on metal surfaces.

To remove discoloration caused by industrial gases, wash the surface with a very diluted solution of acetic or muriatic acid, and rinse well with water.

5.1.3 Disinfecting

To disinfect a surface, wash it with household or industrial disinfectant, following the manufacturer's instructions. Bleach is a good disinfectant for use on hard, nonabsorbent surfaces. Always rinse with water.

5.1.4 Tacking

A tack cloth, which is treated with varnish to make it tacky, is used to wipe dust, lint, and particles from woodwork. It is especially effective at removing sanding dust. When using a tack cloth, turn it frequently to expose a fresh surface.

5.1.5 Pressure Washing And Wet Abrasive Blasting [For SSPC-SP6 (NACE-3) And SSPC-SP7 (NACE-4) Profiling]

Pressure washing uses a low volume of water at high pressure to give the water more cleaning power.

Low-pressure washing ranges up to 5,000 psi (352 kg/cm^2). Pressure washers can spray water with or without detergents; water with or without abrasives; and steam with or without detergents.

The use of pressure washing depends on the surface condition. Its effectiveness depends on the level of pressure and the use of additives. Low-pressure washing or steam cleaning will remove dirt, grime, loose paint, chalking paint, and heavy soil when solvent cleaning is difficult or impractical. It is commonly used to clean steel, brick, concrete, or masonry.

Injecting an abrasive into the water stream gives pressure washing added power to remove tight, intact paint and adherent rust and mill scale. It also provides the correct surface profile. Wet abrasive washing may also be used to satisfy the surface **abrading** requirements of ASTM D-4259, *Standard Practice for Abrading Concrete*. The requirement is that the surface be free of laitance, glaze, efflorescence, form release, and curing compounds. The surface should also have a roughened, textured appearance.

Pressure washing or blasting with abrasives is a good alternative when dry blasting is not possible because of environmental restrictions. It does not create as much dust or any sparks, which can be dangerous. Extra caution is required near electrical equipment. Pressure washing with an abrasive requires a second washing to remove surface sediment. It may also require the use of an inhibitor to prevent flash rusting on steel surfaces. One such inhibitor is water containing 0.32% sodium nitrate and 1.28% by weight secondary ammonium phosphate (dibasic). The coating system must be compatible with whatever inhibitor is used.

Wet abrasive blasting of steel with low-pressure washers is not recommended when an SSPC-SP5 (**NACE**-1) white metal blast profiled or SSPC-SP10 (NACE-2) near-white blast profiled surface is required. It can be used to achieve SSPC-SP6 (NACE-3) commercial blast profiled or SSPC-SP7 (NACE-4) brush-off blast profiled surfaces.

A commercial blast cleaned surface is defined by SSPC-SP6 (NACE-3) as a surface, when viewed without magnification, that is free of all visible oil, grease, dirt, dust, mill scale, rust, paint, oxides, corrosion products, and other foreign matter. Staining shall be limited to no more than 33% of each square inch of surface area and may consist of light shadows, streaks, or discoloration caused by stains of rust, mill scale, or paint. If the surface was pitted, a slight residue of rust and paint may be left in pits.

A brush-off blast cleaned surface is defined by SSPC-SP7 (NACE-4) as a surface, when viewed without magnification, that is free of all visible oil, grease, dirt, loose mill scale, loose rust, and loose paint. Tightly adherent mill scale, rust, and paint may remain on the surface. Material is considered to be tightly adherent if it cannot be removed by lifting with a dull putty knife.

Working with high-pressure water calls for extra safety precautions. Appropriate protective clothing includes a waterproof suit, face shield or helmet with visor, hearing protection, gloves, steel-toed shoes, and rubber boots. Operators should be well trained, and equipment should be inspected regularly to assure good working condition. A deadman control—a safety device that stops the blasting when released—is a must.

5.2.0 HAND TOOL CLEANING (SSPC-SP2)

Cleaning with hand tools (*Figure 36*) is most effective at removing small amounts of loose contaminants from wood, concrete, or steel substrates. Wire brushes, sandpaper, nonwoven abrasive pads, scrapers, knives of various shapes and sizes, chipping hammers, and hammers and chisels are good for cleaning away dirt and loose paint, rust, and mill scale. The best tool to use depends on the type and condition of the surface. One or all of them may be needed.

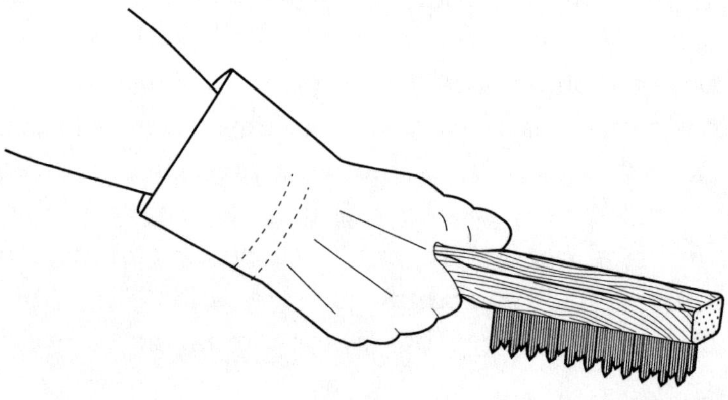

Figure 36. Hand Tool Cleaning With A Wire Brush

Hand tools cannot be expected to satisfactorily remove oil and grease, tightly adhered rust, paint, mill scale, or contaminants stuck in pits or crevices. In those cases, other cleaning methods are required.

When using hand tools, it is important not to scar the surface or raise burrs that may later protrude through the coating and allow a place for rust to begin.

Before cleaning with hand tools, wash the surface with water, detergent, or solvent to remove dirt, oil, and grease. After cleaning the surface, brush, dust, sweep, or blow away loose debris.

Hand tool cleaning can be done in a confined area or near sensitive equipment since it creates little dust. Hand tool cleaning is good when an entire surface does not require cleaning.

For large jobs, it can be slow and costly. Since the quality of cleaning with hand tools may be limited compared to more powerful cleaning methods, coating life expectancy may not be as good. However, some coatings, such as slow drying, oil-based paints with good wetting ability, or epoxy mastics, work well on hand tool cleaned surfaces, making this an attractive option in certain circumstances.

SSPC-SP2 defines hand tool cleaning as the removal of all loose scale, rust, paint, and other loose material (other than adherent).

Working safely with hand tools requires the use of safety goggles, gloves, and a NIOSH-approved cartridge respirator to avoid inhaling dust, rust, and paint chips. Be careful not to create sparks around flammable materials or volatile vapors.

5.3.0 POWER TOOL CLEANING (SSPC-SP3 AND 11; ASTM D-4259)

Power tools are much more effective than hand tools at removing paint, rust, mill scale, and other surface contaminants from wood, concrete, and steel. As discussed previously, there are three types of power tools used for surface preparation: impact tools, rotary/vibrating tools, and rotary impact tools.

- Impact tools, such as scaling hammers, needle guns (*Figure 37*), and power chisels, are used to clean steel and concrete. They can clean a surface by knocking off the contaminants, including heavy layers of rust, mill scale, thick paint, weld flux, slag, and weld spatter.

106F37.EPS

Figure 37. Using A Needle Gun To Clean A Surface

- Rotary/vibrating tools are used on wood as well as concrete and steel. They include sanders, wire brushes, and grinding wheels (*Figure 38*). They brush, grind, or sand away uneven wood or concrete, repair compounds, loose rust, paint, loose mill scale, weld flux, and slag. They are also used to prepare welded surfaces.

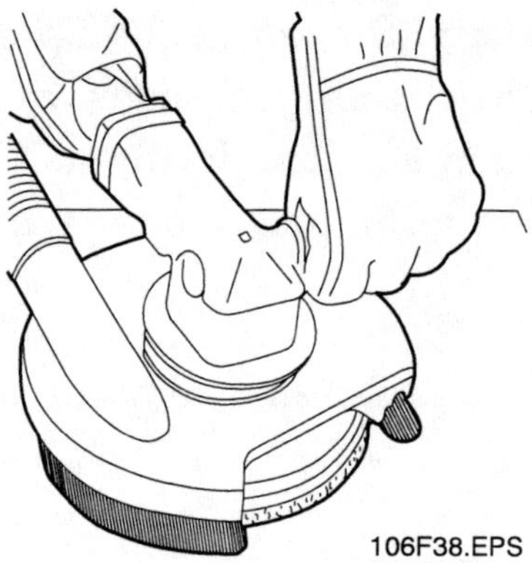

Figure 38. A Nonwoven Abrasive Disc Being Used With A Shrouded, Right-Angle Grinder

- Rotary impact tools (*Figure 39*) combine the working features of both rotary and impact tools. They use rotary action to hit a steel or concrete surface with an abrasive material. The names of these tools (e.g., rotary peeners, rotary flaps, and rotary hammers) reflect how they work to remove old paint, tight mill scale, rust, weld flux, and slag.

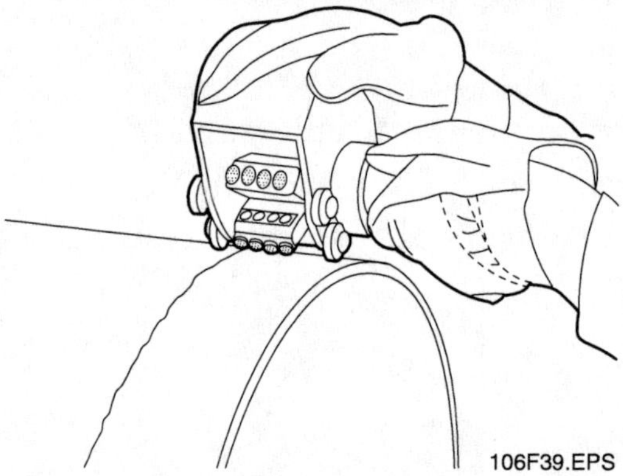

Figure 39. Rotary Peener

SSPC-SP3 defines power tool cleaning as the removal of all loose scale, rust, paint, and other loose material (other than adherent).

ASTM D-4259 (Abrading of Concrete) requires that the surface be free of laitance, glaze, efflorescence, form release, and curing compounds. The surface must also have a roughened and textured appearance. In addition, aggregate may be exposed.

As always, start by washing the surface so that oil and grease are not spread around by the action of the power tools.

Explosion-proof power tools should be used in confined areas, and they are economical for removing limited amounts of most contaminants. They are very effective for maintenance projects in which the painter must spot clean 25% to 30% of the surface. However, cleaning large, heavily contaminated surfaces with power tools can be slow and costly.

To create a profile on steel that is suitable for high-performance coatings, follow SSPC-SP11, *Power Tool Cleaning to Bare Metal*. Power tool cleaning to bare metal is defined by SSPC-SP11 as a surface, when viewed without magnification, that is free of all visible oil, grease, dirt, rust, mill scale, paint, oxides, corrosion products, and other foreign matter. Slight residues of rust or paint may be left in the lower portion of pits if the surface was pitted.

Some power tools are equipped with a vacuum system (*Figure 40*) to recover dust, debris, and surface waste, which is necessary when removing lead-based paint or other hazardous coating material because of the danger to both the worker and the environment.

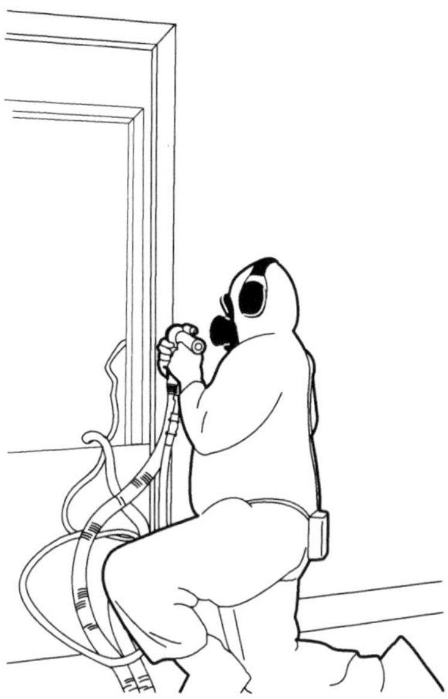

Figure 40. Removing Hazardous Coating Material

WARNING! Electric tools must be properly grounded and must never be used in wet areas. Remove any explosive hazard to avoid ignition from sparks.

Safe use of power tools requires good training, regular inspection of equipment, and appropriate protective gear, including hearing protection and a respirator, especially if removing lead-based paint or any respirable debris.

WARNING! Removing lead-based paint requires strict and careful compliance with the regulations designed to protect the worker and work area from the dangers of contamination.

5.4.0 ETCHING AND NEUTRALIZING (ASTM D-4260 AND D-4262)

While mechanical cleaning is the preferred method, in some situations acid etching of horizontal concrete surfaces is used to remove laitance and give a smooth surface texture, which is required for good coating adhesion. Muriatic acid (which is a diluted form of hydrochloric acid) may be used for this purpose. The surface condition and desired texture will determine the proper concentration of the acid solution. ASTM D-4260, *Standard Practice for Acid Etching Concrete*, outlines the following process.

Precleaning is important because the surface should be free of oil, grease, and dirt that may interfere with the acid action. It also should be moistened, but standing water should not be present. The acid should be spread evenly over the surface; it will foam or bubble as it works. Some surfaces, depending on their original condition, may require more than one treatment. In any case, the finished texture should feel like medium grit sandpaper.

Perhaps even more important than precleaning the surface is rinsing and scrubbing it after etching to remove all acid residue as well as any loose particles. If acid residue is not removed thoroughly, it is likely to interfere with adhesion and cause coating failure.

The pH level of the rinse water and of the rinsed, wet surface should be measured to see if the acid has been neutralized. A litmus paper with a range from 1 to 11 pH units with a capability of measuring in increments of 0.5 pH units should be used.

Two surface readings should be taken for each 500 square feet (46 m^2), or portion thereof, at randomly-selected locations immediately following a final rinse. The pH readings should not be more than 1.0 pH lower or 2.0 pH higher than the pH of the rinse water sampled both before and after the rinse. Rinsing should continue until the readings are acceptable. Ideally, the final pH of the surface should be neutral (7.0 pH). ASTM D-4262, *Standard Test Method for pH of Chemically Cleaned or Etched Concrete Surfaces*, addresses the above procedure.

After the surface has been properly rinsed, allow it to dry thoroughly. Before coating, the moisture should be checked using the plastic sheet method described later in this module.

It is more practical to etch horizontal surfaces than it is to etch vertical surfaces, where the acid cannot be controlled. Also, etching is not as effective as blasting for dense surfaces. Other disadvantages of acid etching include the hazards of working with acid solutions and their fumes, and the need to properly dispose of the used acid and rinse water.

WARNING! In working with acid, avoid contact with skin and exposure to fumes by wearing protective clothing and the appropriate approved respirator. Dispose of the residue from acid etching according to environmental regulations.

Neutralizing is also important in the surface preparation of chemically contaminated concrete, which may occur in an industrial plant such as a paper mill. For instance, a surface contaminated with acid can be treated with an alkaline cleaner, while an alkaline surface can be treated with white vinegar. In either case, the surface should be rinsed thoroughly, and checked for the presence of residual chemicals, pH level, and moisture content before coating.

5.5.0 VACUUMING

An important issue in surface preparation is cleanup of debris, including abrasive, dust, rust, and paint particles from the surface being cleaned. Collecting this debris can eliminate ground and water contamination, and contamination of homes or other indoor areas where the work is being done. Cleanup also prevents the dust and debris from possible recontamination of the clean surface.

When lead-based paint or other hazardous material is being removed, cleanup is not only a practical matter but also an environmental concern, subject to strict regulations, which should be followed carefully to avoid contamination of the worker and the work area.

Vacuum shrouds are also available for many hand-held power tools. Vacuums should be equipped with HEPA filters when hazardous materials are being handled.

5.6.0 CHECKING FOR MOISTURE IN CONCRETE, STUCCO, MASONRY, OR PLASTER SUBSTRATES (ASTM D-4263)

ASTM D-4263 provides a method for indicating capillary moisture in concrete by using a plastic sheet. This is known as the *plastic sheet method*. This method can also be used on other substrates, such as stucco, masonry, or plaster.

The plastic sheet method consists of taping an 18 x 18 inch (457 x 457 mm) square of polyethylene film (about 4 mils thick) securely and tightly to the surface of the substrate using 2 inch (50 mm) wide duct tape or an equivalent. For floors (or all horizontal surfaces) one test area per 500 square feet (46 m^2) or portion thereof is specified. For walls and ceilings combined, one test area per 500 square feet (46 m^2) or portion thereof is specified. In addition, for walls, one test area for each 10 feet (3 m) of vertical rise in all elevations starting within 12 inches (305 mm) of the floor is specified.

All test areas must be shielded from direct sunlight, heat, or damage to the plastic film sheet or the test results will be invalid. Allow the plastic film sheet to remain in place for a minimum of 16 hours. After the minimum amount of time has elapsed, remove the sheet and visually check the underside of the sheet and the substrate surface for the presence of moisture. If moisture is found, the source of the moisture should be located and eliminated, or the surface should be allowed to dry further. In either case, the test will have to be repeated on the problem test areas until no moisture is detected (even though some coatings may be applied with water).

5.7.0 REPAIR/REPLACEMENT OF SUBSTRATES

Any number of things can go wrong with a substrate or surface, but as a general rule, never paint a problem. Fix it first. If you don't, chances are the problem will only get worse and you'll end up redoing the job anyway.

Given the variety of common substrates and the endless possibilities of surface conditions, there is no way to anticipate every problem a painter might encounter. However, here are some general guidelines for repairing and replacing substrates.

- Secure loose surface material to its framework.
- Remove rotted, crumbling, or deteriorated material until reaching solid substrate and replace it with sound material.
- Clean the repair surface thoroughly to remove all dust and debris that may interfere with patching adhesion or the finished appearance.
- Sand or scrape to remove loose material and to give the repair surface the correct profile.
- Choose patching material carefully. Match it as closely as possible to the texture, color, and performance properties of the surface being repaired. Follow the manufacturer's directions for proper application and drying time.
- Moisten the edges of absorbent materials before applying water-based patches to help guarantee a solid bond. Do not moisten nonabsorbent materials; this will interfere with adhesion.
- Patch deep holes or cracks by filling them in layers. Scoring or sanding the patching material between layers may help them bond.
- Recognize when a surface is too badly damaged to be repaired and should be replaced.

The following paragraphs offer specifics for some types of repairs on certain substrates and conditions.

5.7.1 Caulking, Sealing, And Puttying

Detailed information on caulks, sealants, and putties and their applications will be covered in another module.

5.7.2 Spackling

Spackle is easy to use. Simply clean away any loose paint, plaster, or debris around the hole and apply the spackle with a putty or broad knife. For a smooth, even coating, use a knife that is wider than the hole or crack being filled. Allow the spackle to dry thoroughly, then sand lightly and wipe away the dust. If the surface is to be coated with an enamel, prime the spackled spot to prevent flashing or unevenness in the texture and sheen of the finish.

5.7.3 Filling And Sealing Concrete, Masonry, And Stucco Substrates

A common repair material for filling cracks in cement, stucco, and stone is one part portland cement with two or three parts sand. It is mixed with water to an appropriate consistency. Adding a latex vehicle to the water improves its density and working qualities.

A number of commercial grouts and repair materials such as crack injection epoxies, as well as surface preparation treatments such as sealers, fillers, and surfacers are also available for concrete.

Concrete sealers and penetrants are typically clear, silicone-based products that are applied using a brush, roller, or spray gun. These products protect against water penetration by filling the pores and small holes in the surface.

Sealing concrete protects it from moisture as well as acids, alkalis, salts, and other corrosives that can penetrate the surface and result in cracking, flaking, and corrosion of rebars (steel reinforcing bars).

These rebars can be subjected to corrosion for a variety of reasons; for example, if the structure is made of porous, low-quality concrete, or if the bars are not covered with a sufficient amount of concrete material. It may be necessary to remove unsound concrete in order to expose the reinforcing steel, clean away the corrosion, and treat the steel before replacing the concrete. Severe cases may require the reinforcing steel to be replaced.

Heavy-bodied coatings known as *block fillers* are widely used on rough surfaces such as concrete block. These products are not just primers or undercoats. Properly applied, they will seal the surface and fill the voids in the blocks to an almost level surface. All methods of application are acceptable. One of the most popular methods is to spray on the material and roll the wet film to level it.

Surfacers, which are generally made of epoxies, make a concrete surface smooth and resistant to chemical contamination. Topcoats can be applied to fillers and surfacers, but any coating used on concrete should be able to resist its alkaline nature.

5.7.4 Plastic Patching, Repairing, And Preparation

If the plastic type is unknown but is rigid and opaque or will be painted so that the back of the panel is covered, sand the damaged area carefully and apply a polyester automotive body filler. Allow it to dry and use very fine sandpaper to carefully sand it flush with the surface. Fill in any pinholes with automotive spot putty. When dry, wet sand with 400 to 600 grit paper until it is flush with the surface.

If the type of plastic substrate is known, contact a manufacturer of that type of plastic through a wholesale plastics distributor to determine the surface repair kits, pretreatments, primers, and finishes that may be applied. A number of publications are also available from the Society of Plastics Engineers, Inc., Brookfield, CT, covering the joining, pretreatment, priming, and painting of various plastic surfaces:

- *Joining Plastics* (#9653) by K.W. Allen, 1993
- *Decorating and Coating of Polymers* (#9509) by P.J. Robinson, 1993
- *Plastics Surface and Finish* (#953) by W.G. Gordon Simpson, 1994
- *Adhesives*, 8th Edition (#27) by International Plastics Selector, 1995

These publications are relatively technical and are meant primarily for manufacturers. However, they do reveal many types of joining methods, surface preparation techniques, and finishes used for plastics. If you will be extensively involved in repairing, finishing, or refinishing plastic surfaces, they may be worthwhile references.

5.7.5 Wood Patching

If at all possible, damaged or rotted wood should be replaced rather than repaired. If replacement is not possible or feasible, wood consolidants and epoxy fillers may be used to repair damage to wood or small areas of soft and rotting wood. All damaged, soft, or rotted wood must be removed. The remaining wood fibers should be thoroughly soaked with multiple applications of a consolidant. Once the consolidant has cured, voids may be filled with an epoxy filler. When cured, the filler and stabilized wood may be sanded or finished like any other piece of wood.

5.7.6 Weld Patching Of Metal

Welding (*Figure 41*) is one way of repairing deep pitting in corroded metal surfaces, such as the interior of water storage tanks. Welding can be used either to prevent leaking, which can result from spot pitting, or to patch or replace steel that is corroded so severely that it is structurally unsafe.

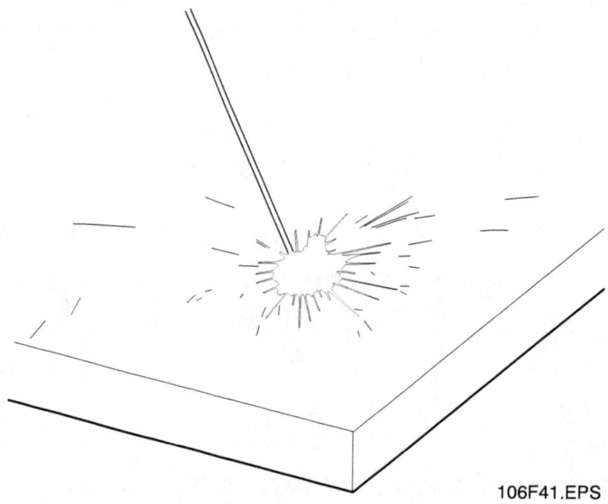

106F41.EPS

Figure 41. Welding As A Surface Preparation Method For Repairing Steel

Among the considerations are the size, depth, and frequency of the pitting. The surface should be welded if the pits are too deep or too small to be cleaned properly with abrasives, and if residual contaminants cannot be removed from the pits in any other way.

Welding can repair all sizes of pits. It is especially effective when the pits are more than 1/4 inch deep. Before welding, the area must be cleaned thoroughly to determine the size and depth of the pits and to remove any surface contamination that may interfere with the weld. The welds should be ground smooth but left unpolished if the surface is to be recoated at the time of the repairs.

5.7.7 Fiberglass Patching Of Metal Or Concrete

Fiberglass has been used extensively since the early 1960s to repair both metal and concrete structures. Fiberglass-reinforced plastic, commonly referred to as *FRP*, reinforces coatings, bridges holes and pits, and acts as a barrier to future internal corrosion. It has proven to be an economical method of making steel tank repairs.

The term FRP is used interchangeably with RPT (reinforced thermoset plastic), RTR (reinforced thermoset resin), and GRP (glass-reinforced plastic) and refers to any system of fiberglass or other fiber-reinforced plastic lining used in combination with a resin coating.

The advantage of fiberglass is its ability to withstand coating shrinkage during curing and drying, as well as its ability to compensate for resin brittleness. The resulting lining is hard due to its high solids content, high tensile strength, and strength-to-weight ratio. Varying the amount of reinforcement in a system changes its physical and chemical characteristics. A higher fiberglass content provides greater physical strength, while a higher resin content creates greater chemical resistance.

Laying down an FRP lining requires surface preparation using hand tools, power tools, or abrasive blast cleaning (the most common). Pitting can be filled with a flexible putty. A primer is often applied last to seal the surface, hold the blast, and improve adhesion of the FRP.

WARNING! Installing an FRP system can be hazardous and every precaution should be taken. Fiberglass should not come in contact with skin and eyes, or be inhaled. Protective clothing is critical, including eye and ear protection, gloves, boots, and a respirator. Ventilation is also important. Inside tanks, explosion-proof exhaust systems equipped with special air quality gauges, and low-voltage lighting are necessary in most cases.

5.7.8 Concrete Or Plaster Patching

To repair a deep hole in a plaster or concrete surface, start by removing any old, loose debris. For plaster, secure lathing if loose, then undercut the concrete or plaster at an angle so the opening is wider at the back than at the surface. This will help form an anchor for the patch. Vacuum out dust and residue. Moisten the inside and edges of the opening with water so the patch will form a good bond with the existing material. Apply the patching material, forcing it into the hole. Smooth the surface after it dries. Apply one or more coats, as needed.

5.7.9 Repairing Drywall

Small cracks, dents, and defects in a drywall surface may be filled easily with a dab of spackle or joint compound applied with a fingertip or a putty knife. Let it dry and sand it smooth, if necessary.

A hole about 2 inches in diameter may be repaired by covering it with a piece of joint tape "buttered" with joint compound or with adhesive-backed mesh tape, as shown in *Figure 42*. Two or three pieces applied in different directions may be needed for extra strength. Apply a coat of compound, feathering the edges. After it dries, apply a skim coat, if needed. Allow to dry and then sand.

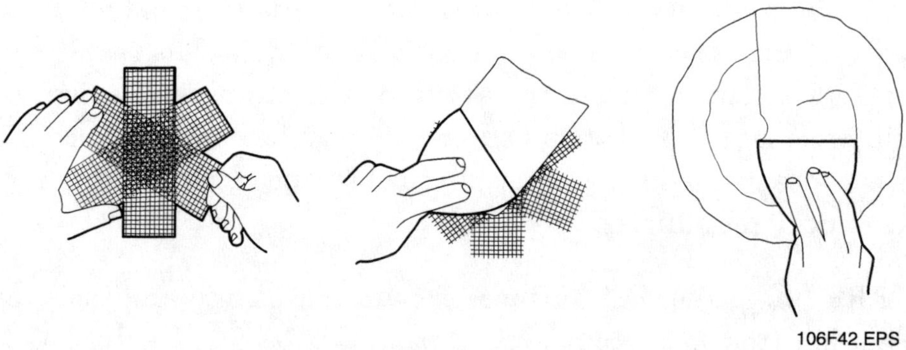

Figure 42. Repairing A Hole

Larger holes should be patched with a piece of drywall. One method is shown in *Figure 43*. First, square off the area around the hole with a saw and then cut a piece of new drywall to fill it. If there is no structural backing behind the hole, it may be necessary to support the patch. One way is to apply adhesive to the ends of a thin, wooden backing strip cut wider than the hole. Feed the strip into the opening and hold it in place across the inside of the hole, or use screws to secure it, until the adhesive begins to set. Several backing strips may be needed depending on the size of the hole. When they are secure, set the drywall patch into place and finish the seams with joint compound, feathering the edges. After it dries, sand it, apply a skim coat, allow that to dry, and then sand again. If the hole is very large, the panel should be cut back to the framing so that the patch has adequate support. Panels that are badly damaged should be removed and replaced with a new panel.

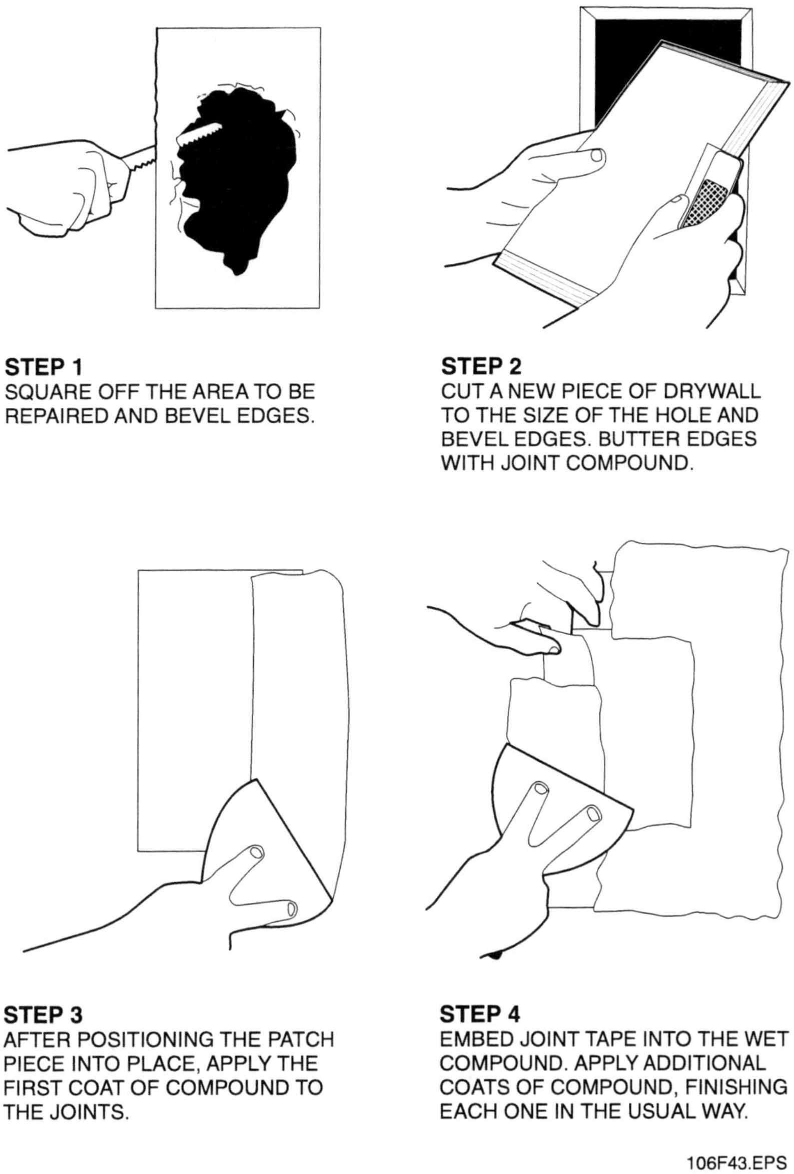

Figure 43. Repairing Drywall

5.7.10 Surface Preparation Standards And Inspection Techniques

The quality of the surface preparation has a direct bearing on how well a coating performs and how long it lasts. It is believed that the majority of all premature coating failures are due, all or in part, to faulty surface preparation and/or coating application. That being the case, it is important to determine how well a surface has been prepared and to catch any deficiencies in the quality of the preparation before they are covered with a coat of paint.

Standards and inspection techniques for surface preparation have been established to help protect against premature coating failures. Generally, the more industrial the application, the more standards there are likely to be. And the more standards there are, the more formal the inspection process is. For instance, there are volumes of standards on proper steel coating practices and ways to be sure they are followed. A contractor hired to coat a nuclear power plant literally would have to meet hundreds of standards in the course of the job and undergo strict, formal inspections following established guidelines and procedures. Can you imagine a residential painter having to do the same thing? Yet in both cases, surface preparation most likely will determine the success of their efforts. The principle is the same whether a painter stops to eyeball his or her own work or it is evaluated by a team of inspectors with calibrated instruments.

The following is a brief overview of some generalized standards for surface preparation of various substrates and how inspectors, such as those shown at work in *Figure 44*, would evaluate them.

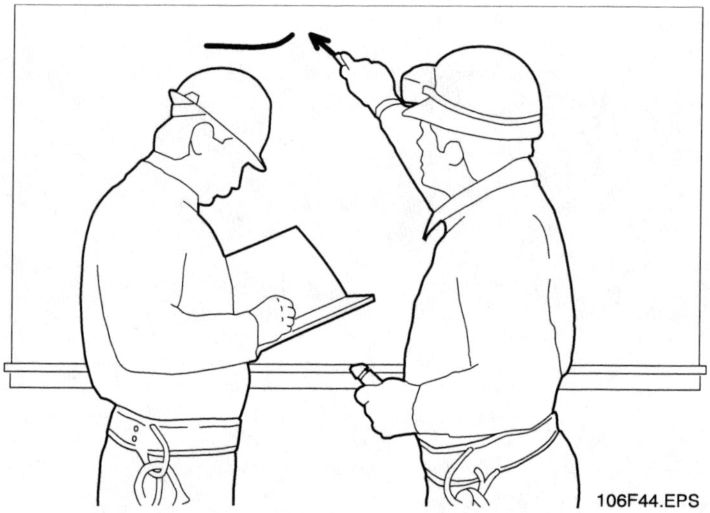

Figure 44. Inspectors At Work In The Field

1. Wood surfaces:
   - The moisture content of new wood should be no more than 12% to 15%, lower if possible.
   - New wood should be primed as soon as possible.
   - Sap spots or resin streaks should be sealed with a pigmented shellac material; knot holes should be filled.
   - Grease, dirt, oil, and other surface contaminants should be cleaned off with detergent or solvent.
   - If the wood has been exposed to sunlight for several weeks, it should be checked for ultraviolet damage. Any loose or discolored surface fibers should be sanded off before the wood is primed.

Most of these standards are easily checked by visual inspection. A moisture meter (*Figure 45*) may be used by inspectors to help determine how wet the wood is. The standards established by the Steel Structures Painting Council for solvent cleaning (SSPC-SP1) can apply to all types of surfaces.

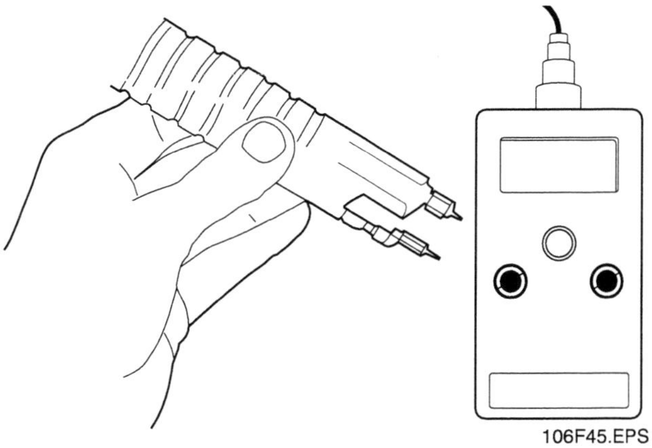

Figure 45. Moisture Meter

2. Masonry and concrete surfaces:
   - Generally, concrete should cure for at least 28 days before coating.
   - The surface should be cleaned with water, detergent, or steam to remove oil, grease, and dirt (*Standard Practice for Cleaning Concrete*, ASTM D-4258).
   - Cracks, holes, and voids in the surface should be filled and sealed.
   - Laitance and efflorescence should be removed, as should form release agents and curing compounds that are not compatible with the coating system. If simple cleaning is not sufficient, more aggressive methods, such as abrasive blasting, power tool cleaning, or water jetting may be used.
   - The surface should be roughened, if necessary, by mechanical abrading, abrasive blasting (ASTM D-4259), or, for horizontal surfaces, acid etching (ASTM D-4260).
   - The surface should be dry and either swept or vacuumed before coating.

Visual inspection is enough to check conformity to some, but not all, of these standards. For instance, surface grease and oil may not be visible without magnification. One way to check for them is to apply water to the surface. If the water beads, it indicates the presence of oil, grease, or some other contaminant that is preventing it from wetting the surface.

A litmus test (ASTM D-4262) will check the pH of the surface if it has been etched with acid. A moisture meter or the plastic sheet method (ASTM D-4263) can be used to check for moisture.

3. Steel surfaces:
   - Slag, flux, weld spatter, and rough edges should be removed.
   - The surface should be washed with solvents to remove grease, dirt, oil, debris, and other loose contaminants.

- The surface should be cleaned of all rust, dirt, mill scale, old paint, or other tightly adhered contaminants according to the specification requirements, using one or more of the following: hand tools, power tools, wet or dry abrasive blast machines, and high-pressure or ultra high-pressure water jetting machines with or without abrasives.
- The surface should be profiled with abrasives to produce the specified anchor pattern.

The two primary areas of concern to inspectors in the surface preparation of steel are cleanliness and profile. SSPC and NACE have developed similar standards for evaluating four degrees of abrasive blast cleaning. Starting at the lowest degree, they are brush-off blast cleaning (SSPC-SP7 or NACE-4), commercial blast cleaning (SSPC-SP6 or NACE-3), near-white blast cleaning (SSPC-SP10 or NACE-2), and white metal blast cleaning (SSPC-SP5 or NACE-1).

As a guide for comparing surface cleanliness to these standards, SSPC has developed reference photographs, and NACE has pieces of metal encased in plastic to illustrate each grade. These visual standards are for steel that has not previously been painted.

SSPC also has visual standards for hand and power tool cleaning (SSPC-SP2, SP3, and SP11) of both unpainted and previously painted steel.

To check surface profile, inspectors use tools, such as a profile comparator (*Figure 46*), which helps them to make a visual analysis of the surface compared to established standards; a profile gauge (*Figure 47*), which measures the depth of the profile; or replica tape (*Figure 48*), which makes an impression of the profile. They then measure the impression to gauge the profile. How clean and profiled a surface needs to be will depend on the type of structure and its location. A bridge in an alkaline desert and a water tank along the coast may need the same types of protection because the alkaline desert and the seacoast are harsh, with salt-laden or alkaline dust-laden air that attacks the paint. As a result, their surfaces are probably prepared to the same specifications.

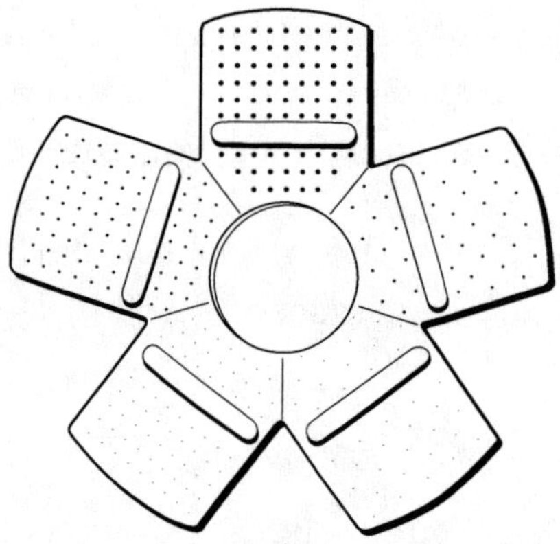

106F46.EPS

Figure 46. Profile Comparator

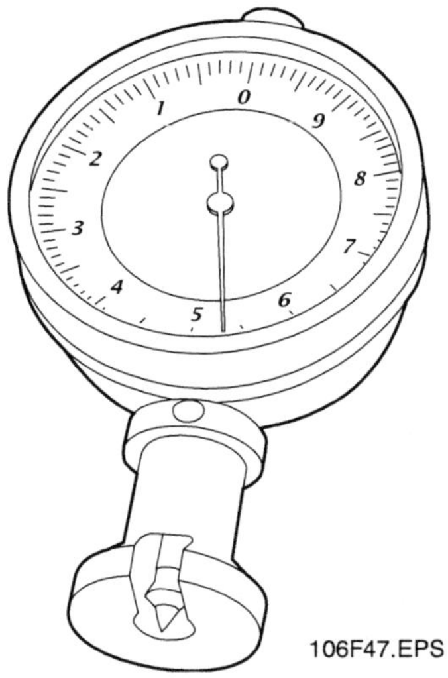

Figure 47. Profile Gauge

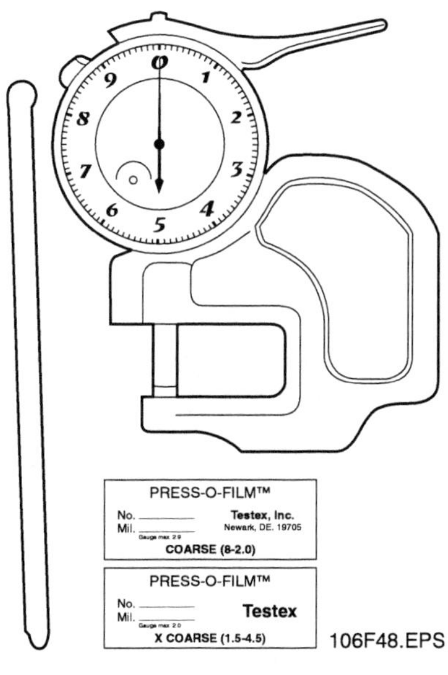

Figure 48. Replica Tape And Micrometer

Other standards for surface preparation of steel concern the cleanliness of air used for blast cleaning, the quality of abrasives, and ambient weather conditions during blasting. During blasting and painting, the temperature of the steel must be at least 5°F above the dewpoint to prevent condensation, which causes flash rusting. The steel temperature and dewpoint can be measured with two simple tools—a surface temperature gauge and a sling psychrometer.

4. Drywall surfaces:
    - The surface should be clean and free of dirt, oil, grease, and other foreign matter.
    - Holes, cracks, voids, or other defects should be filled with spackling or patching compound.
    - Drywall panels should be securely anchored to their support, but the paper facing should not be broken.
    - Nails or screws should be countersunk and covered with spackle or joint compound.
    - Seams should be covered with joint tape and several layers of joint compound.
    - Joint compound should be dried and sanded before coating.
    - The surface should be vacuumed or dusted prior to coating.

All of these practices can be checked by visual inspection.

Drywall surfaces require different levels of finish depending on their location, lighting conditions, decorative treatment, and other factors. Descriptions of five levels of finish for drywall surfaces are outlined in a *Recommended Specification on Levels of Gypsum Board Finish* (GA-214-96) developed jointly by the Painting and Decorating Contractors of America, the Association of Wall and Ceiling Industries – International, the Gypsum Association, and the Ceiling & Interior Systems Construction Association. Details of this specification are included in a later module.

5. Plaster surfaces:
    - Plaster should be sufficiently dry before painting.
    - The surface should be clean and free of dirt, oil, grease, and other foreign matter.
    - Holes, cracks, or defects should be filled with spackle or patching material.
    - Loose or soft plaster should be repaired, replaced, or hardened with a white vinegar wash, as applicable.
    - The moisture content should be 6% or less.
    - The alkaline level should be neutral.
    - The surface should be dusted or vacuumed before painting.

A plastic sheet taped to the plaster surface for 16 hours can be used to check for moisture. A litmus test will determine its alkalinity. Both factors determine the plaster's receptivity to paint. The rest can be evaluated by visual inspection.

6. Existing coatings:

    In evaluating existing coatings, inspectors check the quality of paint adhesion, the film thickness, and the condition of the substrate. Adhesion can be evaluated visually or with the aid of adhesion testing equipment (*Figure 49*) or by the use of a tape test (ASTM D-3359). The condition of the substrate can be checked by removing a little bit of coating and visually inspecting it. Among the problems that inspectors look for are evidence of mill scale or rust on steel and efflorescence on concrete. A variety of gauges are available to measure film thickness, which indicates the quantity of the coating and how well it was applied.

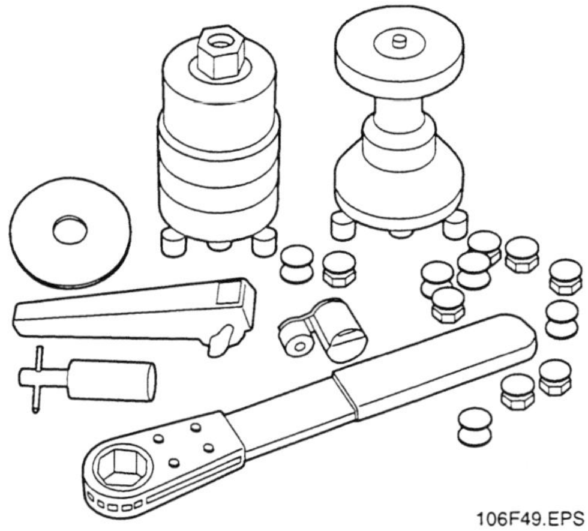

Figure 49. Adhesion Testing Equipment

6.0.0 GENERAL PREPARATION PROCEDURES FOR SPECIFIC SURFACES/SUBSTRATES

No matter what type of surface is involved, the goal of surface preparation is to create proper conditions for a coating to adhere and form a strong, durable, attractive barrier against contamination and deterioration. Each surface type has features that warrant individual consideration of how best to prepare it for coating. Following are general procedures for surface preparation of wood, concrete and masonry, plaster and drywall, steel, other metals, and various other substrates. Depending on the circumstances encountered, the steps of each procedure may not have to be accomplished completely or in the order given. In any case, the specific instructions of the coating manufacturer should be followed in lieu of these procedures, where applicable.

Instructions such as clean, fill, repair, etc. in the following procedures assume that the reader is familiar with the tools, materials, and methods previously covered. Remember, the goal of surface/substrate preparation is to have the surface/substrate clean, dry, and dull.

6.1.0 WOOD SUBSTRATES (INCLUDING PLYWOOD, PARTICLE BOARD, AND FLAKE BOARD)

6.1.1 New Exterior Wood Substrates

Step 1 Sand rough spots smooth.

Step 2 Sand any mill glaze (shiny spots) to open pores.

Step 3 Set any protruding nail heads flush or below the surface.

Step 4 Brush rough-hewn or textured surfaces to remove loose fibers and splinters.

Step 5 Clean the surface of the wood to remove any contaminants using detergent or alkaline cleaners, as required, and allow to dry. Brush or vacuum to clean the surface of dust and debris.

Step 6 Allow the wood to dry if washed or if wetted by rain or dew. Using a moisture meter, check that the moisture content is at least 12% to 15%.

Step 7 Make sure the temperatures will remain above 50°F (10°C) and prime as soon as possible to prevent ultraviolet light damage to the surface.

Step 8 Seal all knots or resin streaks after the primer has dried. Also, fill any cracks and nail holes. Caulk door and window trim.

6.1.2 Previously-Coated Exterior Wood Substrates

WARNING! Paint applied prior to 1978 may contain lead or other toxic heavy metals. Use approved personal and environmental protection equipment and procedures when sanding or removing paint from these surfaces.

Step 1 Clean the surface of any contaminants or chalking using detergent or alkaline cleaners.

WARNING! Never mix bleach with ammonia-based cleaners. A poisonous gas will be released.

Step 2 Remove and kill any mildew (black or gray fungus) by washing with a solution of 1 part household bleach and 3 or 4 parts water. Rinse thoroughly with fresh water.

Step 3 With hand or power tools, remove deteriorated paint (chalking, flaking, etc.). Use appropriate personal protection equipment.

Step 4 Feather edges of bare spots by sanding.

Step 5 Remove any deteriorated caulking. Clean sanded areas and areas where caulking was removed to eliminate any loose dust and debris.

Step 6 Allow the surface to dry if washed or wetted by rain or dew. Using a moisture meter, check any bare wood areas to see if the moisture content is at least 12% to 15%.

Step 7 Make sure temperatures will remain above 50°F (10°C). Spot prime all bare areas. Seal knots or resin streaks after the primer has dried. Also, fill any cracks and nail holes and replace any removed caulking with new caulk.

Note If mildew was a problem on the existing coating, a mildewcide paint, or if recommended by the paint manufacturer, a paint additive for compatible paint may be required.

6.1.3 Aged (Uncoated) Exterior Wood Substrates

Step 1 Clean the surface of any contaminants using detergent or an alkaline wash, as required, and allow to dry.

Step 2 Remove and kill any mildew by washing with a solution of 1 part household bleach to 3 or 4 parts water. Rinse thoroughly with fresh water.

Step 3 With hand or power tools, sand the surface to fresh wood. Use appropriate personal protection equipment.

Step 4 Remove and replace or patch any rotted or deteriorated wood.

Step 5 Clean the surface of loose dust and debris.

Step 6 Make sure that temperatures will remain above 50°F (10°C). Prime the surface as soon as possible to prevent ultraviolet light damage to the surface and allow it to dry.

Note If mildew was present on the wood, a mildewcide paint or, if recommended by the manufacturer, a paint additive for compatible paint may be needed.

6.1.4 Pressure-Treated Wood Substrates

WARNING! The chromated copper-arsenate (CCA), ammoniacal copper-arsenate (ACA), or ammoniacal copper zinc-arsenate (ACAZA) compound used in pressure-treated wood is toxic. Proper respiratory protection and dust collection equipment should be used when sanding surfaces.

New, previously coated or aged pressure-treated wood is prepared for cooling the same as normal wood. Finishes can be stains, wood toners, paint, or water repellants. A number of products on the market are specifically formulated for use on pressure-treated wood. The ideal finish will be water repellant, contain pigments or UV absorbers, and will be fortified with mildewcides. Horizontal surfaces, such as decks, should not generally be coated with a paint, varnish, or solid color stain. Moisture content of new pressure-treated wood is a problem and must be checked before any coating is applied. When preparing aged wood, wood brighteners, bleaches, or restorers may be required.

6.1.5 Fire-Retardant Treated Wood Substrates

Fire-retardant treated wood substrates are prepared for coating the same as normal wood. However, they must be finished with fire-retardant coatings in accordance with the coating manufacturer's instructions. Finishing with other than a fire-retardant coating will defeat the fire-retardant properties of the treated wood.

6.1.6 New Interior Wood Substrates

Step 1 Make sure the wood is clean and dry. The temperature in the rooms should be above 55°F (12°C).

Note All plaster or drywall should have been dried prior to application of wood trim.

Step 2 If the trim was not already factory filled and primed, fill all voids and open-grained wood with wood filler if wood is to be stained, or vinyl spackle if it is to be painted with latex. Soft wood may require the application of a clear wash sealer prior to staining to prevent uneven stain absorption.

Step 3 With hand or power tools and appropriate personal protective equipment, sand all wood surfaces smooth. Then brush, vacuum, or wipe the surfaces to remove all dust and debris.

Step 4 Prime or stain the surfaces and allow them to dry. If wood is to be painted, seal any knots or resin streaks.

6.1.7 Previously-Coated Interior Wood Substrates

WARNING! Paint applied prior to 1978 may contain lead or other toxic heavy metals. Use approved personal and environmental protection equipment and procedures when sanding or removing paint from these surfaces.

Step 1 Clean surfaces with detergent or alkaline cleaner, rinse thoroughly, then allow to dry.

Step 2 Remove any chipped, flaking, chalking, or damaged paint or coating using hand or power tools. Use appropriate personal protective equipment. Sand the surfaces and feather the edges of the damaged areas.

Step 3 If the old finish was gloss or semi-gloss, scuff sand the entire surface or dull it with a chemical deglosser.

Step 4 Fill any voids or cracks with wood filler if staining, or vinyl spackle if latex paint will be applied. Sand smooth.

Step 5 Brush, vacuum, or wipe the surface to remove dust and debris.

Step 6 With the temperature at 55°F (12°C) or above, prime or stain the repaired areas.

Note If painting over previously-stained wood, first apply a test patch of primer to an inconspicuous spot of stain to see if the stain will bleed through or if the old finish will lift off. If staining is observed, seal the finish before priming with a stain sealer. If the old finish lifts off, it will have to be completely removed before painting.

6.2.0 MASONRY AND CONCRETE SUBSTRATES

6.2.1 New Concrete Substrates

Step 1 Allow surfaces to cure for 28 days or more at temperatures exceeding 75°F (24°C). Clean surfaces with detergent or alkaline cleaners to remove any contaminants, including curing compounds, form release agents, oil, grease, etc.

Note In some cases, a curing agent may be added to concrete to accelerate the set and curing time, or special masonry conditioners may be used on concrete that has not cured for the 28 day time period.

Step 2 Using appropriate personal protective equipment, abrade the surface with hand or power tools or wet abrasive washing to remove efflorescence, fins, mortar, splatter, and laitance and to provide a proper profile. Acid etching may also be used on concrete floors to remove laitance and provide the correct profile. Thoroughly rinse the surface and properly dispose of rinse water.

Step 3 Check the pH of the surface after cleaning, abrading, or etching. Use an alkaline or vinegar wash and water rinse to establish a neutral pH near a value of 7. Properly dispose of rinse water.

Step 4 Allow the surface to dry, then check the moisture content using the plastic sheet method. If the moisture level is unacceptable, allow it to dry further, or find and remedy the source of the water.

Step 5 Fill any holes, voids, and cracks with grout or patching compound and allow to dry.

Step 6 Apply primer and finish coatings in temperatures of at least 50°F (10°C) or as recommended by the coating manufacturer.

6.2.2 Previously-Coated Concrete Substrates

WARNING! Paints applied prior to 1978 may contain lead or other toxic heavy metals. Use approved personal and environmental protection equipment and procedures when sanding or removing paint from these surfaces.

Step 1 If necessary, remove surface contaminants such as oil, grease, etc., by spot cleaning with a solvent.

Step 2 Using appropriate personal protection equipment, abrade the surface using hand or power tools or wet abrasive washing to remove fins, protrusions, old paint, etc. Acid etching and neutralization may also be used on concrete floors. Properly dispose of acid and rinse water.

Step 3 Clean the entire surface with detergent or alkaline cleaners. Then rinse thoroughly and check for a neutral pH value of about 7. Properly dispose of the rinse water.

Step 4 Allow the surface to dry, then use the plastic sheet method to check for moisture. If the moisture content is unacceptable, find and correct the source of the water.

Step 5 If the surface is chalky or dusty, a masonry conditioner may be applied before finishing.

Step 6 Apply finish coatings in temperatures of 55°F (12°C) or above.

6.2.3 New Concrete Block Masonry, Stucco, Stone, Or Gunnite Substrates

Step 1 Allow surfaces to cure for 28 days or more at temperatures exceeding 75°F (24°C). Using appropriate personal protection equipment, abrade the surface, as necessary, using hand or power tools or wet abrasive washing to remove fins, mortar splatter, protrusions, etc. Acid etching is not recommended.

Note In some cases, a curing agent may be added to concrete to accelerate the set and curing time, or special masonry conditioners may be used on concrete that has not cured for the 28 day time period.

Step 2 Clean surfaces with detergent or alkaline cleaners. After rinsing, check that the pH is about 7. Properly dispose of cleaning and rinse water.

Step 3 Allow the surfaces to dry, then check for moisture using the plastic sheet method. If the moisture content is unacceptable, allow it to dry further or find and correct the source of the water.

Step 4 Fill voids and cracks with grout or patching compound. Use block filler to fill voids in concrete block masonry.

Step 5 Apply primer and finish coatings in temperatures of at least 50°F (10°C) or as recommended by the coating manufacturer.

6.2.4 Previously-Coated Concrete Block Masonry, Stucco, Stone, Or Gunnite Substrates

---

WARNING! Paints applied prior to 1978 may contain lead or other toxic heavy metals. Use approved personal and environmental protection equipment and procedures when sanding or removing paint from these surfaces.

Step 1 If necessary, spot clean oil, grease, etc. with solvent.

Step 2 Using appropriate personal protection equipment, abrade the surface using hand or power tools or wet abrasive washing as necessary to remove loose paint, efflorescence, disintegrated mortar, etc.

Step 3 Clean surfaces with detergent or alkaline cleaners and check that the pH is about 7 after rinsing. Properly dispose of cleaning and rinse water.

Step 4 Allow surfaces to dry, then check for moisture using the plastic sheet method. If the moisture content is unacceptable, find and correct the source of the water.

Step 5 Fill voids, cracks, holes, etc. with grout or patching compound.

Step 6 If surface is chalky or dusty, a masonry conditioner may be applied before finishing.

Step 7 Apply primer and finish coatings in temperatures of at least 50°F (10°C) or as recommended by the coating manufacturer.

6.2.5 New Brick Or Previously-Coated Brick Substrates

New brick should be allowed to weather for one year. It is then prepared the same as previously-coated concrete block masonry, stucco, stone, or gunnite substrates.

Previously-coated brick is prepared the same as previously-coated concrete block masonry, stucco, stone, or gunnite substrates.

6.2.6 Asbestos Or Fiber-Cement Board, Fire-Stop Board, Flexboard, And Transite Substrates

WARNING! Do not abrade substrates containing asbestos unless proper permits, special protective equipment, and dust containment procedures are implemented. If the surface requires abrading, submit a sample for laboratory analysis, or determine by other means if the substrate contains asbestos.

Step 1 If an existing coating is flaking badly or is deteriorated, abrade the surface with hand or power tools, taking all necessary precautions.

Step 2 Clean the surface with detergent or alkaline cleaners and rinse thoroughly. After rinsing, check that the pH of the surface is approximately 7. Properly dispose of cleaning and rinse water. Allow the substrate to dry.

Step 3 If the substrate contains asbestos, coat with special primers for encapsulating asbestos fibers. If the substrate does not contain asbestos, apply a masonry conditioner to the surface before priming.

Step 4 Apply priming and finishing coatings in temperatures of at least 50°F (10°C) or as recommended by the coating manufacturer.

6.3.0 PLASTER AND DRYWALL SUBSTRATES

6.3.1 New Plaster Or Drywall Substrates

WARNING! Sanding dust from plaster, plaster patch, or drywall joint compound contains silicates and other compounds that require approved respiratory protection equipment.

Step 1 Repair any surface damage to the plaster or drywall surface.

Step 2 For drywall, make sure that all nails are set below the surface and are covered by joint compound. Make sure that all seams are taped and covered by joint compound. Using appropriate personal protective equipment, sand joint compound flush with drywall surface.

Step 3 Allow the drywall to dry at least four days and the new plaster to dry for at least 30 days.

Step 4 New plaster moisture content should be checked using the plastic sheet method. If moisture is detected, additional drying time is required.

Step 5 Thoroughly clean all sanding dust from surfaces of plaster or drywall by brushing or vacuuming and wiping with a tack cloth.

Step 6 Check that the pH of new plaster is approximately 7. If the pH is below 7, an alkaline-resistant primer sealer should be applied.

Step 7 With the temperatures above 55°F (12°C), prime the plaster or drywall and check for dull spots or uneven color and gloss after the primer dries. Reprime the problem spots, or reprime the entire surface until the surface is uniform in appearance.

6.3.2 Previously-Coated Plaster Or Drywall Substrates

WARNING! Older plaster, drywall ceilings, or walls may have asbestos fiber in the plaster itself, in plaster patch material, or in the drywall joint compound. In addition, some surfaces of the plaster or drywall may have been coated with a texture finish or a sprayed flocked finish that contains asbestos fiber. Before abrading, repairing, or removing these substrates and finishes, they should be laboratory tested for the presence of asbestos. If the surface/substrate contains asbestos, proper permits, special personal protection equipment, and dust containment equipment must be employed if the material is to be abraded, removed, or repaired. Special asbestos-fiber encapsulating paints are also available for use on these surfaces.

WARNING! Sanding dust from plaster, plaster patch, spackle, and joint compound contains silicates that require wearing approved respiratory equipment.

Paints applied prior to 1978 may contain lead or other toxic heavy metals. Use approved personal and environmental protection equipment and procedures when sanding or removing paint from these surfaces.

Step 1 If covered with a wallcovering that is to be removed, use a commercial wallcovering remover.

Step 2 Check for the presence of a calcamine finish by wiping the painted surface with a wet cloth. If the finish comes off on the cloth, all traces of the finish must be removed.

Step 3 Except for flocked-finished surfaces, use detergent or alkaline cleaners to clean the surface of all contaminants. Rinse thoroughly and check that the pH of the surface is approximately 7.

Step 4 Using the proper personal protection equipment, remove any loose or flaking paint with a dull putty knife. Feather the edges by sanding, or fill with spackle or joint compound. When dry, sand smooth.

Step 5 Fill or repair all holes, cracks, and soft or damaged plaster or drywall. Soft plaster may also be hardened with a white vinegar wash. When dry, sand repairs smooth using proper personal protection equipment.

Note If water damage is evident, find and eliminate the source of the water or the damage will recur. Use stain killing primer to recoat the damaged area.

Step 6 If the old surface was a gloss or semi-gloss finish, scuff sand or use a chemical deglosser to dull the surface. Then wipe, brush, or vacuum all loose dust and dirt from the surface.

Step 7 With the temperature at 55°F (12°C) or above, spot prime all repair areas with an appropriate primer/sealer. If wallcovering was removed, prime the entire surface. When dry, check that the primed surface is uniform in gloss and color. If not, apply primer to dull spots or reprime the entire surface.

6.4.0 METAL SUBSTRATES

6.4.1 New Steel Substrate

Step 1 Clean heavy deposits of oil, grease, and dirt from the surface using a solvent cleaner.

Step 2 Clean any remaining oil, grease, dirt, or other contaminants from the surface using detergent or alkaline cleaners. Rinse thoroughly with water to neutralize the surface.

Step 3 If required, determine the rust grade of the steel using SSPC-VIS 1-89 as a visual guide.

Rust Grade A – Covered with adherent mill scale, little or no rust visible.
Rust Grade B – Covered with both mill scale and rust.
Rust Grade C – Completely covered with rust; little or no pitting visible.
Rust Grade D – Completely covered with rust; pitting is visible.

Step 4 If required, determine the degree of cleaning required by using the SSPC-VIS 1-89 cleaned surface comparison guide.

Step 5 Wearing proper personal protection equipment, use the hand tool method, power tool cleaning to bare metal method, or wet abrasive low-pressure wash method for brush-off blast cleaning or commercial blast cleaning to achieve the desired degree of cleaning and profiling.

Note Near-white or white metal blast cleaning and profiling will be covered in a later module.

Step 6 Using proper personal protection equipment, remove weld spatter, rough welds, protrusions, and sharp edges with power tools.

WARNING! Many industrial coatings contain lead or other toxic heavy metals or solvents. Check product Material Safety Data Sheets for information. Wear approved respiratory protection when applying such coatings.

Step 7 With the temperature at least 50°F (10°C) or as recommended by the coating manufacturer and using proper personal protection equipment, apply any required conversion coatings and primers as soon as possible after cleaning the surface. Make sure any rust-back is removed before coatings are applied.

6.4.2 Previously-Coated Steel Substrate

WARNING! Many industrial coatings contain lead or other toxic heavy metals or solvents. Use approved personal and environmental protection equipment and procedures when sanding or removing paint from these surfaces. Wear approved respiratory protection when applying such coatings. Check product Material Safety Data Sheets for information.

Step 1 Clean heavy deposits of oil, grease, and dirt from the surface using a solvent cleaner.

Step 2 Clean remaining oil, grease, dirt, or other contamination from the surface using detergent or alkaline cleaners. Rinse thoroughly with water to neutralize the surface.

Step 3 Repair any cracks, holes, or deteriorated metal by welding.

Step 4 If required, determine the rust grade of steel in the same way as for new steel (substituting paint for mill scale).

Step 5 If required, determine the degree of cleaning in the same way as for new steel (substituting paint for mill scale).

Step 6 Wearing proper personal protection equipment, use the same methods as for new steel to achieve the desired degree of cleaning, profiling, and removal of protrusions and sharp edges, weld spatter, and rough welds.

Step 7 With the temperature at least 50°F (10°C) or as recommended by the coating manufacturer and using proper personal protection equipment, apply any required conversion coatings and primers as soon as possible after cleaning the surface. Make sure that any rust-back is removed before coatings are applied.

6.4.3 Aluminum Substrate

Note Aluminum does not require painting unless located near a seacoast, an alkaline environment, or when it is necessary for decorative purposes.

Step 1 If new, aged, or previously painted, clean any heavy deposits of oil, grease, or dirt with a solvent. Then wash with a detergent or alkaline cleaner and rinse thoroughly to neutralize the surface.

Step 2 Repair any cracks or holes by welding.

Note It is not necessary to remove any tightly-adhering oxides.

WARNING!	Many industrial coatings contain lead or other toxic heavy metals or solvents. Use approved personal and environmental protection equipment and procedures when sanding or removing paint from these surfaces. Wear approved respiratory protection when applying such coatings. Check product Material Safety Data Sheets for information.

Step 3 Clean loose oxides from the surface using hand or power tools and proper personal protection. Wipe the surface to clean any loose dust.

Step 4 Using proper personal protection and at temperatures of 40°F (6°C) or above, apply zinc chromate primer or other primer(s) to bare surfaces as directed by the manufacturer of the selected final coating.

6.4.4 Galvanized Steel Substrate

Step 1 If new, determine if the surface was treated with factory-applied stabilizer to prevent "white rusting" during shipment and storage. Test a small area with a commercially-available chrome test kit.

Step 2 If new, aged, or previously coated, clean any heavy deposits of oil, grease, or dirt with a solvent. Then wash with a detergent or alkaline cleaner and rinse thoroughly to neutralize the surface.

Step 3 If new, allow any exterior surface to weather for six months, if possible; otherwise, etch using a prepackaged etching solution or a solution of vinegar and water. Rinse thoroughly with water to neutralize the surface.

WARNING!	Many industrial coatings contain lead or other toxic heavy metals or solvents. Use approved personal and environmental protection equipment and procedures when sanding or removing paint from these surfaces. Wear approved respiratory protection when applying such coatings. Check product Material Safety Data Sheets for information.

Step 4 If aged or previously coated, repair any cracks, holes, or deteriorated metal by welding. Using proper personal protection equipment, remove any cracked or peeling paint, rust, oxides, or other contaminants with hand or power tools. Wipe the surface to clean any loose dust.

Step 5 Using proper personal protection equipment and at temperatures of 40°F (6°C) or above, apply zinc chromate or zinc dust primer or other primer(s) to the bare surfaces as directed by the manufacturer of the selected final coating.

6.4.5 Prefinished (Factory-Finished) Metal Substrate

Step 1 Clean off any heavy deposits of oil, grease, or dirt with a solvent. Then wash with a detergent or alkaline cleaner.

Step 2 Repair any cracks, holes, or deteriorated metal by welding.

WARNING! Many industrial coatings contain lead or other toxic heavy metals or solvents. Use approved personal and environmental protection equipment and procedures when sanding or removing paint from these surfaces. Wear approved respiratory protection when applying such coatings. Check product Material Safety Data Sheets for information.

Step 3 Using proper personal protection equipment, dull the surface using hand tools or power tools.

Note Metal finishes composed of silicon, teflon, or other fluorocarbons are difficult to paint and require brush blasting and/or special bonding primers to promote adhesion.

Step 4 At recommended temperatures and using proper personal protection equipment, apply primers as recommended by the manufacturer of the selected final coating.

6.4.6 Copper, Brass, Bronze, And Zinc Substrates

Step 1 For new bright zinc, brass, bronze, or copper, remove any deposits of oil, grease, or dirt with solvent. Then wash with detergent or alkaline cleaner.

Step 2 For new, aged, or previously-coated zinc, prepare and prime in the same manner as new, aged, or previously-coated galvanized steel.

Step 3 For new copper, brass, or bronze, dull the surface with steel wool or fine sandpaper.

WARNING! Many industrial coatings contain lead or other toxic heavy metals or solvents. Use approved personal and environmental protection equipment and procedures when sanding or removing paint from these surfaces. Wear approved respiratory protection when applying such coatings. Check product Material Safety Data Sheets for information.

Step 4 For aged or previously-coated copper, brass, or bronze, using proper personal protection equipment, remove any deteriorated finish and heavy oxide with hand tools, power tools, and deoxidizers. Then wash with diluted nitric acid and rinse thoroughly.

Step 5 At recommended temperatures and using proper personal protection equipment, apply the appropriate primer as recommended by the manufacturers of the selected final coating.

6.5.0 SYNTHETIC SUBSTRATES

6.5.1 Acoustic Tile Substrate

WARNING! Some older acoustic tile may contain asbestos fiber. Do not abrade the surface. Special asbestos-fiber encapsulating coatings are available for use over these types of surfaces.

Step 1 Vacuum the surface and then clean with a damp cleaning sponge or the equivalent. Heavily-soiled surfaces may be cleaned with a spray-and-wipe alkaline cleaner, followed by a clear-water wipe.

Step 2 Apply non-bridging coatings. Heavy coats of normal latex could destroy the sound-absorbent quality of the tile.

6.5.2 Pipe Or Duct Insulating Wrap Substrates

WARNING! Most older pipe or duct insulating wrap consisted of asbestos tape or asbestos with a cemented and plastered asbestos or canvas cover. Unless lab tests confirm the absence of asbestos, do not abrade or attempt to repair these wrappings. Special asbestos-fiber encapsulating coatings are available for use over these types of wraps.

Step 1 Vacuum the surface and then clean with a damp cleaning sponge or the equivalent. Heavily-soiled surfaces may be cleaned with a spray-and-wipe alkaline cleaner, followed by a clear-water wipe.

Step 2 Apply the appropriate primer; allow to dry.

6.5.3 New Or Previously-Coated Fiberglass Substrates

Step 1 Wash with a detergent or alkaline cleaner.

Step 2 Clean with a wax and silicone wax remover solvent.

Step 3 Lightly dull the surface with fine sandpaper.

Step 4 Repair or patch any surface or substrate damage. There should be no exposed fibers.

Step 5 Prime with the appropriate primer/surfacer following the coating manufacturer's instructions. Sand the primer/surfacer smooth.

6.5.4 Plastic Substrate

Step 1 Wash with a detergent or alkaline cleaner.

Step 2 Clean with a wax and silicon wax remover solvent after testing a small area to be sure that the solvent does not dissolve the plastic.

Step 3 If possible, identify the type of substrate (PVC, ABS, etc.).

Step 4 Contact the manufacturer of that type of substrate and determine the repair methods and coatings that are recommended. Use the manufacturer's information to repair and coat the surface.

Step 5 If unable to identify the plastic type and the plastic is relatively rigid, use polyester autobody filler or spot putty to repair any surface damage. Apply a primer/surfacer to spot repairs and wet sand with 400 grit sandpaper.

Step 6 Dull the surface with fine sandpaper.

Step 7 Apply the coatings recommended by the manufacturer, or if the plastic is unidentified, apply latex-based paints.

6.5.5 Wallcovering Substrate

Note Painting wallcoverings is not recommended and they should be removed. Wallpapers that are flocked or have a raised pattern cannot be painted without the pattern or flocking showing through.

Step 1 Check that the wallcovering and all edges are tightly bonded to the wall. The wallcovering will have to be removed if bubbles exist under portions of the wallcovering or the edges or other sections of the wallcovering are loose or poorly bonded. The wall should then be thoroughly cleaned and treated as a new plaster or drywall substrate.

Step 2 If the adherence of the wall covering is satisfactory, it should be damp wiped with a detergent solution, then sponged with clear water. Allow the surface to dry and recheck the adherence of the wallcovering.

Step 3 Prime the surface with one or two coats of acrylic primer/sealer. Make sure that no color bleeding occurs through the sealer.

Step 4 Coat with a latex-based paint, as required.

SUMMARY

No matter what the surface or substrate is, three actions may be required to prepare it for coating:

- Cleaning
- Repairing, if necessary
- Pretreating, if necessary

Not all of these actions are required every time and, depending on the coating manufacturer's instructions, they may not have to be done in a specific order. They may or may not have to be repeated to obtain the desired result, which is a surface or substrate that is clean, dry, dull, and ready for coating.

References

For advanced study of topics covered in this task module, the following books are suggested:

Adhesives, 8th Edition, Society of Plastics Engineers, Brookfield, Connecticut.

Decorating and Coating of Polymers, Society of Plastics Engineers, Brookfield, Connecticut.

Joining Plastics, Society of Plastics Engineers, Brookfield, Connecticut.

Painting and Coating Systems, The Sherwin-Williams Company, Cleveland, Ohio.

Painting and Decorating Craftsman's Manual and Textbook, 8th Edition, Painting and Decorating Contractors of America, Fairfax, Virginia.

Painting and Decorating Encyclopedia, The Goodheart-Willcox Company, Inc., Tinley Park, Illinois.

Plastics Surface and Finish, Society of Plastics Engineers, Brookfield, Connecticut.

REVIEW/PRACTICE QUESTIONS

1. There are _____ basic actions involved in surface preparation.
   a. two
   b. three
   c. four
   d. five

2. Detergents are used to _____ a contaminant.
   a. dissolve
   b. saponify
   c. emulsify and disperse
   d. chemically convert

3. Solvents are used to _____ a contaminant.
   a. dissolve
   b. saponify
   c. emulsify and disperse
   d. chemically convert

4. Alkaline cleaners are used to _____ a contaminant.
   a. dissolve
   b. saponify
   c. emulsify and disperse
   d. chemically convert

5. A consolidant is used to:
   a. concentrate solvents.
   b. modify primers.
   c. restore wood.
   d. accelerate acid activity.

6. A type of joint tape that is adhesive-backed is usually a:
   a. paper tape.
   b. cloth tape.
   c. mesh tape.
   d. perforated tape.

7. The metal butt of a putty or broad knife is used to:
   a. remove putty can lids.
   b. indent drywall seams.
   c. score drywall.
   d. hammer nailheads.

8. Power tool cleaning to bare metal is defined in:
   a. SSPC-SP11.
   b. SSPC-SP3.
   c. SSPC-SP7.
   d. SSPC-SP10.

9. Half-sheet sanders operate primarily as _____ sanders.
   a. oscillating
   b. belt
   c. straight-line or oscillating
   d. straight-line

10. Needle guns are good for cleaning:
    a. irregular areas.
    b. flat areas.
    c. aluminum.
    d. plastic.

11. Rotary impact power tools commonly use _____ types of abrasive hubs.
    a. two
    b. three
    c. four
    d. five

12. Water blast cleaning equipment is classified by _____ pressure ranges.
    a. two
    b. three
    c. four
    d. five

13. Low-pressure wet abrasive washing can be effectively used to achieve the finish of _____ standards.
    a. SSPC-SP5 and SP6
    b. SSPC-SP6 and SP10
    c. SSPC-SP5 and SP7
    d. SSPC-SP6 and SP7

14. Impact tools are primarily used to clean:
    a. steel.
    b. all metals.
    c. concrete.
    d. concrete and steel.

15. Muriatic acid is used to etch _____ for the proper profile.
    a. old plaster surfaces
    b. concrete or masonry floors
    c. brick
    d. steel

16. The neutral value of pH is:
    a. 7.
    b. 8.
    c. 6.
    d. 5.

17. The plastic sheet method uses a _____ sheet.
    a. 15 x 15 inch
    b. 18 x 18 inch
    c. 16 x 18 inch
    d. 24 x 24 inch

18. The moisture content of new exterior wood should be between _____ or lower before coating.
    a. 10% and 25%
    b. 3% and 5%
    c. 15% and 20%
    d. 12% and 15%

19. Preferably, new pressure-treated wood should be:
    a. allowed to weather for six months.
    b. prepared and finished like other wood once it is dry.
    c. stained only with solid color stains.
    d. all of the above.

20. New soft woods may require _____ before staining.
    a. a wash sealer
    b. priming
    c. washing
    d. planing

ACKNOWLEDGMENTS

Figures 1 and 2 courtesy of Knights Marketing Corporation.
Figure 3 courtesy of Painting and Decorating Contractors of America.
Figure 13 courtesy of Porter-Cable Company.
Appendix A courtesy of Sherwin-Williams Company.

APPENDIX A

Summary of SSPC-SP Series Standards

Steel (structural, plate, etc.) should be cleaned by one or more of the nine surface preparations described below. These methods were originally established by the Steel Structures Painting Council in 1952, and are used throughout the world for describing methods for cleaning structural steel. Visual standards are available through the Steel Structures Painting Council; ask for SSPC-Vis 1-89. A brief description of these standards together with numbers by which they can be specified follow.

- *Solvent Cleaning (SSPC-SP1)* – Solvent cleaning is used to remove oil, grease, dirt, soil, drawing compounds, and other contaminants by use of solvents, emulsions, cleaning compounds, or steam cleaning. Solvent cleaning does not remove rust or mill scale. Change rags and cleaning solution frequently so that deposits of oil and grease are not spread over additional areas in the cleaning process. Be sure to follow manufacturer's safety precautions. Do not use hydrocarbon solvents to clean grease, oil, or waxes; hydrocarbon solvents do not dissolve these, they merely spread them around.
- *Hand Tool Cleaning (SSPC-SP2)* – Hand tool cleaning is the method of removing loose rust, loose mill scale, and other loose contaminants from the surface after cleaning the surface in accordance with SSPC-SP1, Solvent Cleaning. It does not require the removal of intact rust or mill scale.
- *Power Tool Cleaning (SSPC-SP3)* – Power tool cleaning provides for the removal of loose rust, loose mill scale, and other loose contaminants from the surface after cleaning the surface in accordance with SSPC-SP1, Solvent Cleaning. It does not require the removal of intact rust or mill scale.
- *White Metal Blast Cleaning (SSPC-SP5 or NACE-1)* – White metal blast cleaning is used when a totally cleaned surface is required. This method of cleaning is defined as a blast cleaned surface of a gray-white uniform metallic color. The surface must be free of all oil, grease, dirt, mill scale, rust, corrosion products, oxides, paint, stains, streaks, or any other contaminants across 100% of every square inch. The surface must first be cleaned in accordance with SSPC-SP1, Solvent Cleaning.
- *Commercial Blast Cleaning (SSPC-SP6 or NACE-3)* – Commercial blast cleaning creates a surface on which all grease, dirt, mill scale, rust, and oil paint have been completely removed from the surface except that light streaks or discolorations caused by rust stain, mill scale oxides, or slight, tightly adhering residues of paint or coating remain. If the surface is pitted, slight residues of rust or paint may remain at the bottom of the pits. The slight discolorations are limited to one-third of every square inch. The surface must first be cleaned in accordance with SSPC-SP1, Solvent Cleaning.

- *Brush-Off Blast Cleaning (SSPC-SP7 or NACE-4)* – Brush-off blast cleaning creates a surface from which all oil, grease, dirt, loose mill scale, loose rust, and loose paint or coatings are completely removed. Tight mill scale and tightly adhering rust and paint or coatings may remain as long as the entire surface has been exposed to abrasive blasting. The surface must first be cleaned in accordance with SSPC-SP1, Solvent Cleaning.
- *Power Tool Cleaning to Bare Metal (SSPC-SP11)* – Power tool cleaning to bare metal is defined as cleaning metallic surfaces to such a degree that, when viewed without magnification, the surface is free of all visible oil, grease, dirt, dust, mill scale, rust, paint, oxide corrosion products, and other contaminants. Slight residues of rust and paint may be left in the lower portions of the pits if the original surface is pitted. The surface must be roughened to a degree suitable for the specified paint system. The surface profile must be no less than 1 mil. The surface must first be cleaned in accordance with SSPC-SP1, Solvent Cleaning.
- *Near-White Blast Cleaning (SSPC-SP10 or NACE-2)* – Near-white blast cleaning provide a surface from which all oil, grease, dirt, mill scale, rust, corrosion products, paint, or other contaminant have been completely removed from the surface. Very light shadows, very light streaks or discolorations caused by rust stain, mill scale oxides, or slight, tight residues of paint or coatings are permitted to remain. Only 5% of every square inch may contain the slight discolorations mentioned above. The surface must first be cleaned in accordance with SSPC-SP1 Solvent Cleaning.

NCCER CRAFT TRAINING USER UPDATES

The NCCER makes every effort to keep these manuals up-to-date and free of technical errors. We appreciate your help in this process. If you have an idea for improving this manual, or if you find an error, a typographical mistake, or an inaccuracy in the NCCER's Craft Training Manuals, please write us, using this form or a photocopy. Be sure to include the exact module number, page number, a description of the problem, and the correction, if possible. Your input will be brought to the attention of the Technical Review Committee. Thank you for your assistance.

Instructors – If you found that additional materials were necessary in order to teach this module effectively, please let us know so that we may include them in the Equipment/Materials list in the Instructor's Guide.

Write: Curriculum and Revision Department
National Center for Construction Education and Research
P.O. Box 141104
Gainesville, FL 32614-1104

Fax: 352-334-0932

Craft ____________________ Module Name ____________________

Module Number ____________________ Page Number(s) ____________________

Description of Problem

____

____

____

____

(Optional) Correction of Problem

____

____

____

(Optional) Your Name and Address

____

____

Sealants and Repair/Fillers

Module 07107

Painting Trainee Task Module 07107

SEALANTS AND REPAIR/FILLERS

PAINTING AND DECORATING CONTRACTORS OF AMERICA

NATIONAL CENTER FOR CONSTRUCTION EDUCATION AND RESEARCH

Objectives

Upon completion of this module, the trainee will be able to:

1. Describe the composition and function of various sealants and fillers.
2. Select an appropriate product for a given application and surface, stating the important properties for product selection.
3. Describe tools and additional materials required for sealant application.
4. Apply and smooth suitable products to the following joints using appropriate tools:
   - Fixed inside corner joint between two types of substrate
   - Fixed outside corner joint between two pieces of the same substrate
   - Expansion joint with 25%-50% expected movement
   - Long, wide, deep crevice in substrate
5. Based on application conditions, judge whether a post-primer sealant or filler will adhere properly after curing.

Prerequisites

Successful completion of the following Task Modules is required before beginning study of this Task Module: Core Curricula, Painting Level 1, Modules 07101 through 07106.

Required Trainee Material

1. Trainee Module
2. Appropriate Personal Protective Equipment

Course Map

This course map shows all of the *Wheels of Learning* task modules in the first level of the Painting curricula. The suggested training order begins at the bottom and proceeds up. Skill levels increase as a trainee advances on the course map. The training order may be adjusted by the local Training Program Sponsor.

Course Map: Painting, Level 1

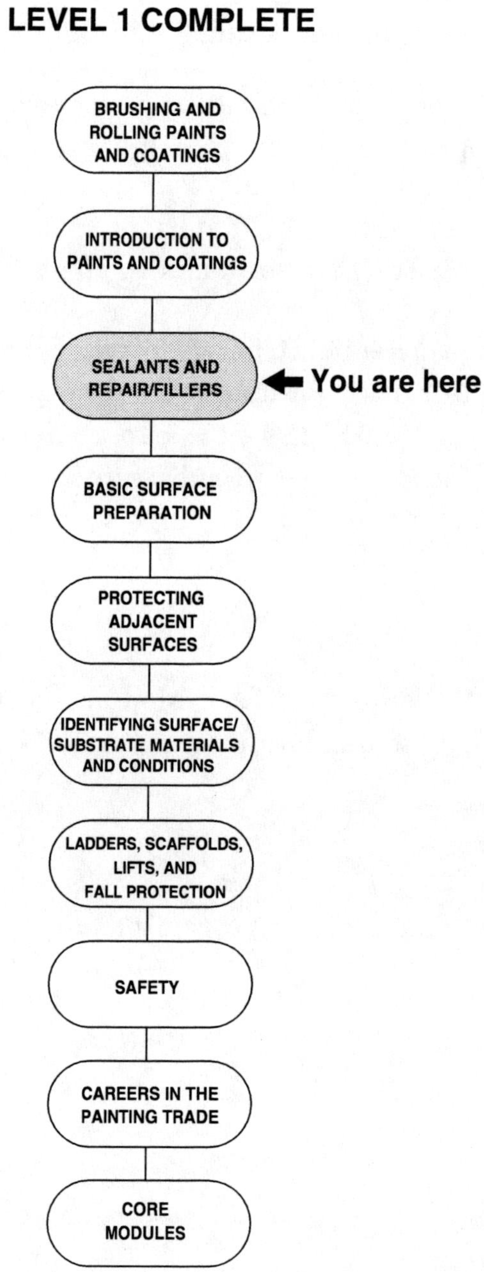

CMAP107.EPS

TABLE OF CONTENTS

Trade Terms Introduced In This Module

Base: The liquid that holds or binds the solids together in a sealant or filler. Bases are similar to the binders in paint coatings.

Bead: A string of material squeezed from the tip of a cartridge nozzle onto a substrate.

Cartridge: A hollow cardboard or plastic tube with a nozzle on one end, filled with sealant or filler. Cartridges are available in 6, 10, or 32 ounce sizes and are usually disposable.

Caulk: Sealants applied as a bead on construction joints which have little or no movement. *Caulk* is often used as an overall term for all sealants and fillers available in cartridges.

Elastomeric: Having the properties of excellent flexibility and elongation. Products can be slightly to highly elastomeric.

Elongation: The distance a cured bead can be stretched without breaking. It is expressed as a percentage of the original length of the cured bead.

Filler: Products applied to substrates to fill joints and repair cracks, gouges, tears, and other uneven spots in substrate surfaces.

Flexibility: A measure of a sealant or filler's ability to maintain its sealing quality after squeezing. It is expressed as excellent, good, or poor. Elastomeric products have excellent flexibility.

Glazing: Sealant (usually painter's putty) specifically used to seal the joint between a windowpane and its primed sash.

Gun grade: Sealants and fillers that are thicker than pour grade. Most are dispensed from a cartridge. Most gun grade products are the consistency of toothpaste.

Knife grade: Sealants and fillers of stiffer consistency than gun grade. Usually applied with knives or similar tools. May resemble the consistency of stiff frosting.

Paintability: Refers to how well paint coatings adhere to cured sealants.

Pour grade: Sealants and fillers that are of thin enough consistency to be poured from a bucket.

Putty: Used to describe some knife grade consistency products that are often used as sealants and fillers. Some can be thinned to gun or pour grades.

Sealant: Products applied to substrates as a barrier to moisture, air, and heat.

Shelf life: The amount of time a product remains usable.

Shrinkage: The amount of volume lost as a product dries and cures. Minimal shrinkage is desired for sealants and fillers.

Siliconized: A manufacturer's term for the addition of silicon particles to resin-based sealants. Identical to the phrase "with silicone added."

Solids content: The percentage of solid particles in the base.

Vacuumized: The process of removing air from a sealant or filler during packaging to reduce shrinkage.

1.0.0 INTRODUCTION

Your bathtub has them. Windows and log cabins have them. Even parking garages and bridges have them. All of these have sealants and fillers applied to their joints. Sealing and filling cracks and joints is an important step in substrate preparation, and if not done properly, can result in coating and substrate failure. For example, building materials exposed to moisture will deteriorate. Building materials contract and expand due to temperature changes, impacts, or vibrations. Joints with movement can collapse when joint flexibility fails. Fuel costs increase when heat is lost. Surfaces appear unsightly when they are pitted, gouged, dented, or uneven. Sealants and fillers are applied to prevent and fix these problems. They conform to odd-shaped openings, and create a smooth, continuous, moisture-resistant barrier on interior and exterior substrates.

2.0.0 PURPOSES/USES

Sealants are applied to substrates as a barrier to moisture, air, and heat. **Fillers** are used to fill holes, cracks, tears, and other spaces between substrates for cosmetic reasons, planned structural reinforcement, or unplanned structural reinforcement (such as repairing tears or cracks). The term *caulking* is often used to describe any sealant or filler that is available in **cartridges**. **Caulks** and **putties** are actually those sealants and fillers used on construction joints which have little or no movement. A single product can be both a sealant and a filler. A tube of caulk purchased at a local building supplier can serve more than one purpose. Manufacturers list the uses for each product on the product label or data sheet.

2.1.0 SUBSTRATES/USES

Sealants and fillers are used on a wide variety of substrates including wood, metal, glass, ceramic, plastic, and some types of foam. All sealants and fillers discussed here are post-primer types, meaning that they are used after primer has been applied to a substrate and before the final coating is applied. Keep in mind that some of these sealants and fillers can also be used before priming. Always read the product label and its data sheet.

Sealants are used on two types of joints: fixed or flexible. Fixed joints are those with little or no anticipated substrate movement. A drywall joint and a tile seam are examples of fixed joints (*Figure 1*). Flexible joints have considerable movement due to temperature changes, vibrations, or impacts (*Figure 2*). Expansion joints in concrete, metal, or masonry are examples.

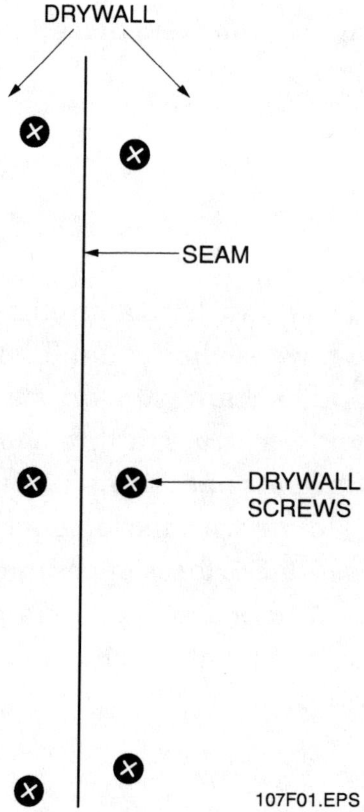

Figure 1. Fixed Joint – No Expected Movement

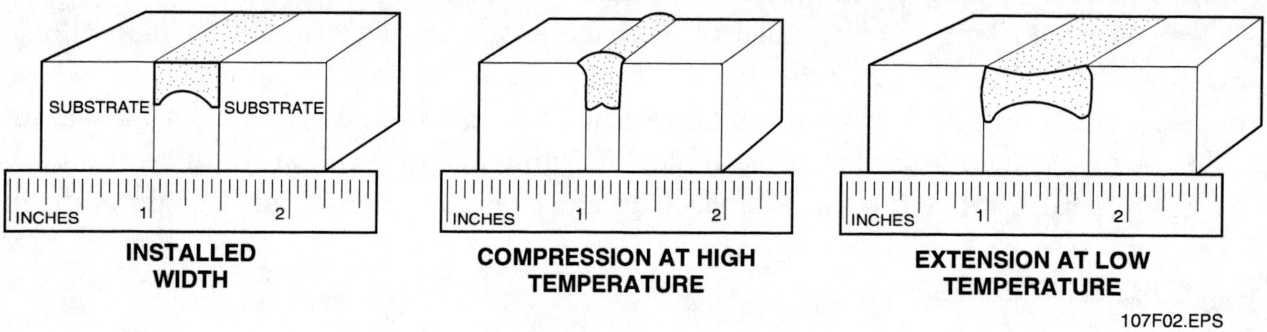

Figure 2. Flexible Joint – Expected Movement

Fillers are used when sealed joints have unplanned or excess movement, rupturing the existing sealant or filler. Fillers are also used to restore a smooth, even, continuous surface on substrates that have been gouged, cracked, dented, or damaged. Sealants and fillers are best suited to particular substrates based on their ingredients and properties.

3.0.0 INGREDIENTS

Sealants and fillers have two basic ingredients: a base and solids.

3.1.0 BASE

The **base** serves as the binder for the other ingredients (the solids). Sealants and fillers can be resin-based, silicone-based, or oil-based. Almost all resin-based products are acrylic latex and clean up with soap and water. Oil and silicone bases are usually grouped together as solvent bases, since paint thinners or other solvents must be used for cleanup. Linseed oil, fish oil, and petroleum are types of oil bases. Polyurethane, polysulfide, and urethane are types of petroleum oil bases. Drying oils, alcohols, and other liquids are often added to the base to affect drying and curing.

3.2.0 SOLIDS

The **solids content** includes fillers, thickeners, dispersing agents, and colorants. At least 80% of the weight is usually solids. Fillers and thickeners increase volume and stiffen the consistency. Clay, talc, mica, and calcium carbonate are common fillers. Dispersing agents disperse, or spread out the solids and base evenly in the product. Colorants create color. Color is not important when the sealant or filler will be painted. Some base and solids mixtures are not paintable, such as 100% silicone sealant. Matching the substrate color is important when using these sealants and fillers.

You may see sealants and fillers referred to as *one-part* or *two-part* products. In a one-part product, the base and solids are premixed and ready to use when purchased. Two-part sealants and fillers come with the base and solids in separate containers that you must mix and then use immediately. Two-part products are almost exclusively solvent-based.

4.0.0 PROPERTIES

Different combinations of ingredients create certain properties. Properties describe the way a product behaves when applied. They are important to consider when selecting a sealant or filler for a particular job.

There are several common properties used to select a sealant or filler:

- Cohesion
- Adhesion
- Shrinkage
- Elongation
- Flexibility
- Curing time
- Paintability

- Cleanup
- Life expectancy
- Consistency and texture

In addition, storage and application temperatures as well as possible staining and bleeding are important things to consider. *Figures 3* and *4* are examples of typical product data sheets showing ingredients and properties.

WHAT IT IS:

WEPCO (BUTYL) CAULK IS A SINGLE COMPONENT SYNTHETIC RUBBER JOINT COMPOUND THAT IS PRE-MIXED AND READY FOR USE DIRECTLY FROM THE CONTAINER.

WEPCO (BUTYL) CAULK ELIMINATES COSTLY "ON THE JOB" MIXING PROBLEMS AND APPLICATION TIME LIMITS. SHELF AND STORAGE LIFE IS UNLIMITED.

WEPCO (BUTYL) CAULK CHARACTERISTICS:

COMPOSITION:	PURE BUTYL, RUBBER CO-POLYMER, STABILIZERS, PLASTICIZERS AND PIGMENTS.
SOLIDS CONTENTS:	KNIFE GRADE 95% APPROX. GUN GRADE 90% APPROX.
SAG OR SLUMP:	NONE (COMPLIES WITH FED. SPEC. TT-C-589)
SHRINKAGE:	0 TO 5%
TENSILE ADHESIVENESS:	88 LBS. (COMPLIES WITH FED. SPEC. TT-P-781)
SHEARING ADHESIVENESS:	22 LBS. (COMPLIES WITH FED. SPEC. TT-P-781)
APPEARANCE AFTER DRYING:	NO CRACKING (COMPLIES WITH FED. SPEC. TT-C-598)
ELONGATION AT RUPTURE:	155%
CHEMICAL RESISTANCE:	EXCELLENT RESISTANCE TO: OZONE, SUNRAYS, WATER, ALCOHOL, ALKALIES AND ACIDS

USES:

WEPCO (BUTYL) CAULK IS HIGHLY RECOMMENDED AS A SEALING AND JOINT COMPOUND WHERE AN EXCELLENT ADHESION AND BOND IS REQUIRED ON CONCRETE, MASONRY, GLASS, WOOD, METAL, CERAMIC AND STONE.

WEPCO (BUTYL) CAULK ELIMINATES THE USE OF A PRIMER AND IS EXCELLENT FOR SEALING GLASS TO METAL, METAL SASH TO CONCRETE, AND FOR GENERAL GLAZING PURPOSES.

WEPCO (BUTYL) CAULK MAY BE USED AS A SEALANT FOR COPING STONES, PIPE JOINTS, AND FOR PROVIDING A FLEXIBLE SEAL IN EXPANSION JOINTS.

WEPCO (BUTYL) CAULK IS RECOMMENDED FOR ALUMINUM, STEEL, OR CERAMIC CURTAIN WALL CONSTRUCTION.

WEPCO (BUTYL) CAULK IS HIGHLY RECOMMENDED AS AN ALL-PURPOSE SEALANT WHERE A LASTING, ELASTIC WEATHERTIGHT JOINT IS DESIRED.

WHAT IT DOES:

WEPCO (BUTYL) CAULK IS EASY TO APPLY—NO PRE-MIXING OR PRIMING IS REQUIRED TO OBTAIN A TIGHT, FLEXIBLE RUBBERSEAL WHEN APPLIED TO VERTICAL OR HORIZONTAL SURFACES. HAVING EXCELLENT ADHESION, IT IS RESISTANT TO AGING AND POSSESSES EXCELLENT DURABILITY.

WEPCO (BUTYL) CAULK EXCEEDS THE PERFORMANCE OF CONVENTIONAL CAULKING MATERIALS DUE TO ITS ABILITY TO WITHSTAND SURFACE CONTRACTION AND EXPANSION. HAVING A VERY LOW PERMEABILITY TO MOISTURE, VAPOR, AND GASES, IT FORMS AN AIRTIGHT SEAL.

WEPCO (BUTYL) CAULK WILL NOT STAIN OR BLEED INTO ANY SURFACE.

DIRECTIONS FOR APPLICATION:

THE USE OF A PRIMER ON ANY SURFACE IS NOT REQUIRED. SURFACE TO BE CAULKED SHOULD BE FREE OF DUST, OIL, GREASE OR OTHER FOREIGN MATTER. WEPCO (BUTYL) CAULK MAY BE APPLIED IN THE CONVENTIONAL MANNER EITHER BY CAULKING GUN, PUTTY KNIFE, OR PRESSURE-TYPE CAULKING GUN. APPLICATORS OR TOOLS MAY BE CLEANED WITH ORDINARY PAINT THINNERS OR TURPENTINE.

COVERAGE:

ESTIMATED QUANTITY

LINEAR FEET PER ONE (1) GALLON OR 231 CUBIC INCHES

DEPTH OF JOINT	WIDTH OF OPENING						
	1/4	3/8	1/2	5/8	3/4	7/8	1
1/4	308	205.3	154.6	122.6	102.6	88	77.3
3/8		136.	102.6	82.6	68	58.6	50.6
1/2			77.3	61.3	50.6	44	38.6
5/8				49.3	41.3	34.6	30.6
3/4					34.6	29.3	25.3
7/8						25.3	22.6
1							18.7

COLORS:

WEPCO (BUTYL) CAULK IS AVAILABLE IN BLACK, WHITE, ALUMINUM AND GRAY. OTHER COLORS ARE AVAILABLE ON SPECIAL ORDER IN QUANTITIES OF 50 GALLONS OR MORE.

PACKAGING:

WEPCO (BUTYL) CAULK IS AVAILABLE IN ONE AND FIVE GALLON CANS AND IN ONE LB. CARTRIDGES.

107F03.EPS

Figure 3. Butyl Sealant Product Data Sheet

PRODUCT	ELMER'S WHITE ACRYLIC LATEX CAULK — PAINTABLE
DESCRIPTION	A SUPERIOR GENERAL PURPOSE INDOOR-OUTDOOR CAULK SUITABLE FOR USE WITH ALL LATEX AND OIL-BASED PAINTS. ITS DRY FILM IS TOUGH BUT FLEXIBLE, HAS EXCELLENT WEATHER RESISTANCE, IS MILDEW RESISTANT, AND CONTAINS NO EXTERNAL PLASTICIZER THAT COULD SOFTEN, BLEED THROUGH, AND DISCOLOR PAINT. FAST-SETTING FORMULA ALLOWS FOR PAINTING AROUND WINDOWS, DOORS, TUBS, SINKS, BASEBOARDS, AND FOR SEALING WOOD, METAL, BRICK, AND STONE. IT CAN BE APPLIED TO SLIGHTLY DAMP SURFACES AND ALLOWS FOR EASY CLEANUP WITH WATER BEFORE DRYING.

OTHER FACTS ABOUT ELMER'S WHITE ACRYLIC LATEX CAULK

TYPICAL PROPERTIES

TYPE	ACRYLIC LATEX
% SOLIDS	84.0
CONSISTENCY	GUN
SLUMP RESISTANCE	EXCELLENT
(INCHES – CHANNEL METHOD)	0
pH	8.0
WEIGHT/GALLON	12.4 LBS.
FLASH POINT	NONE, NON-FLAMMABLE
FLEXIBILITY	EXCELLENT
% ELONGATION	MAXIMUM STRESS 450 BREAKING POINT 620
LOW TEMPERATURE FLEXIBILITY @ -15°F 1/4" SLAB ON ALUMINUM, 180° BEND	NO CRACKING

STAINING OR SOFTENING DUE TO PLASTICIZER MIGRATION

AT ELEVATED TEMPERATURE OF 200°F FOR 2 HRS. – NONE
RESISTANCE TO DIRT PICKUP – EXCELLENT

RESISTANCE TO DISCOLORATION AND DEGRADATION ON AGING – EXCELLENT

INITIAL	WHITE/FLEXIBLE
AFTER 300 HRS. IN FADE-OMETER	NO COLOR CHANGE
AFTER 300 HRS. IN WEATHER-OMETER	NO COLOR CHANGE – REMAINS FLEXIBLE

PAINTABILITY – EXCELLENT

WITH ALL OIL-BASED AND LATEX TYPES OF PAINT

DRYING TIME

TACK FREE IN APPROX. 30 MINUTES. UNDER NORMAL CONDITIONS, MAY PAINT AFTER 30 MINUTES. ALLOW MORE TIME FOR HIGH HUMIDITY, LOW TEMPERATURE, AND HEAVY APPLICATION ON NON-POROUS BASE. CURES THROUGHOUT IN APPROX. 3-5 DAYS. DO NOT APPLY WHERE CURED BEAD WILL BE CONTINUALLY SUBMERGED IN WATER. APPLICATION MAY BE DAMAGED BY HEAVY WIND-DRIVEN RAINFALL DURING THE FIRST 10-16 HOURS AFTER APPLICATION.

107F04.EPS

Figure 4. Acrylic Latex Caulk Product Data Sheet

4.1.0 COHESION

Cohesion refers to how well solids hold together or stick to each other. High cohesion improves a product's resistance to weather, chemicals, and age. Highly cohesive sealants and fillers are especially important in places such as food preparation plants or exterior joint seals that are exposed to salt water.

4.2.0 ADHESION

Adhesion refers to how well the sealant or filler sticks to the substrate(s) to which it has been applied. Adhesion may be decreased by high humidity, improper substrate preparation, or the wrong product for the joint type or substrate. Adhesion is expressed as excellent, good, and fair for various substrates. Most sealants and fillers can be used on glass, wood, metal, ceramic, plastic, masonry, and stone. Product labels list substrate applications specific to the product.

4.3.0 SHRINKAGE

Shrinkage is the volume a sealant or filler contracts, or shrinks, after application. All sealants and fillers will shrink to some degree as they cure and dry. Shrinkage is expressed as a percentage of the original volume, such as 10%. This means the volume of cured sealant will be 10% smaller than it was when first applied. Shrinkage occurs as the liquid portion of the base evaporates during curing. A higher solids content means less shrinkage. Low shrinkage is desired, especially for exterior substrates. Too much shrinkage may crack a paint coating, expose a joint, or cause pockets in a smoothly filled surface.

4.4.0 ELONGATION

Elongation is the distance a cured bead of sealant or filler can be stretched without breaking. Elastomeric sealants and fillers have high elongation. **Elastomeric** sealants can also return to their original shape after some amount of elongation. Elongation is expressed as a percentage of the original length. If the length was doubled before breakage occurred, the elongation is 100%. If the length tripled, the elongation is 200%. Elongation is important to consider in flexible joints since it indicates how much a sealant or filler can be stretched before the seal is broken (*Figure 5*).

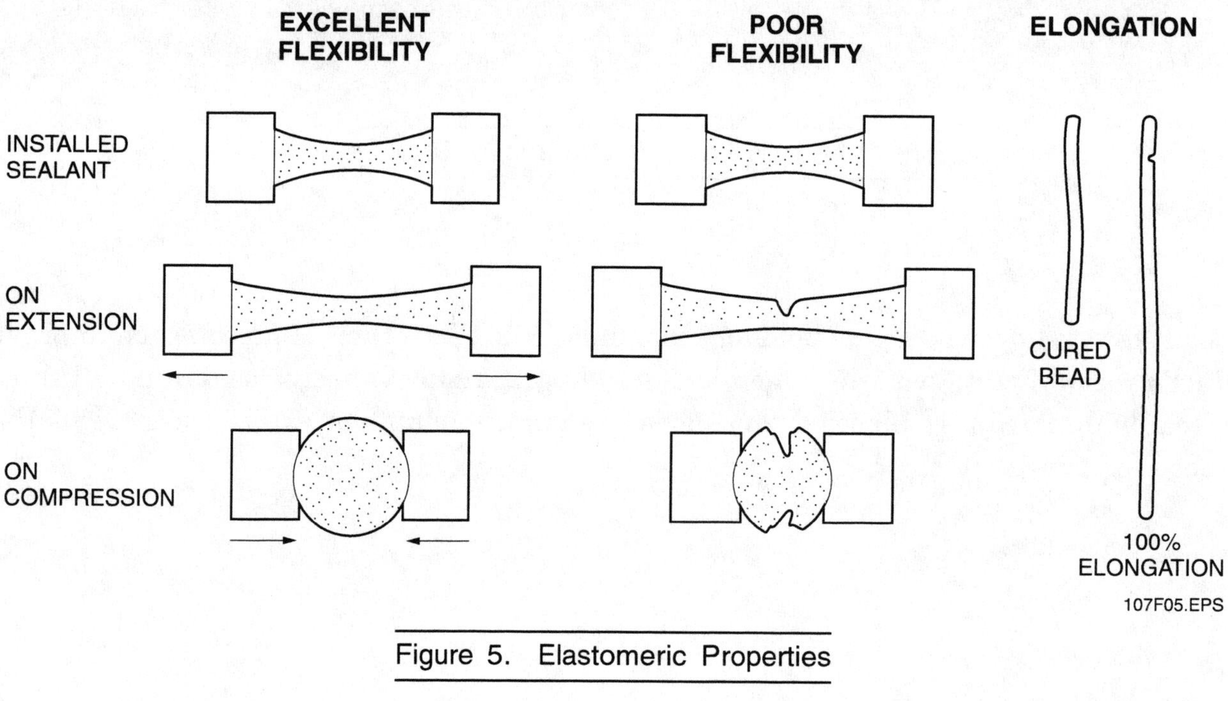

Figure 5. Elastomeric Properties

4.5.0 FLEXIBILITY

Flexibility is a measure of a sealant or filler's ability to withstand squeezing and maintain its sealing quality. It is expressed as excellent, good, or poor. Elastomeric products have excellent flexibility (see *Figure 5*). Flexibility is important to consider in flexible joints. The greater the expected movement in the joint, the more important the flexibility and elongation properties. Butyl rubber, polyurethane, polysulfide, silicone, and even some acrylic latex sealants have excellent flexibility.

4.6.0 CURING TIME

Curing time refers to the time it takes a product to harden completely. Drying time is slightly different, as it refers to the time required for the exterior or skin to harden. Tack-free time indicates when a sealant or filler is firm enough to paint or finish. Times can vary from minutes to hours, depending on the product, temperature, and humidity after application.

Curing and drying occur in two ways. In resin-based and some solvent-based products, liquids in the base evaporate into the air, leaving the dry solids. Resin-based products dry quickly. In solvent bases such as polysulfide, silicone, and polyurethane, chemical curing occurs. Solvent-based products dry slowly. The product basically "cooks" in moisture from the air until it is hardened. The exposed surface hardens slowly (hours), and the center remains pliable and sticky. Resin-based products skin quickly (sometimes in minutes) and are paintable with latex paint coatings once the skin has formed. Most resin-based products must cure before oil paint coatings are applied.

4.7.0 PAINTABILITY

Paintability refers to how well paint coatings will stick to dry or cured sealants and fillers.

Resin-based products can generally be painted with both oil and latex paint coatings. Some can be painted as soon as 30 minutes after application, but often curing is recommended before painting. Some resin-based products can *only* be painted with latex paint. Poor paintability means paint will peel off or repel the sealant or filler. Some sealants, such as butyl rubber, 100% silicone, and polyurethane cannot be painted and must contain a colorant in order to match the final paint coating. Always check the label or data sheet for paintability.

4.8.0 CLEANUP

Cleanup refers to the liquids and methods required to remove unwanted sealant. Almost all resin-based products are acrylic latex and can be removed from tools and substrates using a damp cloth or soap and water. Solvent-based products and a few resin-based products require the use of paint thinners or other solvents to clean up tools and excess product. Product labels and data sheets will state any special cleanup instructions.

4.9.0 LIFE EXPECTANCY OR DURABILITY

Life expectancy is the number of years a cured sealant or filler is expected to maintain its properties. Life expectancy for sealants and fillers is based on solids content, cohesion, adhesion, and proper application. Resin-based products have the longest life expectancy, ranging from 10 to 50 years. Solvent-based products last at least a couple of years before they require replacing. There are also seasonal and removable sealants on the market for use when temporary seals are required.

4.10.0 CONSISTENCY AND TEXTURE

Consistency refers to how thick a product is, which is based primarily on solids content. **Pour grade** refers to a pourable consistency. **Gun grade** refers to a toothpaste-like consistency which can be delivered using a caulking gun. **Knife grade** refers to stiffer consistency putties and pastes that are applied with hand-held tubes, putty knives, and similar tools. Some products can be thinned. Keep in mind that thinning a product can affect such properties as shrinkage, cohesion, flexibility, and life expectancy.

Texture refers to the surface smoothness of the applied product. A smooth texture is desirable for paintability and as a cosmetic concern. Acrylic latex products have superior smoothness.

4.11.0 STAINING AND BLEEDING

Staining and bleeding are undesirable properties of sealants and fillers. This refers to a discoloration of the surrounding substrate due to contact with the product. Stains and bleeding usually occur due to the use of certain colorant solids in the product, improper priming of the substrate, or using unsuitable products on properly primed substrates. Follow the manufacturer's directions to reduce the risk of staining or bleeding. If in doubt, and the appearance is important, apply a sample of the sealant or filler to a scrap of substrate to test it before final application.

4.12.0 STORAGE

Most sealants and fillers should be stored at temperatures between 40°F and 120°F. Each product has a recommended application and storage temperature to maintain its properties and life expectancy. Always check the **shelf life**. The shelf life is the period of time a sealant or filler is useful. Pre-mixed sealants and fillers generally have the longest shelf life, as long as they are in an airtight container. Two-part sealants and fillers have long shelf lives before they are mixed. Once mixed, their shelf life varies from hours to days, depending on the particular product.

5.0.0 PACKAGING

Sealant and filler packaging depends on the consistency of the product and its use (*Figure 6*). Gun grade sealants and fillers are usually sold as "caulks" and are packaged in 6, 12, or 32 ounce disposable cartridges. Some products can be purchased in one gallon or five gallon pails for high-volume applications. Softer consistency products can be purchased in tubes. Stiffer consistency products are usually purchased in cans or tubs. Coverage is expressed in lineal feet of a certain diameter bead. For example, one gallon of polysulfide sealant covers 1,500 lineal feet with a 1/8 inch bead. Coverage will be listed on product labels or data sheets.

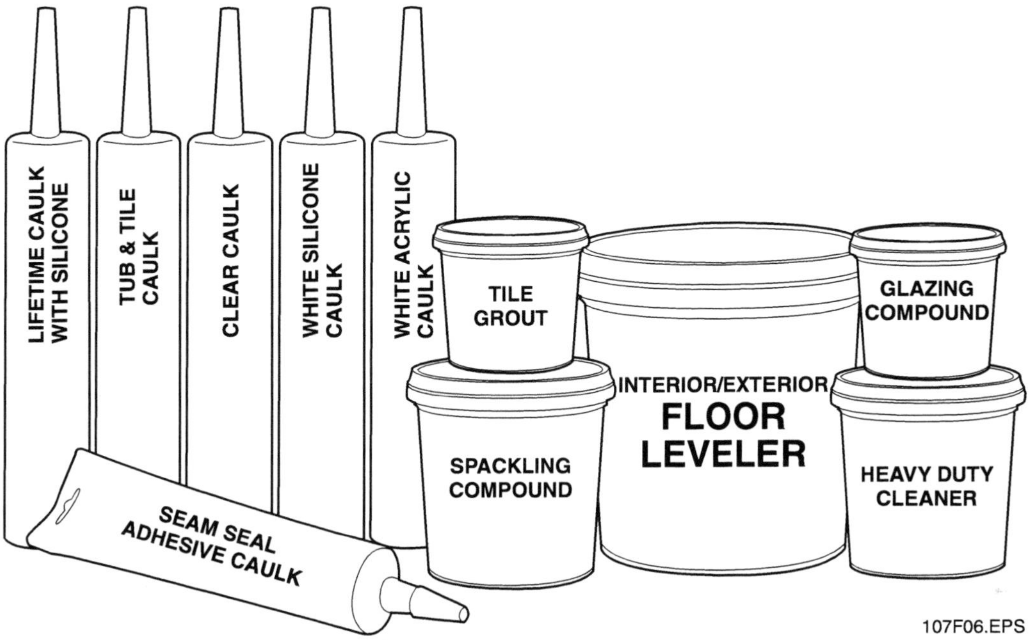

Figure 6. Typical Sealant And Filler Packaging

Some labels will state that the product has been **vacuumized**. Vacuumizing is the removal of air bubbles; less air means less shrinkage. However, even products that don't say vacuumized are still processed to remove air. Acrylic latex caulk that is **siliconized** and acrylic latex caulk "with silicone added" are the same type of sealant. Both refer to resin-based caulk that has silicone particles as part of the solids content. However, the *amount* of silicone particles mixed in may be different (to increase flexibility and life expectancy). Both "vacuumized" and "siliconized" are words used to sell products and make them look special.

Another thing to watch for is products that are labeled as *putties*. Some products have been called putties since their creation at the beginning of this century. So even now, when they may be available in other consistencies and for other uses, they are still called putty due to tradition. An example of this is painter's putty, which is available in both knife grade and gun grade in cans and cartridges. It can be thinned and is used for sealing and filling purposes.

Products labeled as **glazing** are actually putty-consistency products intended specifically for sealing window glass to window sashes. Always read the label to make sure you're getting the properties you require for the job.

6.0.0 ADDITIONAL MATERIALS

When sealing or filling joints, deep cracks, and damage, there are several materials commonly used to fill most of the space before applying the sealant or filler product (*Figure 7*). Using these along with sealant is generally less expensive than filling a large space with sealant alone. There are three main materials used to take up space in deep joints:

- Bondbreaker tape
- Backer rod
- Polyethylene foam

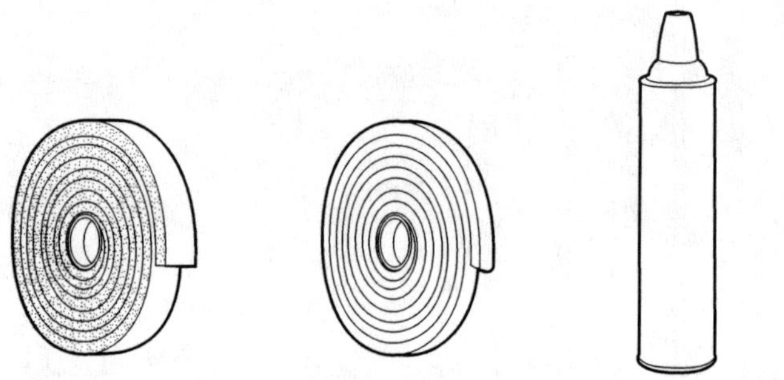

Figure 7. Additional Sealing And Filling Materials

Bondbreaker tape is made of polyethylene foam or similar materials. Bondbreaker tape fits into the back of the joint, preventing adhesion of the sealant to the back of the joint. This is especially useful in joints with considerable movement that can potentially be greater than the flexibility of the sealant. Bondbreaker tape allows the sealant to move with the joint, preventing tears in the sealant (*Figure 8*).

Backer rod is foam material sold in rolls from 3/8 inch to 3 inches wide. It can be used in place of bondbreaker tape, since it also prevents complete three-sided adhesion. It is also an economical alternative to filling a large joint completely with sealant or filler product.

Polyethylene foam is sold in cans. It expands and hardens into a stiff foam. It is useful in large fixed joints where it would be expensive to fill the complete joint with sealant or filler.

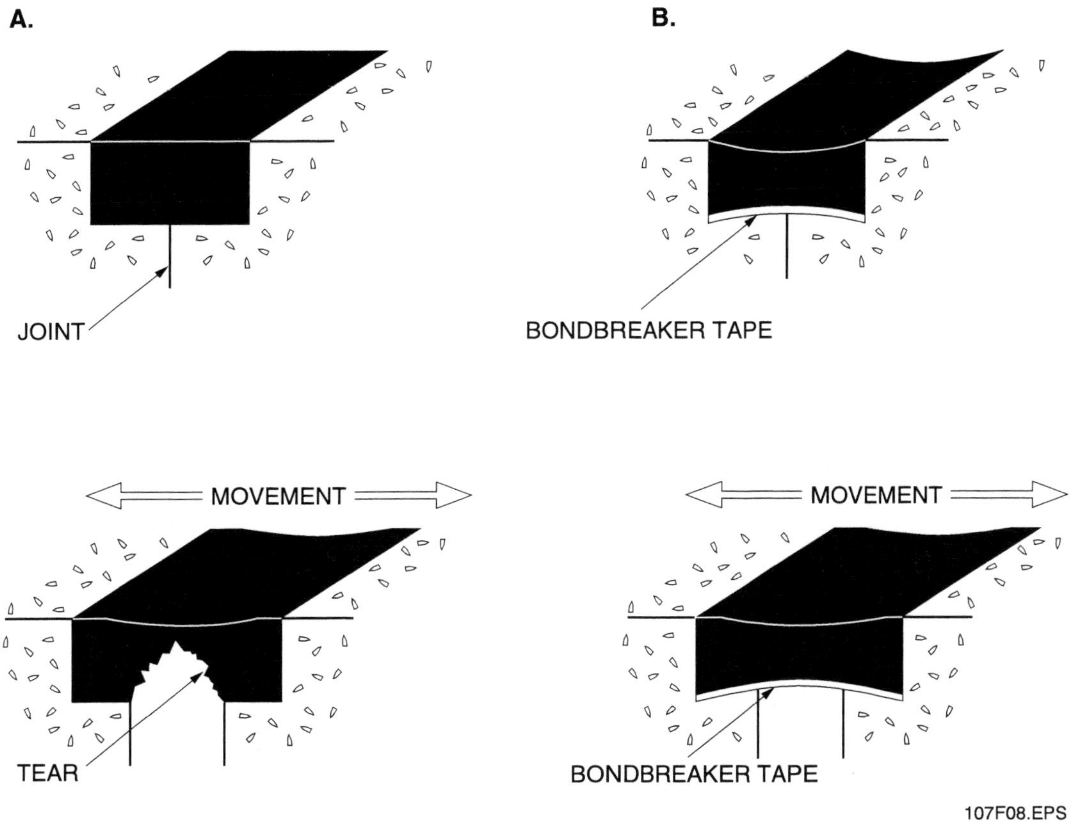

Figure 8. Flexible Joint (A) Without Bondbreaker Tape And (B) With Bondbreaker Tape

7.0.0 APPLICATION TOOLS

There are many tools for applying sealants and fillers. Perhaps the most common tools are caulking guns. A gun is a metal or plastic tool with a handle, plunger, and trigger that uses hand or air power to squeeze sealants and fillers out of cartridges through a nozzle opening to form a bead. There are hand-operated, battery-operated, and pneumatic (air-powered) guns (*Figures 9* and *10*). Almost every gun has a dripless feature, which backs off the pressure on the plunger when the trigger is released to prevent oozing. Hand-operated guns require you to repeatedly squeeze a trigger. Each time you squeeze the trigger, a little more sealant or filler is extruded from the nozzle. Battery-operated guns do not require repetitive hand motions to advance the plunger. This prevents hand fatigue, and supplies product at an even rate, making it easier to apply a smooth bead. Pneumatic guns are usually sold as compressor attachments that can apply large volumes of product at high rates of speed. Attachments for guns that can deliver products from a pail instead of a cartridge are also available for large jobs.

Knife grade products in cans and tubs are usually applied with putty knives or similar tools in much the same way that drywall joint compound is applied. The difference is that sealants and fillers are applied *after* the primer, and drywall joint compound is applied *before* the primer.

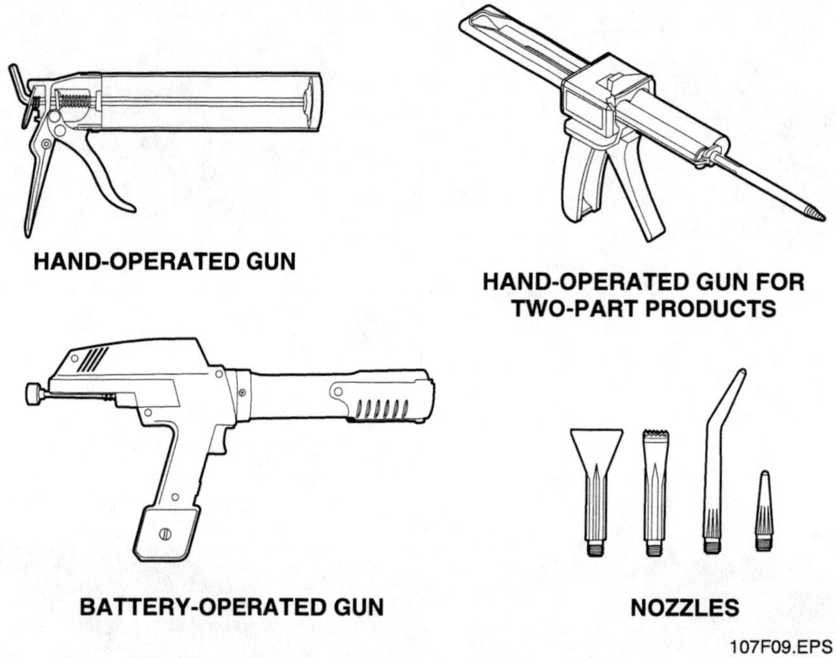

Figure 9. Gun-Type Application Tools

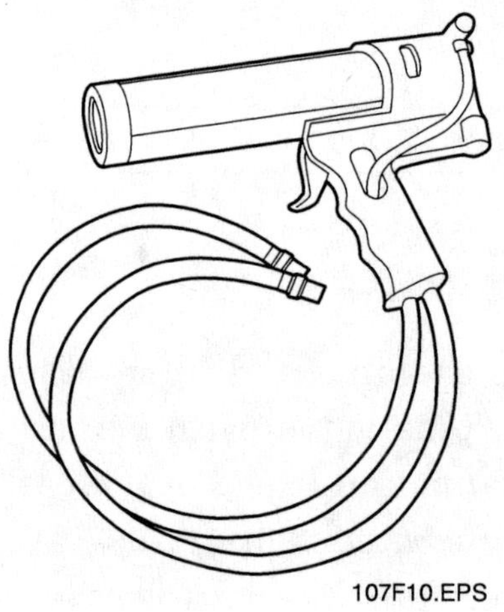

Figure 10. Pneumatic Gun

Pour grade products can be applied with spray applicators, such as those used for paint coatings, except with a larger nozzle. When using this equipment, use a positive pressure, air-supplied respirator or an air line respirator.

Always select your tools and equipment according to the job. Small jobs require few tools, such as a hand-operated gun and 10-ounce cartridges, cleanup supplies, and personal protection equipment, when necessary. Large jobs may require an air-powered gun, five gallon pails of sealant, cleanup supplies, and personal protection equipment for several people.

8.0.0 SELECTING A SEALANT OR FILLER

To select a sealant or filler, you will need the following information:

- Substrate type (masonry, metal, glass, wood, plastic, ceramic tile, etc.)
- Primer type (latex, oil, etc.)
- Joint type (fixed or flexible, shallow or deep)
- Environment for application (damp substrate, cold weather, etc.)
- Intended paint coating

After you have this information, determine the most important sealant properties for the job. For example, a damp exterior wall to window joint seal requires sealant properties such as excellent paintability, short curing time, adhesion to damp surfaces, and smooth texture. Some flexibility is necessary, but is not nearly as crucial as for sealant used in concrete expansion joints. In the window example, a good choice would be an acrylic latex sealant with silicone added.

WARNING! Always obtain and read the Material Safety Data Sheets (MSDSs) for the products being applied. Observe all safety precautions to avoid health and safety hazards to yourself and others.

Always use the product labels and data sheets to help you select the product with the best properties for the job conditions. Some manufacturers provide charts of properties and recommended uses for easy product selection. *Figure 11* is a comparison chart of general product types; manufacturers offer specific product comparison charts.

9.0.0 APPLYING SEALANTS AND FILLERS

Following are some general guidelines and procedures for applying sealants and fillers.

There are five basic steps in applying any sealant or filler:

- Preparing the substrate
- Preparing the sealant and tools
- Applying the sealant or filler to the joint
- Smoothing the sealant/filler surface
- Cleaning up and checking for adhesion

TYPE	PAINTABLE	CLEANUP	FLEXIBILITY	DURABILITY	ADHESION TO:			
					WOOD	GLASS	METAL	MASONRY
CLEAR SILICONE NEUTRAL CURE	NO	MINERAL SPIRITS	4	4	3	4	4	4
CLEAR SILICONE ACETOXY	NO	MINERAL SPIRITS	4	3	2	4	4	0
TWO-PART BLADE SILICONE NEUTRAL CURE	NO	SOLVENT	4	4	3.5	3	4	4
COLORED INDUSTRIAL POLY-URETHANE (COMBUSTIBLE)	YES	SOLVENT	4	4	4	3	4	4
WHITE URETHANE ACRYLIC	YES	SOAP & WATER	4	4	4	3.5	3.5	3.5
CLEAR SOLVENT (FLAMMABLE)	YES	SOAP & WATER	3.5	2.5	3	3.5	3.5	3.5
WHITE 50-YEAR ACRYLIC LATEX WITH SILICONE	YES	SOAP & WATER	4	4	4	3.5	3	3.5
WHITE 35-YEAR ACRYLIC LATEX WITH SILICONE	YES	SOAP & WATER	4	3	4	3	3	3
WHITE 25-YEAR SILICONIZED ACRYLIC LATEX	YES	SOAP & WATER	3	3	3	3	2.5	2
WHITE 5-YEAR LATEX	YES	SOAP & WATER	1	1	2	1	1	1
PAINTER'S WHITE 25-YEAR ACRYLIC LATEX	YES	SOAP & WATER	3	2.5	3	2.5	2	2.5
PAINTER'S WHITE 15-YEAR ACRYLIC LATEX	YES	SOAP & WATER	2	2	2.5	2	1.5	2

RATING SCALE: EXCELLENT 4/BEST 3/BETTER 2/GOOD 1/NR – NOT RECOMMENDED

107F11.EPS

Figure 11. Typical Comparison Chart For Selecting Products

9.1.0 PREPARING THE SUBSTRATE

Poor substrate preparation results in poor quality seals that can fail long before their life expectancy has expired. Preparing the substrate involves providing a clean, dry, solid surface for the best sealant adhesion. *Table 1* lists specific preparation recommendations for particular substrates.

TYPE OF CONSTRUCTION MATERIAL	SURFACE PREPARATION
CONCRETE AND MASONRY BRICK CONCRETE BLOCK POURED CONCRETE PRECAST CONCRETE TILT-UP CONCRETE MORTAR GROUTS CEMENT-ASBESTOS (FACTORY MADE)	SURFACES MUST BE DRY, SOUND, DUST FREE AND FREE OF FORM OILS OR TREATMENTS THAT PREVENT ADHESION. STIFF BRISTLE BRUSHING, GRINDING, OR SANDBLASTING MAY BE REQUIRED. SURFACES VARY GREATLY FROM JOB TO JOB AND TEST APPLICATIONS ARE RECOMMENDED TO DETERMINE DEGREE OF PREPARATION REQUIRED. IN SOME CASES, PRIMER MAY BE REQUIRED FOR BEST RESULTS.
STONE GRANITE MARBLE LIMESTONE	SURFACES MUST BE DRY AND DUST-FREE. PRIMER MAY BE REQUIRED. TEST APPLICATIONS ARE RECOMMENDED TO DETERMINE SURFACE PREPARATION, PRIMER NEED AND STAINING POTENTIAL.
GLASS/PORCELAIN SHEET, FLOAT, OR PLATE GLASS TINTED GLASS REFLECTIVE GLASS GLAZED CERAMIC TILE PORCELAIN COATED METAL VITRIFIED SURFACES	SURFACES MUST BE DRY AND FREE OF DUST, OIL, OR OTHER CONTAMINANTS. DETERGENT OR SOAP AND WATER TREATMENTS ARE NOT RECOMMENDED. SURFACE SHOULD BE CLEANED WITH ALCOHOL OR OTHER SUITABLE SOLVENT. THE SOLVENT USED SHOULD BE CHECKED FOR COMPATIBILITY WITH ADJACENT MATERIALS THAT IT WILL CONTACT.
PAINTS ACRYLIC LATEX ACRYLIC-THERMOSET ALKYD LATEX ALKYD ENAMEL SILICONE ALKYD SILICONE ACRYLIC POLYURETHANE POLYVINYL CHLORIDE POLYVINYL FLUORIDE RESIN-BASED PAINTS	SURFACES MUST BE CLEAN AND DRY. AN ISOPROPYL ALCOHOL WIPE WITH CLEAN RAG IS RECOMMENDED. SINCE FORMULATION MAY CHANGE, TEST APPLICATION OF MATERIAL TO BE USED ON THE JOB IS RECOMMENDED.
PLASTICS ACRYLIC SHEET POLYESTER/FIBERGLASS EPOXY MATRIX PANELS POLYESTER MATRIX PANELS POLYSTYRENE POLYVINYL CHLORIDE POLYVINYL FLUORIDE POLYCARBONATE SHEET	SURFACES MUST BE CLEAN AND DRY. ALCOHOL WIPE WITH CLEAN RAGS.
RUBBER BUTYL E.P.D.M. NEOPRENE POLYURETHANE POLYSULFIDE ACRYLIC	RUBBER FORMULATIONS VARY GREATLY AND MAY CONTAIN MATERIALS THAT CAUSE STAINING OR CHEMICAL REACTION. TEST APPLICATIONS ARE REQUIRED IN EACH CASE TO DETERMINE COMPATIBILITY.
SILICONE SEALANTS ALL	ALCOHOL WIPE TO CLEAN SURFACE.
METALS ALUMINUM - MILL FINISH ALUMINUM - ANODIZED ALUMINUM - LACQUERED COPPER LEAD STEEL - RED LEAD PRIMED STEEL - BRIGHT/CLEAN STEEL - WEATHERED STEEL - STAINLESS STEEL - GALVANIZED	SURFACES MUST BE OIL AND DUST-FREE. ALCOHOL WIPE-REMOVE OILS AND DUST. ALCOHOL WIPE-REMOVE OILS AND DUST. SAND OFF OXIDE-ALCOHOL WIPE. ALCOHOL WIPE-REMOVE OILS AND DUST. ALCOHOL WIPE-REMOVE DUST. ALCOHOL WIPE-REMOVE OILS AND DUST. WIRE BRUSH-ALCOHOL WIPE. SURFACES MUST BE OIL AND DUST-FREE. ALCOHOL WIPE-REMOVE OILS AND DUST.
WOODS UNFINISHED FINISHED	UNFINISHED WOOD IS NOT A PROPER SURFACE FOR SEALANT ADHESION. WOOD SURFACES SHOULD BE SEALED WITH PAINT OR OTHER COATING. SURFACES MUST BE CLEAN, DRY, AND DEGLOSSED.

"THESE RECOMMENDATIONS AND DETAILS WERE DEVELOPED BY GE SILICONES FOR USE WITH GE SILICONES CONSTRUCTION SEALANTS. PERFORMANCE CHARACTERISTICS MAY VARY BY MANUFACTURER, THEREFORE CONSULT THE MANUFACTURER'S RECOMMENDATIONS FOR SPECIFIC PRODUCT USE."

GENERAL ELECTRIC COMPANY, U.S.A. IS NOT CONNECTED WITH THE ENGLISH COMPANY OF A SIMILAR NAME. FOR MORE INFORMATION ON GE SILICONES CONSTRUCTION SEALANTS, PLEASE CONTACT 1-800-332-3390.

107TBL1.EPS

Table 1. Typical Surface Preparation By Substrate

Figure 12 illustrates the first three steps of the following general substrate preparation procedure:

Step 1 Remove loose mortar, coatings, dirt, moisture, and old sealant.

Step 2 Clean with soap and water or solvent, then let dry.

Step 3 Prime the surface (as directed in the surface preparation module) and let dry.

Step 4 Insert foam, bondbreaker tape, or backer rod in the back of the joint (if necessary).

Figure 12. Preparing The Substrate

Poor preparation such as loose dirt left on the surface, incomplete washing, wet or damp substrates, wet or missing primer, or incompatible primers can cause sealants to fail. Without primer, the substrate can soak up the base in the sealant, not allowing it to cure properly and staining the substrate. *Do not rush preparation.* It is a lot harder to go back to a job site, remove poorly adhered sealant, and reapply new sealant, than it is to do it right the first time.

9.2.0 PREPARING THE SEALANT AND TOOLS

If the sealant is a two-part sealant, you may need to mix it in a container like the one shown in *Figure 13*. The mixed sealant is then put in a cartridge, bucket, or pail, depending on the consistency grade. Some guns will accept two cartridges and mix two-part sealant as it is applied. Insert the mixed or pre-mixed cartridge(s) into the gun. Then cut the nozzle tip off at a 30° or 45° angle (*Figure 13*), leaving an opening slightly larger than the width of the joint opening. Finally, pierce the inner seal inside the nozzle with a nail or other sharp object.

If you are using a battery-operated gun, you may need to replace the batteries or recharge the battery pack. If you use a pneumatic gun, you'll need to connect the hoses, check for pressure, and either load a cartridge or attach a hose to the bucket of sealant.

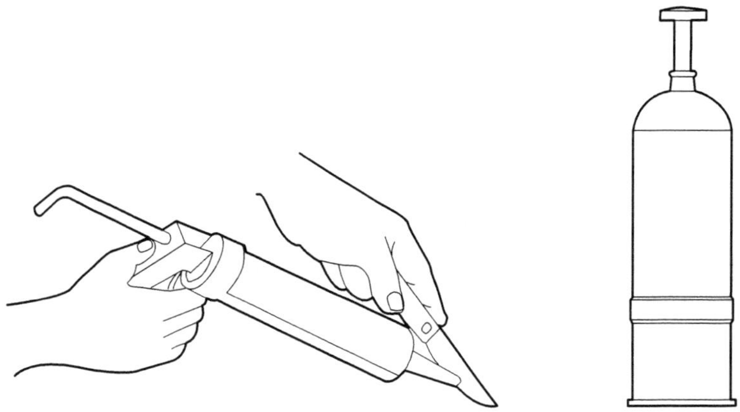

Figure 13. Preparing Sealants

For pour grade sealant, load the liquid into the application tool; choose a container you can carry easily (sometimes this is just a bucket). For knife grade sealant, make sure a putty knife or other application and smoothing tool(s) are available.

For resin-based products without solvents, prepare a bucket of clean, soapy water, clean cloths, and a clean latex glove for smoothing and cleanup. For oil-based products and other solvent-based products, have a solvent cleaner, clean cloths, solvent-resistant gloves, and the proper disposal container ready in a well-ventilated area. Use the appropriate safety equipment.

WARNING! Use proper ventilation when applying sealants and fillers. Even resin-based products can contain solvents that give off toxic fumes and can cause eye, lung, and skin irritation. Avoid eye and skin contact. Wear gloves and remove contaminated clothing as soon as possible.

9.3.0 APPLYING THE SEALANT OR FILLER TO THE JOINT

Smoothly applying sealants and fillers requires practice. It will be messy at first, but with each application your results will be more accurate. Most manufacturers recommend applying a maximum $^1/_2$-inch sealant depth to any joint. Joints larger than $^1/_2$-inch deep or those that will allow sealant or filler to seep through the back of the joint should be filled with foam or backer rod. Always allow a minimum $^1/_8$-inch depth of sealant.

CAUTION Have adequate lighting in the work area. Poor lighting causes eye fatigue.

9.3.1 Applying Gun Grade Sealants Or Fillers

Squeeze a little bit of material from the nozzle onto a piece of paper to get the feel of the flow. Place the tip of the nozzle in the joint. Hold the flat side of the nozzle tip parallel to the joint/substrate surface. This is critical. If you hold the gun too high, the nozzle actually scrapes material out of the joint as it is supplied, thinning the middle of the seal and thickening the edges. If you hold the gun too low, you get a lumpy seal that skips areas. Use your free hand to guide the nozzle end of the cartridge along the joint. Now draw a **bead** (*Figures 14* and *15*).

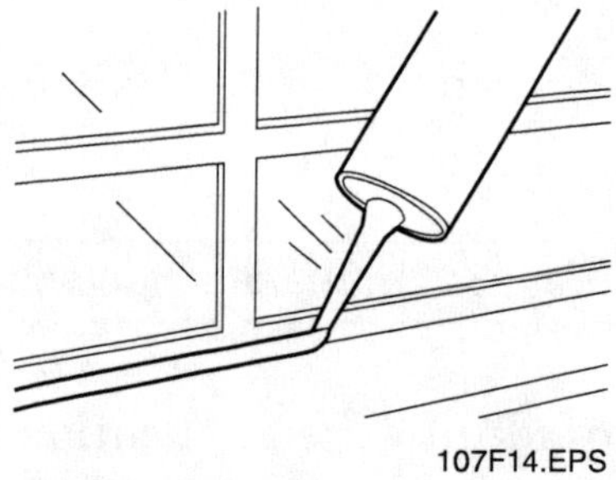

Figure 14. Drawing A Bead Below Waist Height

Figure 15. Drawing A Bead Above Waist Height

Drawing usually has two motions: squeezing the trigger and moving the nozzle tip along the joint. Firmly hold the gun cartridge while depressing the trigger to avoid jarring the bead. Pull the cartridge along the joint rather than pushing it. Draw the bead the length of the joint in a continuous, smooth motion. The bead should touch each side of the joint and be centered along the opening. Drawing too quickly will produce a thin, stretched bead. Drawing a bead too slowly produces a thick, lumpy bead.

To stop the bead flow at the end of the joint, release the pressure on the plunger. Most guns have a dripless feature that backs off the pressure and stops the bead from extruding. If your gun doesn't have this feature, you'll need to relieve the gun pressure by rotating the handle about 90° and pulling it out.

9.3.2 Applying Knife Grade Sealants Or Fillers

Follow the same procedures for applying knife grade sealants and fillers that you learned for applying drywall joint compound. Press a little bit of product onto a piece of paper to get the feel of it. Push the product firmly into the crevice with the blade of a putty knife (*Figure 16*). For hard to reach areas, make a roll with gloved hands and push the roll into the crevice.

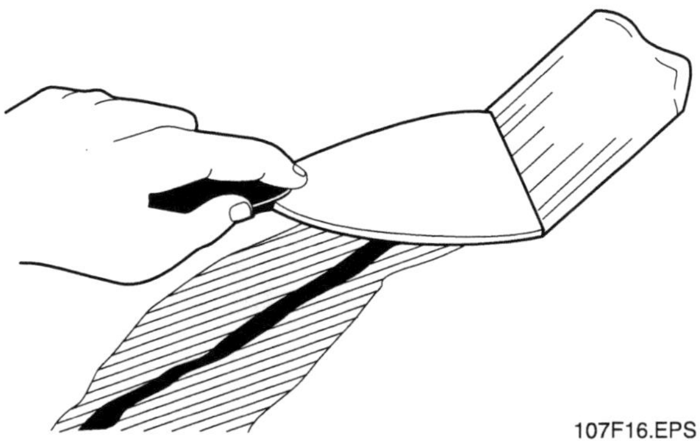

Figure 16. Applying Filler With A Knife

9.3.3 Applying Pour Grade Sealants Or Fillers

Wait for low wind conditions. Spray or pour a little bit of product onto a piece of paper to get the feel of the flow. Operate the equipment to cover the area with a continuous, even layer of sealant. Move from one end of the joint to the other, using a slow and steady pace to get even, complete coverage.

WARNING! For spray applications, wear a positive pressure, air-supplied respirator or an air line respirator. Wear gloves, goggles, and protective clothing to avoid contact with airborne sealant that can cause skin, eye, and lung irritation, breathing problems, and possible blindness. Always check the MSDS for the product you are using and observe all safety precautions listed.

9.4.0 SMOOTHING THE SEALANT/FILLER SURFACE

Sealants and fillers are not sandable as a general rule, so you must make as smooth a surface as possible before the sealant or filler dries. Always leave at least 1/8 inch of sealant depth for proper coverage.

For non-solvent-based beads, use a gloved finger dipped in clean, soapy water. Float your finger lightly over lumps and other irregularities to smooth the bead surface (*Figure 17*). As necessary, wipe the excess off your gloved finger with a cloth and redip in the soapy water to get the best finish.

Figure 17. Several Smoothing Techniques

For solvent-based beads, make a point out of a dry cloth and dip it in solvent. Wrap the point over your gloved index finger and float the cloth over the bead to smooth it out. Make a new point as the excess sealant is picked up on the cloth. You can also smooth out a bead with a rounded knife.

For knife grade sealant, use a putty knife to smooth out large areas. Slight mounding of knife grade products is not necessary for sealant surfaces that must be flush with the substrate surface. Most fillers and sealants have far less shrinkage than products such as joint compound.

Excess product can also be abraded away or trimmed with a knife once the sealant or filler has cured. *Figure 18* shows cross-sections of several different types of smoothed beads.

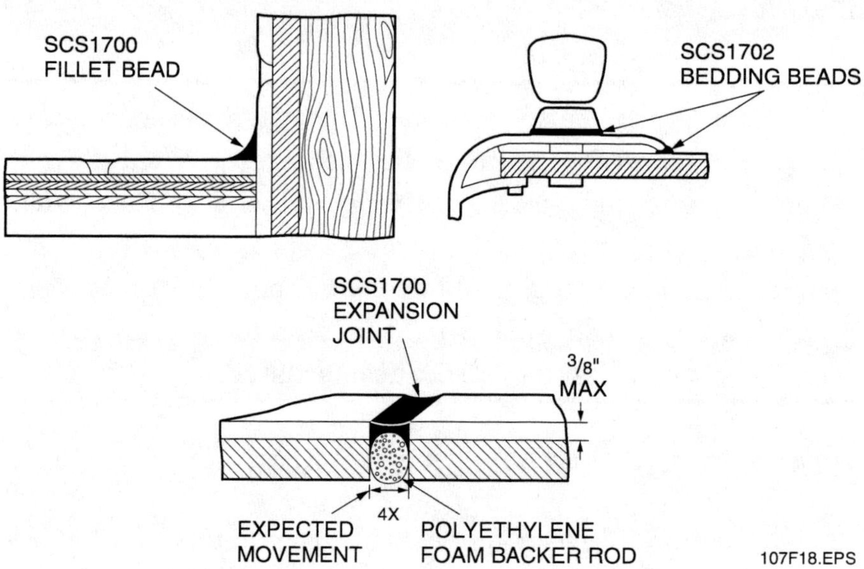

Figure 18. Smoothed Beads In Tub And Tile Applications

9.5.0 CLEANUP AND ADHESION CHECK

Seal partially-used cartridges by inserting a nail in the nozzle and wrapping the tip with tape or plastic wrap. Clean off bucket and can rims and replace the lids. Clean all tools and equipment with soapy water or solvent, depending on the product. Dispose of solvent-soaked rags and empty cartridges, containers, etc.

WARNING! Use special care with products and cleaning solvents containing xylol and toluol, which are harsh solvents. Toxic fumes can cause lung and eye irritation. Always check the MSDS for the product being used.

It is always a good idea to check for proper adhesion to avoid future problems. One way is to do a hand pull test once the sealant or filler has completely cured (*Figure 19*). Make a knife cut through the sealant all the way across the joint. Starting at the first cut, make two cuts, each about two inches long, along the edges of the joint. Grasp the three-sided piece of sealant firmly between your fingers and pull down at a 90° angle or more. Try to pull the uncut sealant out of the joint. If adhesion is proper, the sealant should tear cohesively (in itself) before releasing adhesively (from the substrate), *or* elongate at least 100% before releasing adhesively. Sealant is replaced in the test area by cutting off the two-inch piece plus any other damaged pieces. Wipe the cut area to remove any contaminants. Original sealant surfaces need to be clean and dry. Apply more sealant in the same manner (assuming good adhesion was obtained). Make sure the new sealant is in contact with the cured sealant.

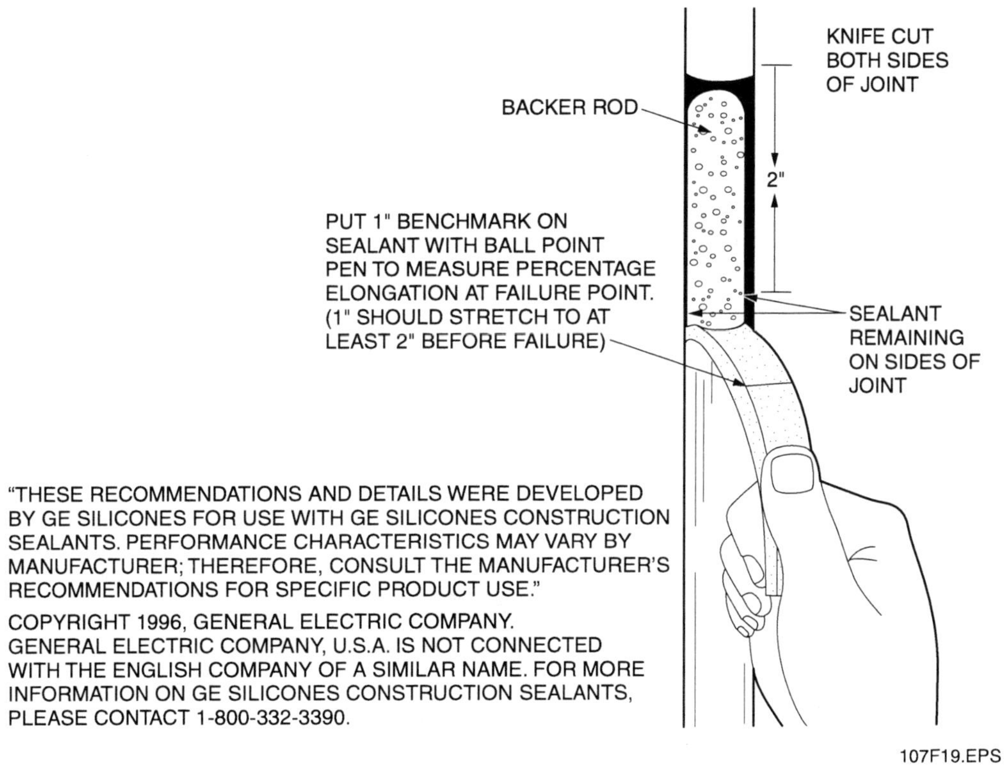

Figure 19. Testing For Adhesion

SUMMARY

Sealing construction joints and filling/repairing substrate surface damage provides protection from moisture damage and heat loss. It also provides cosmetically pleasing surfaces. Remember to apply sealant when you need a barrier to air, moisture, or heat, and use filler when you need to repair damage to substrate surfaces or establish smooth, even surfaces for appearance or coating.

References

For advanced study of topics covered in this task module, the following books are suggested:

Building Sealants: Materials, Properties, and Performance, ASTM Special Technical Publication, Philadelphia, PA.

Caulking and Weatherstripping, First Edition, 1985, U.S. Department of Energy, Conservation, and Renewable Energy Inquiry and Referral Service.

Proper Painting and Surface Preparation, National Park Service, U.S. Department of the Interior, Washington, D.C.

REVIEW/PRACTICE QUESTIONS

1. Sealants are typically used:
   a. for priming substrates.
   b. to keep painted surfaces from yellowing.
   c. as a barrier to moisture, heat loss, and substrate damage.
   d. to repair large holes in a substrate.

2. Fillers may be used:
   a. for priming substrates.
   b. to keep painted surfaces from yellowing.
   c. as a barrier to moisture, heat loss, and substrate damage.
   d. to repair holes in a substrate.

3. When there are two types of substrate at a joint, the most critical property is sealant:
   a. adhesion to both substrates.
   b. cohesion.
   c. colorant to match substrates.
   d. consistency.

4. You are applying paintable filler to a large area of primed wall on the exterior of a building that many people will be walking past. An important property to consider when choosing the filler is:
   a. consistency.
   b. shrinkage.
   c. colorant.
   d. primer color.

5. The type of curing (chemical or evaporating) depends on the:
   a. type of solids in the product.
   b. weather.
   c. type of base in the product.
   d. primer on the substrate.

6. If paintability is important, which sealant would you choose?
   a. Silicone
   b. Acrylic latex
   c. Butyl rubber
   d. Asphalt-based

7. Soap and water can be used to clean up:
   a. oil-based sealants and fillers.
   b. solvent-based sealants and fillers.
   c. chemically curing sealants and fillers.
   d. resin-based sealants and fillers that do not contain solvents.

8. The life expectancy of properly applied resin-based sealants and fillers is usually:
   a. 10 to 50 years.
   b. seasonal.
   c. 1 to 2 years.
   d. 5 to 6 years.

9. At a job site, you are handed cartridges of unfamiliar filler material and asked to complete the job. The first thing you do is:
   a. get the caulking gun.
   b. put on a respirator.
   c. read the label safety precautions.
   d. pierce the tip and check the consistency with your finger.

10. If primer is not applied to the substrate, the sealant or filler:
    a. adhesion will be poor.
    b. cohesion will be poor.
    c. paintability will be the same.
    d. will cure perfectly.

11. After preparing the substrate joint by scrubbing, cleaning, and priming:
    a. check joint depth and insert backer rod or foam, if necessary.
    b. smooth the bead of sealant.
    c. paint the substrate.
    d. spray pour grade sealant into the cracks.

12. Drawing a bead too quickly will result in a:
    a. thin, stretched bead that leaves gaps in the joint.
    b. thick, lumpy, uneven bead.
    c. smooth but thick bead.
    d. thin bead that tends to curl up at the edges.

13. Drawing a bead too slowly will result in a:
    a. thin, stretched bead that leaves gaps in the joint.
    b. thick, lumpy, uneven bead.
    c. smooth but thick bead.
    d. thin bead that tends to curl up at the edges.

14. If the gun is held too high when applying gun grade products, the:
    a. tip will scrape away the bead you just applied.
    b. bead will ooze out the side of the nozzle.
    c. bead will stretch and break.
    d. tip will smooth the bead as it goes.

15. The minimum depth of sealant that should be present after smoothing is:
    a. $^{1}/_{8}$ inch.
    b. $^{1}/_{4}$ inch.
    c. $^{1}/_{2}$ inch.
    d. 2 inches.

ACKNOWLEDGMENTS

Figure 3 courtesy of Weatherguard Products Corporation.

Figures 4 and 13 courtesy of Borden Chemical, a Division of Borden.

Figure 8 courtesy of Woodmont Products, Inc.

Figure 11 courtesy of Macklanburg-Duncan, a Division of Dow Chemical Company.

Figures 18 and 19 courtesy of G.E. Silicones, a Division of General Electric Company.

NCCER CRAFT TRAINING USER UPDATES

The NCCER makes every effort to keep these manuals up-to-date and free of technical errors. We appreciate your help in this process. If you have an idea for improving this manual, or if you find an error, a typographical mistake, or an inaccuracy in the NCCER's Craft Training Manuals, please write us, using this form or a photocopy. Be sure to include the exact module number, page number, a description of the problem, and the correction, if possible. Your input will be brought to the attention of the Technical Review Committee. Thank you for your assistance.

Instructors – If you found that additional materials were necessary in order to teach this module effectively, please let us know so that we may include them in the Equipment/Materials list in the Instructor's Guide.

Write: Curriculum and Revision Department
National Center for Construction Education and Research
P.O. Box 141104
Gainesville, FL 32614-1104

Fax: 352-334-0932

Craft ____________ Module Name ____________

Module Number ____________ Page Number(s) ____________

Description of Problem ____________

____

____

____

(Optional) Correction of Problem ____________

____

____

(Optional) Your Name and Address ____________

____

Introduction to Paints and Coatings

Module 07108

Painting Trainee Task Module 07108

INTRODUCTION TO PAINTS AND COATINGS

PAINTING AND DECORATING CONTRACTORS OF AMERICA

NATIONAL CENTER FOR CONSTRUCTION EDUCATION AND RESEARCH

Objectives

Upon completion of this module, the trainee will be able to:

1. Explain the function(s) of pigments, resins, solvents, and additives.
2. Describe the basic differences between water-based and oil-based paints and coatings, including the film-forming mechanisms, advantages, and disadvantages of both types.
3. Use manufacturer's literature and/or product labels to identify coating(s) recommended for use with various substrates (wood, metal, etc.) and exposure conditions. Also identify the recommended method of surface preparation for each coating.
4. Describe the properties and/or functions of paints or coatings.
   - Properties:
     - Alkyd
     - Latex
     - Epoxy
     - Urethane (polyurethane)
   - Functions:
     - Primers/undercoats
     - Tie coats
     - Finish coats
     - Sealers
     - Shellacs, varnishes, and lacquers
     - Stains
     - Special purpose coatings
5. Demonstrate and/or explain the general methods used for the cleanup and disposal of water-based and oil-based paints.

Prerequisites

Successful completion of the following Task Modules is required before beginning study of this Task Module: Core Curricula, Painting Level 1, Modules 07101 through 07107.

Required Trainee Material

1. Trainee Module
2. Appropriate Personal Protective Equipment

Course Map

This course map shows all of the *Wheels of Learning* task modules in the first level of the Painting curricula. The suggested training order begins at the bottom and proceeds up. Skill levels increase as a trainee advances on the course map. The training order may be adjusted by the local Training Program Sponsor.

Course Map: Painting, Level 1

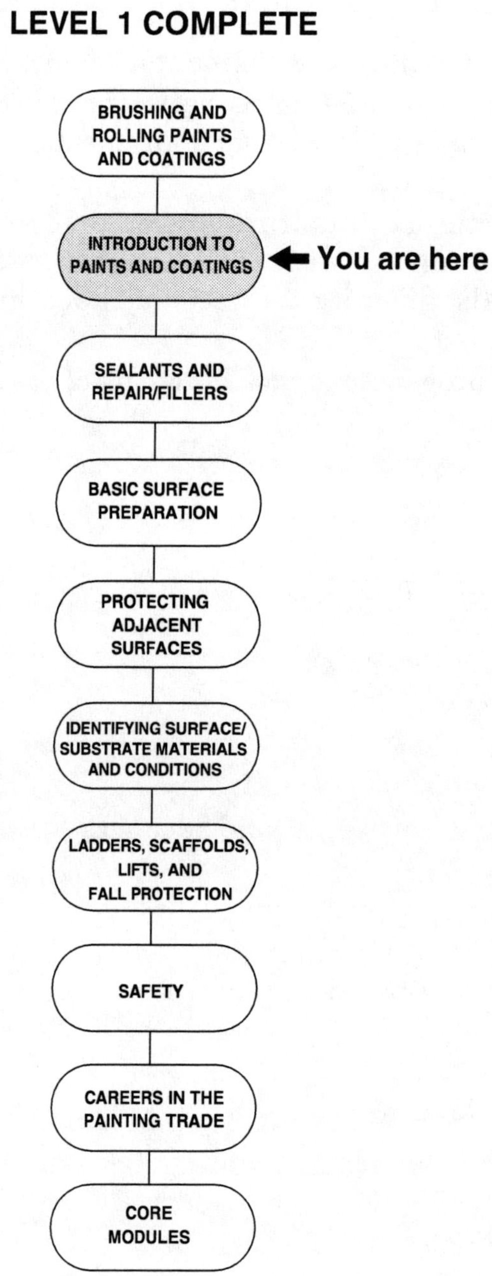

CMAP108.EPS

TABLE OF CONTENTS

Table of Contents

Trade Terms Introduced In This Module

Additive: A substance added to a coating material to adjust, enhance, or improve emulsion, suspension, drying, application, weathering, or other properties.

Alkyd: An oil synthesized from organic sources such as plant fiber, hydrocarbons, and petroleum-based chemicals.

Binder: The nonvolatile portion of a paint which serves to bind or cement the pigment particles together. Oils and varnishes are examples of binders.

Coalescing solvents: Solvents in a paint mixture that allow resin particles to fuse together (coalesce) forming a continuous paint coating.

Coating: Any liquid, liquefiable, or mastic composition that, when applied as a thin layer to a surface, is converted into an solid, adhesive film for the purpose of protecting, decorating, and/or providing other functional properties to the surface.

Coating system: A protective film consisting of one or more coats applied in a specific order using recommended methods.

Crosslinking: A method by which chemicals unite or link to form polymer films.

Curing agent: An additive, sometimes called a *hardener* or *promoter*, that helps a coating film cure by chemical reaction.

Drying oils: Oils such as linseed oil, tung oil, etc., that convert to a solid film when exposed to oxygen in the air.

Emulsifier: A material added to dissimilar materials, such as oil and water, to produce a stable, uniform emulsion.

Emulsion: A liquid preparation in which tiny, immiscible (non-soluble) particles of one liquid are suspended in another, such as oil in water or vice versa.

Hot spot: An area of improperly mixed plaster. If paint is applied over a hot spot, a chemical reaction causes *burning*, which damages the paint film.

Opacity: A paint's ability to cover or hide the underlying surface.

Paint: Any pigmented liquid, liquefiable, or mastic composition designed for application to a substrate in a thin layer that is converted to an opaque, solid film after application. It is used for protection, decoration, identification, or to serve some other purpose.

Pigment: Material in the form of fine powder which is dispersed in a liquid vehicle to make paint. It imparts color and other essential properties to a paint such as opacity, hardness, durability, and corrosion resistance.

Resin: General term applied to a wide variety of transparent and fusible products. They may vary widely in color. Higher molecular weight synthetic resins are more generally referred to as *polymers*. In a broad sense, the term is used to designate any polymer that is a basic material for coatings and plastics.

Sheen: The amount of luster or gloss that a finish paint coat has when fully cured.

Solvent: A liquid, usually volatile, which is used in the manufacture of paint to dissolve other materials. It evaporates during drying and does not become part of the dried film. Solvents are also used to dissolve and remove oil, dirt, grease, soil, and waxes from various surfaces.

Surfactant: A surface active agent that is used to break down the surface tension of liquids, such as oil and water, to enable pigments and binders to readily mix with the liquid.

Thermoplastic: A material that becomes soft when heated and hard when cooled.

Vehicle: The liquid portion of the paint. It is composed of the binder and solvent in which the pigment is dispersed.

Water-based paint: Latex paint and paints/coatings containing a water-soluble binder.

1.0.0 INTRODUCTION

As a result of the Clean Air Act and many other laws and regulations passed from the 1970s to the 1990s, paint manufacturers have stepped up research and development for both water-based and oil-based paints. Paints and coatings have been re-engineered to eliminate or reduce the use of volatile organic compounds (VOCs) and lead. Today's paints are quite different in their composition and performance than the best paints of just a few years ago. Even the terms *latex* and *oil-based* paint have taken on new meanings. In both cases, improved synthetic resins have yielded better performance, especially in water-based paints. As a result of these efforts to make paints and coatings comply with VOC regulations, many new water-based paints have been developed that have properties equal to, or better than, oil-based paints.

This module is the first in a series of modules covering paints and coatings that you will study throughout your training program. The first part of this module introduces the basic ingredients and film-forming processes common to all paints and coatings. The remainder of the module describes paint systems and the functional categories of paints and coatings with an emphasis placed on water-based and oil-based alkyd paints and coatings.

2.0.0 PAINT AND COATING MATERIALS

Before going any further, let's clarify the terms **paint** and **coating**. These terms mean different things to different people. To some, paint refers to oil-based products while coating refers to newer, resin-based materials. To others, the difference between paints and coatings is that paints always contain pigments but many coatings do not. Also, some coatings are much thicker than paints and do not harden completely. The terms are used interchangeably in the trade and throughout the remainder of this module.

Coatings are applied to substrates for the purpose of protection, decoration, or to serve a functional purpose such as weatherproofing, etc. All coatings contain some or all of the four basic ingredients listed below and shown in *Figure 1*.

- Solvents
- Resins
- Pigments
- Additives

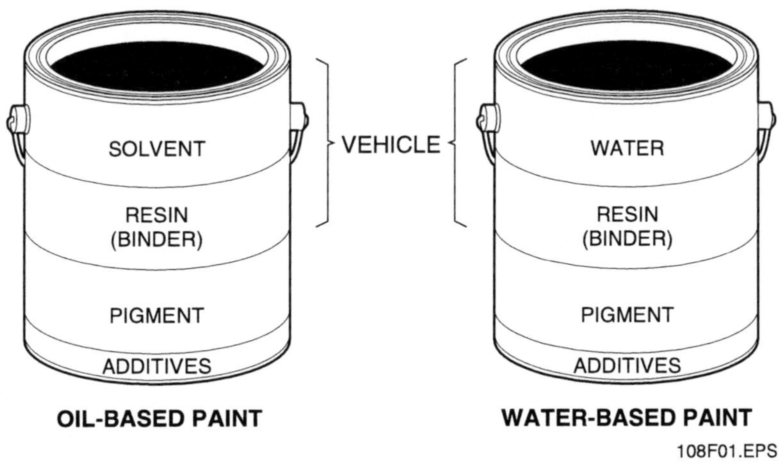

Figure 1. Basic Components Of Coatings

Many tradespeople and texts prefer to describe paints as consisting of **pigments** and a **vehicle**. The vehicle consists of the **resin** or **binder** and a suitable **solvent** (oil-based paint) or water (water-based paint) to make it liquid. Pigment particles, along with those of any additives, are dispersed and mixed into the vehicle and the paint is packaged for distribution and sale. Some water-based paint may also contain a relatively small amount of solvent in the vehicle.

Depending on the specific combination of the four basic ingredients, paints and coatings can be classified into four general categories: paints, stains, clear coatings, and special coatings.

2.1.0 PIGMENTS

Pigments consist of fine solid particles dispersed in the liquid portion of the paint. They may be white or colored, opaque or translucent. Pigments give a paint its color, level of sheen or gloss, and opacity. **Opacity** is a paint's ability to cover or hide the underlying surface. Generally, the more pigment in a coating, the better its opacity. Also, within limits, the finer the pigment is ground, the better the hiding power of the paint. Fine grinding is also used to provide smoothness in paint films. Most pigment particles used in commercial paints range in diameter from 0.0001 to 0.006 millimeters. The addition of special pigments to paints can provide extra protection against ultraviolet (UV) rays, moisture, and airborne contaminants. They also help the paint to resist heat, abrasion, acid, or alkali.

Most pigments consist of two or more elements combined to form an organic or inorganic compound. Basically, inorganic pigments are chemical compounds made from minerals that contain no carbon. They can be made of simple oxides of metal such as iron, zinc, or chromium, or they can be made of a complex combination of elements. Most white and translucent pigments are examples of inorganic pigments. Titanium dioxide is the most popular inorganic white pigment because of its good hiding power and ability to reflect light (refractive index). Rutile titanium, another inorganic pigment, has the highest refractive index and the best coloring power (tinting strength) of all pigments.

Organic pigments are chemical compounds based on the element carbon. They are typically derived from coal tar and petroleum. When elements such as copper, sodium, barium, or calcium are added to organic pigment compounds, they provide a wide range of colors. Compared to inorganic pigments, most organic pigments are generally brighter in color. However, they are usually not as permanent because they are more susceptible to attack from the sun's UV rays, heat, and airborne contaminants.

Pigments may be classified by their primary function (*Figure 2*). These include:

- Color pigments
- Hiding pigments
- Extender pigments
- Barrier pigments
- Inhibitive pigments
- Sacrificial pigments

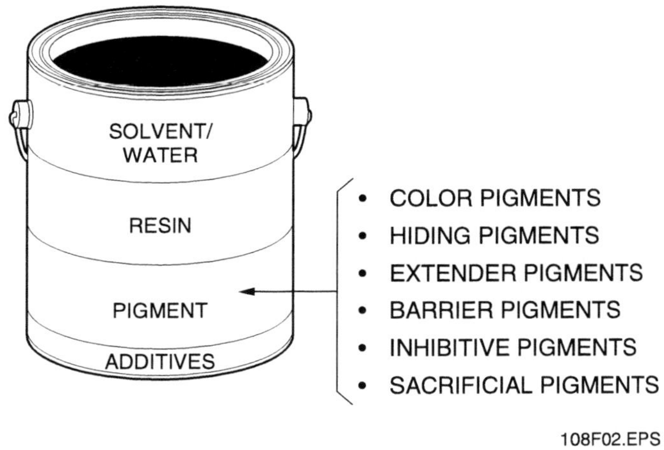

Figure 2. Functional Categories Of Pigments Used In Coatings

2.1.1 Color Pigments

Color pigments are used alone or in combination to add color to paint. Normally, color pigments are used with white pigments. The source of color pigments can be grouped into two general categories: earth pigments and precipitated pigments.

Earth pigments are natural or mineral pigments that are mined like minerals. Common examples of earth pigments are ochres, red iron oxides, siennas, and umbers.

Precipitated pigments are so named because they are made by a process that separates them out (precipitates them) from a chemical solution. They can be organic or inorganic. Chrome yellows and iron blue pigments are examples of inorganic precipitated pigments. Organic precipitated pigments include phthalocyanine blue and green pigments.

Color pigments impart the appearance of color because they absorb a portion of light which falls on them and reflect or return to the eye the remaining light waves of the spectrum. The stimulation and response of the retina in the human eye to these light rays creates colors as we see them. Note that white, black, and gray are often called *colors*, although black is caused by the complete absorption of light rays; white by the reflection of all the rays that produce color; and gray is a mixture of black and white resulting from an imperfect absorption of all these rays.

2.1.2 Hiding Pigments

Hiding pigments are pigments with a high light-refractive index, typically between 1.94 and 2.7 (the higher the index, the greater the opacity). Rutile titanium and titanium dioxide are two common hiding pigments.

2.1.3 Extender Pigments

Extender pigments act to reinforce the paint film or make the wet paint thicker (more viscous). Used alone or in combination, they stabilize pigment suspension, improve leveling, control flow, build film, lower gloss, etc. They affect film properties such as washability and appearance. Extender pigments usually have a low light-refractive index (1.45 to 1.7), resulting in poor opacity. They are sometimes used in excessive quantities in cheaper paints. This is because they are less costly than most hiding pigments. Use of too much extender pigment results in excess chalking and poor durability.

2.1.4 Barrier, Inhibitive, And Sacrificial Pigments

Barrier, inhibitive, and sacrificial pigments all help to protect the substrate. Barrier pigments are fairly large pigments that, when used in paint, act to prevent moisture from reaching the substrate. This is because any moisture that enters the paint film must work itself around and between these larger pigment particles, which increases the distance the moisture must travel to reach the substrate.

Pigments of chromate, molybdate, and borate are inhibitors that provide corrosion protection for metal substrates, usually steel or aluminum. These pigments are slightly water soluble. When the paint film gets wet, moisture enters the film and partially dissolves the pigment, carrying it to the underlying metal surface. There, the dissolved pigment reacts with the metal to form a film that helps make the substrate chemically inert, thus reducing the rate of corrosion. In recent years, the use of chromate pigment in paint has been severely limited because it has been identified as a health hazard.

Sacrificial pigments also provide corrosion protection for steel substrates. High purity zinc dust pigments are used as a sacrificial metal layer to protect the underlying steel substrate. This happens because zinc is a more "active" metal than steel. When exposed to chemicals or corrosive agents, the zinc will corrode (be sacrificed) to protect the steel substrate. This is the same process that allows hot dipped galvanizing to minimize corrosion on steel. In order for this protection to occur, there must be direct contact between the zinc pigment particles and the steel substrate.

2.2.0 RESINS

Resins or binders are fusible materials that bind the solid, non-evaporating (nonvolatile) particles of pigment and additives together and adhere them to the substrate, creating a strong and durable bond. The type of resin used contributes to the useful life of the paint. It also determines most of the properties of the paint. By varying the resins, many different types of paint can be produced.

Resins are typically classified by their origin: natural or synthetic. Natural resins are **thermoplastic**, meaning that they become soft when heated and hard when cooled. They are flammable and can be dissolved in some organic solvents. They are not dissolvable in water. Before the development of synthetic resins, natural resins were the only source of material for making coatings with high gloss and exceptional hardness. Some common natural resins are: copal, lac, rosin, dammon, and manila. Use of natural resins has decreased as a result of the development of improved synthetic resins.

Synthetic resins are man-made resins usually derived from byproducts of chemical refining or manufacturing processes. Synthetic resins can be used in both water-based and oil-based coatings. When compared to natural resins, they have better resistance to moisture, chemicals, and UV light. For this reason, they are widely used in paints made to protect substrates in harsh environments.

Resins can be used alone or in combination. The resins commonly used in modern paints and coatings (*Figure 3*) and their properties are briefly described below.

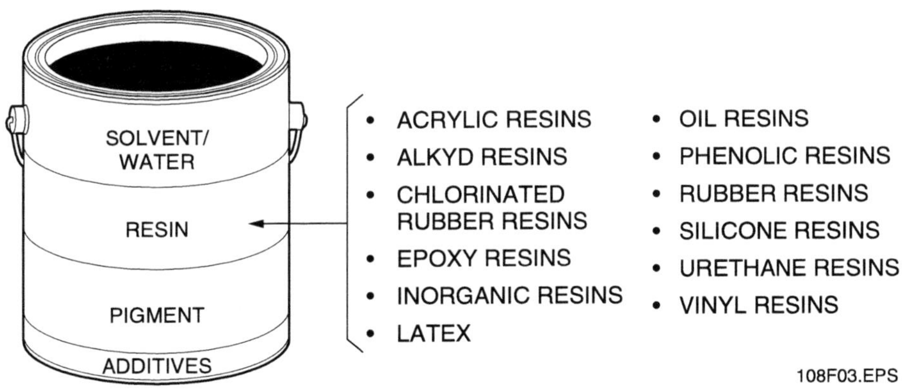

Figure 3. Common Resins Used In Coatings

2.2.1 Acrylic Resins

Acrylic resins are synthetic resins made from derivatives of acrylic acid. They are used in both water-based and solvent-based paints. Acrylic resin is a durable binder and is generally used in combination with latex, vinyl, and epoxy resins. Acrylic resins are commonly used to provide color retention, protection against weathering, and durability in exterior paints.

2.2.2 Alkyd Resins

Alkyd resins are synthetic resins derived as a condensation product usually involving phthalic acid or phthalic anhydride, glycerol, and fatty acids from vegetable oils. Alkyd resins modified by more than 60% oil by weight are called *long oil alkyds*. Those modified with 40% to 60% oil are called *medium oil alkyds*, and those modified with less than 40% oil are called *short oil alkyds*. The amount of oil used as a modifying agent determines the main properties of the alkyd resin. Paints and coatings made with long oil alkyd resins are usually slower drying, tougher, and more elastic than those made with short oil alkyds. Alkyd resins provide for good color and gloss retention and have excellent durability. Alkyd resins are the most commonly used resins in oil-based paints and coatings.

2.2.3 Chlorinated Rubber Resins

Chlorinated rubber resins are synthetic resins made by chlorinating rubber or other polymers. They are soluble in hydrocarbon-type solvents and turpentine, and insoluble in lacquer solvents and alcohol. Paints made from chlorinated rubber are typically used in maintenance and industrial applications. Other uses include fire-retardant paints, marine coatings, paint for masonry, and swimming pools.

2.2.4 Epoxy Resins

Epoxy resins are made from bisphenol. They have outstanding chemical resistance and toughness. Epoxy resins are used in high-end, high-pac, high-performance tile-like finishes. They are also used in catalyzed coatings for concrete, masonry, and structural steel in corrosive and high-use areas.

2.2.5 Inorganic Resins

Inorganic resins are chemical compounds containing no carbon. They are used mainly to react with zinc-powdered pigments to form very hard and corrosion-resistant surfaces. The most common inorganic resins are sodium, potassium, lithium, and ethyl silicates. Inorganic resins are typically used in paints made for marine applications.

2.2.6 Latex Resins

Latex resins are rubber-like natural or synthetic resins used for many years as a binder in water-based emulsion paints. When used alone, latex resins tend to adhere poorly and eventually turn brittle. Normally, they are combined with acrylic or vinyl resins to make a paint with enhanced properties.

2.2.7 Oil Resins

Oil resins are the oldest resins in use. Linseed oil is used the most often. Tung oil, caster oil, fish oil, and soya (soybean) oil are also common. Paints made using only oil resins are typically used to paint exterior wood and metal surfaces. They provide good durability in normal environments. However, they are not very hard or resistant to abrasion, chemicals, or strong solvents. Today, most oil resins are modified with synthetic resins to make paints with improved properties. These modified resins are called *oleoresinuous resins*.

2.2.8 Phenolic Resins

Phenolic resins are synthetic resins made by condensing phenol or a similar compound with formaldehyde. They are usually used in combination with a drying oil to produce paints with excellent resistance to abrasion, water, and mild chemicals. Such paints are frequently used as topcoats for metal in very humid conditions. **Drying oils** are oils such as linseed oil, tung oil, etc., that convert to a solid film when exposed to oxygen in the air. Phenolic resins are often used to modify alkyd resins (phenolic alkyd resins) in alkyd paints to make them suitable for water immersion.

2.2.9 Rubber Resins

Rubber resins are lacquer-type resins, not to be confused with latex-emulsion resins. They dry and cure solely by evaporation of the solvent and not oxidation. Paints made with rubber resins are highly resistant to water and mild chemicals. These paints are commonly used on exterior masonry and in wet and humid areas such as swimming pools, shower rooms, kitchens, and laundry rooms. When combined with silicone resins, they provide a paint with good resistance to heat.

2.2.10 Silicone Resins

Silicone resins are a group of resins containing large amounts of silicone. Paints made with silicone resins are used mainly for waterproofing and heat resistance. Silicone resins are often used to modify the alkyd resins (silicone alkyd resins) in alkyd paints to improve their durability, gloss retention, and heat resistance.

2.2.11 Urethane Resins

Urethane resins are synthetic resins which add abrasion resistance, toughness, and overall performance to paint. Typically used in paints for floors and other high-use surfaces, urethane resins are also often used to modify alkyd resins (uralkyd resins) in alkyd paints to improve their resistance to chemicals, moisture, weathering, and abrasion.

2.2.12 Vinyl Resins

Vinyl resins are synthetic resins made from vinyl compounds. Vinyl acetate is commonly used in water-based (latex) paints. Polyvinyl chloride is used in some solvent-thinned coatings where high chemical resistance is needed. Vinyl resins are often used to modify alkyd resins (vinyl alkyd resins) in alkyd paints to improve their recoatability, moisture resistance, and chemical resistance.

2.3.0 SOLVENTS

Solvents are easily evaporated (volatile) liquids used to dissolve the solid paint pigment and additive particles in liquid oil-based paint. Note that the water used in water-based paints (emulsion paints) acts not to dissolve, but to disperse the pigments and additive particles in the paint. Most paints and coatings use a blend of solvents to achieve the desired properties. Solvents reduce the viscosity of oil-based paint so that it can be readily applied using a brush, roller, etc. Depending on the particular solvents, they can also affect a paint's film formation, consistency, leveling, drying, adhesion, and durability.

Most paint solvents are volatile organic compounds (VOCs). After application of a paint to a substrate, the solvents evaporate from the drying paint film. Upon evaporation into the atmosphere, VOCs react with nitrous oxides (combustion compounds from automotive emissions and the burning of fuels) and sunlight to form ozone and air pollutants. Ozone is an unstable form of oxygen that is highly reactive. When trapped within the Earth's troposphere, it contributes to smog that can be hazardous to humans, animals, and plant life. VOCs have also been identified as one of many sources contributing to poor indoor air quality in buildings. To be compliant with increasingly restrictive VOC legislation, paint manufacturers are working to replace high-VOC solvents with low-VOC solvents and/or are reducing or eliminating the amount of VOC solvents in their paints.

Solvents can be grouped into five broad classes (*Figure 4*):

- Hydrocarbon solvents
- Turpentines
- Ketones
- Esters
- Alcohols

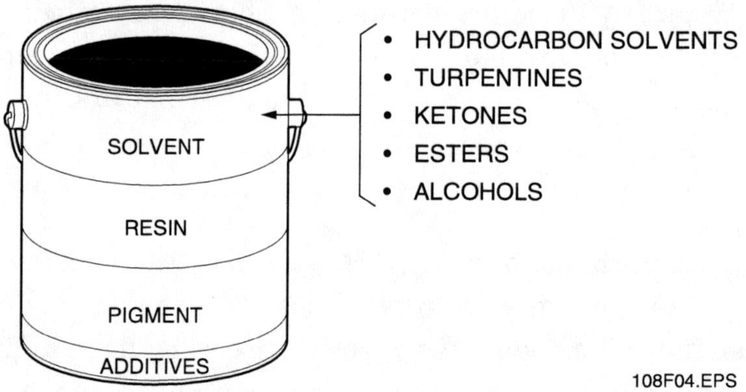

Figure 4. Classes Of Solvents Used In Coatings

2.3.1 Hydrocarbon Solvents

Hydrocarbon solvents are made from petroleum or coal tar sources. These solvents are classified as either aliphatic or aromatic solvents. Aliphatic solvents are made from paraffin-based crude oil. Some common aliphatic solvents are mineral spirits, varnish makers and painters (VM&P), naphtha, hexade, and hapten. Aromatic solvents are better at dissolving another substance than are aliphatic solvents. Common aromatic solvents are xylene (xylol), toluene (toluol), and high-flash naphtha. Aliphatic solvents are generally the least expensive of all solvents and are considered the least toxic.

2.3.2 Turpentines

Turpentines are hydrocarbon solvents derived from pine trees. Gum turpentine is made by distilling crude gum rosins tapped from the trees. Wood turpentine is obtained by distilling the crude rosin from wood ground from the trees. Turpentine solvents are used mainly for thinning oil-based paints and varnishes. Turpentines are seldom used by paint manufacturers in their paint and coating products because they are relatively expensive and have a slow evaporation rate.

2.3.3 Ketones

Ketone solvents are organic solvents used mainly in vinyl, urethane, epoxy, and acrylic paint products. Ketone solvents in common use are acetone, MEK (methyl ethyl ketone), and MIBK (methyl isobutyl ketone).

2.3.4 Esters

Esters are organic solvents made from an alcohol and an organic acid by eliminating the water. Common ester solvents are ethyl acetate, isobutyl acetate, ethylene glycol, and monoethyl ether acetate.

2.3.5 Alcohols

Alcohol solvents are miscible with (soluble in) water. Common alcohol solvents are ethyl alcohol as a shellac solvent and methyl alcohol or wood alcohol used in paint and varnish removers.

2.4.0 ADDITIVES

Additives are special purpose ingredients added in small quantities to a paint to improve or enhance its performance. Additives are classified by the function they perform (*Figure 5*). These functions are:

- Driers
- Antiskinning agents
- Fungicides

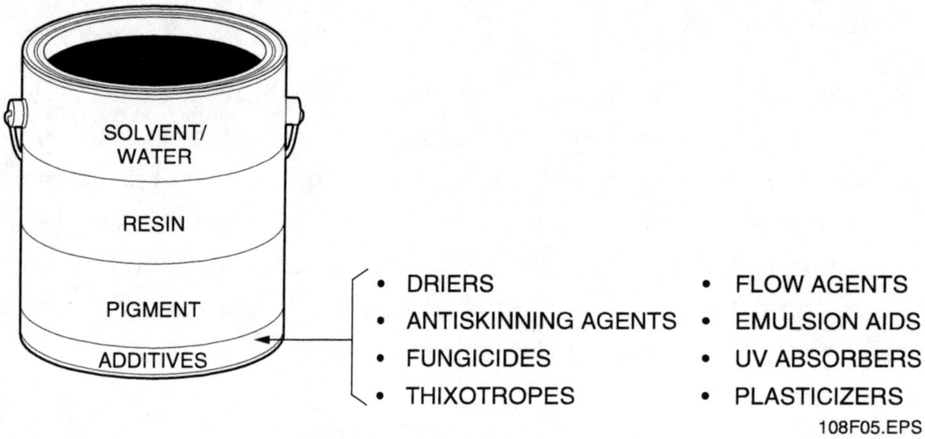

Figure 5. Functional Categories Of Additives Used In Coatings

- Thixotropes
- Flow agents
- Emulsion aids
- UV absorbers
- Plasticizers

2.4.1 Driers

Driers are added to an oil paint to speed up the drying time by boosting the ability of the paint to absorb oxygen. Driers are compounds made of oil and metals such as cobalt, zinc, or manganese. There are two classes of driers: surface driers and through driers. Normally, they are used together. Surface driers are compounds that cause the surface of a drying oil paint to set rapidly and turn almost solid while the underlying paint remains softer. Surface driers are compounds typically made of cobalt and manganese. Through driers are more active than surface driers. When used in conjunction with surface driers, they act to achieve uniform drying throughout the paint film. Through driers are compounds typically made from cadmium, zinc, or zirconium naphthenate.

2.4.2 Antiskinning Agents

Antiskinning agents are compounds used to prevent a tough layer or "skin" from forming on the surface of paint in a container. This skin forms as a result of the paint being exposed to air.

2.4.3 Fungicides

Fungicides, also known as *mildewcides*, are chemical agents added to paint to prevent or retard the growth of fungi such as mold and mildew. Fungicides are often used in water-based paints and other coatings that are especially susceptible to mold and mildew in damp conditions.

2.4.4 Thixotropes

Thixotropes are additives that make paint *thixotropic*. This means that the paint becomes liquid or fluid when stirred or otherwise agitated, but it quickly thickens again when left undisturbed. This property allows the paint to be free flowing during application, but when left undisturbed, it quickly sets to a gel. The advantage of this is that the paint is less likely to drip from the brush and will have less roller splatter. Another advantage is that it can be applied in thicker films without running or sagging.

2.4.5 Flow Agents

Flow agents improve the flow of the paint or coating after application. This helps eliminate brush marks, roller patterns, orange peel from spraying, and other film unevenness and defects.

2.4.6 Emulsion Aids

An **emulsion** is a liquid preparation in which tiny, immiscible (non-soluble) particles of one liquid are suspended in another liquid such as oil in water or vice versa. Emulsion aids, sometimes called **surfactants**, are additives used in oil and water emulsions that break down the surface tension of the liquid. This allows ordinarily dissimilar and immiscible materials such as oil and water to readily mix and disperse in order to produce a stable emulsion. They also enable a water-based emulsion to remain in suspension, keeping it from settling and packing at the bottom of the can during prolonged storage.

2.4.7 UV Absorbers

Ultraviolet light (UV) absorbers are added to exterior paints and other coatings to help prevent their deterioration from the sun's UV rays. UV rays are invisible, short-length light rays that have an active chemical effect on paints and other materials. Excess exposure to UV light normally results in a paint chalking or becoming brittle.

2.4.8 Plasticizers

Plasticizers are additives used in certain plastics and protective paints to make them more flexible and to improve their adhesion. They also act to decrease the material's brittleness, hardness, and glass transition temperature. The *transition temperature* is the temperature at which a material changes from a flexible, rubber-like material to one that is hard and brittle.

3.0.0 FILM-FORMING MECHANISMS OF PAINTS AND COATINGS

Almost immediately after liquid paints and coatings are applied to a surface, they begin a film-forming process to become a dry, stable, and solid surface. The process by which a paint dries and cures is determined mainly by the characteristics of its resins. Paints and coatings dry and cure by one of three processes: oxidation, evaporation, or chemical crosslinking.

3.1.0 FILM FORMING BY OXIDATION

Film forming on certain kinds of alkyd and oil-based paints and coatings happens as a result of these materials being exposed to the oxygen in the atmosphere. After application, the coating initially dries by the evaporation of the solvent (*Figure 6*). Following this, the oxygen in the atmosphere reacts with coating's resins, causing them to crosslink, cure, and harden further. **Crosslinking** is a method by which chemical reactions cause molecules to unite or link to form polymer films. This process, called *oxidation*, occurs rapidly after application of the wet paint or coating, then continues throughout the life of the paint or coating at a much slower rate. Drying oil and oil-based alkyd paints and coatings are two examples of coatings that cure in this manner.

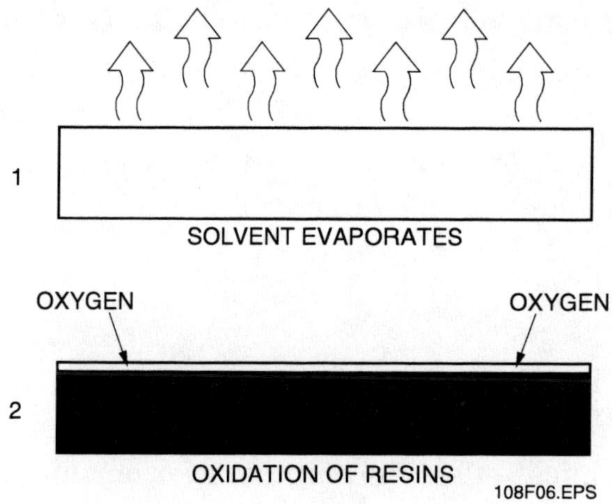

Figure 6. Film Forming By Oxidation

3.2.0 FILM FORMING BY EVAPORATION

Paints and coatings in this group are made by dissolving the resins in a solvent or emulsifying them in water, and mixing in the pigments and additives. After application, these coatings are converted to a solid film by evaporation of the solvent or water. No chemical change occurs to the resins. For solvent-based coatings that cure in this manner, the process is called *solvent evaporation*. For water-based coatings, the process is called *coalescence* (*Figure 7*). The use of solvent-based coatings that cure by evaporation is severely restricted due to VOC regulations. Acrylic, vinyl, epoxy, and polyurethane emulsions are examples of water-based coatings in this category.

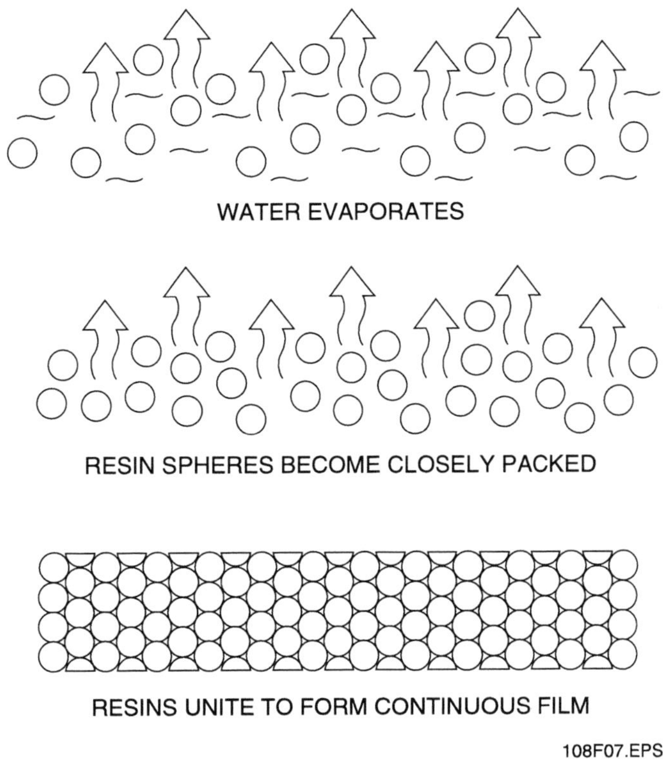

Figure 7. Water-Based Paint Film Formation (Coalescence)

3.3.0 FILM FORMING BY CHEMICAL CROSSLINKING

Coatings in this category harden and cure as a result of a chemical reaction that causes small molecules to be combined to form large molecules. This process, by which the molecules are joined together or crosslinked, is called *polymerization*. Examples of coatings in this group are epoxies, polyesters, urethanes, silicones, and phenolics.

3.4.0 THERMOSET AND THERMOPLASTIC COATINGS

Paints and coatings are also classified as being either thermoplastic or thermoset coatings, depending on the way they cure. Coatings that cure by oxidation or chemical crosslinking are classified as *thermoset coatings*. This is because they set and harden when cured and do not soften or deform upon heating. Coatings that form by solvent evaporation or coalescence are called *thermoplastic coatings*. This is because they become hard when cooled and turn soft and pliable when heated.

4.0.0 COATING SYSTEMS AND COMPONENTS

Paint and coating manufacturers group their products into coating systems. A **coating system** is defined as a protective film consisting of one or more coats applied in a specific order using recommended methods. Most coating systems are based on the type of substrate to which a group of paints or coatings can be applied and under what conditions. The paints or coatings grouped into a particular coating system can have different properties and functions, but they are all designed to be compatible when used with one another.

Figure 8 shows an example of the coating system data given by one major manufacturer. The method or layout used by different manufacturers will vary but all contain the same basic information. It includes:

- The types of paints or coatings for use with different substrates (wood, metal, etc.) and surrounding conditions.
- The correct surface preparation method that should be used prior to application of the paint or coating.
- The sequence and number of coats recommended for application of the paint or coating.

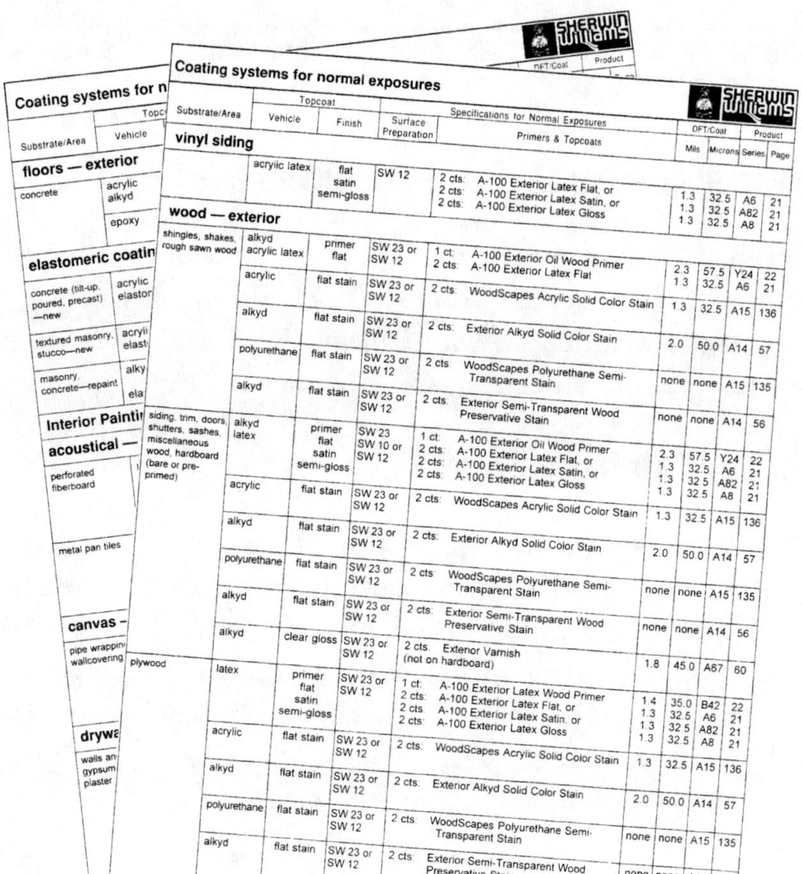

Coating systems for normal exposures

Substrate/Area	Topcoat Vehicle	Topcoat Finish	Surface Preparation	Primers & Topcoats	DFT/Coat Mils	DFT/Coat Microns	Product Series	Product Page
vinyl siding								
	acrylic latex	flat satin semi-gloss	SW 12	2 cts: A-100 Exterior Latex Flat, or 2 cts: A-100 Exterior Latex Satin, or 2 cts: A-100 Exterior Latex Gloss	1.3 1.3 1.3	32.5 32.5 32.5	A6 A82 A8	21 21 21
wood — exterior								
shingles, shakes, rough sawn wood	alkyd acrylic latex	primer flat	SW 23 or SW 12	1 ct: A-100 Exterior Oil Wood Primer 2 cts: A-100 Exterior Latex Flat	2.3 1.3	57.5 32.5	Y24 A6	22 21
	acrylic	flat stain	SW 23 or SW 12	2 cts: WoodScapes Acrylic Solid Color Stain	1.3	32.5	A15	136
	alkyd	flat stain	SW 23 or SW 12	2 cts: Exterior Alkyd Solid Color Stain	2.0	50.0	A14	57
	polyurethane	flat stain	SW 23 or SW 12	2 cts: WoodScapes Polyurethane Semi-Transparent Stain	none	none	A15	135
	alkyd	flat stain	SW 23 or SW 12	2 cts: Exterior Semi-Transparent Wood Preservative Stain	none	none	A14	56
siding, trim, doors, shutters, sashes, miscellaneous wood, hardboard (bare or pre-primed)	alkyd latex	primer flat satin semi-gloss	SW 23 SW 10 or SW 12	1 ct: A-100 Exterior Oil Wood Primer 2 cts: A-100 Exterior Latex Flat, or 2 cts: A-100 Exterior Latex Satin, or 2 cts: A-100 Exterior Latex Gloss	2.3 1.3 1.3 1.3	57.5 32.5 32.5 32.5	Y24 A6 A82 A8	22 21 21 21
	acrylic	flat stain	SW 23 or SW 12	2 cts: WoodScapes Acrylic Solid Color Stain	1.3	32.5	A15	136
	alkyd	flat stain	SW 23 or SW 12	2 cts: Exterior Alkyd Solid Color Stain	2.0	50.0	A14	57
	polyurethane	flat stain	SW 23 or SW 12	2 cts: WoodScapes Polyurethane Semi-Transparent Stain	none	none	A15	135
	alkyd	flat stain	SW 23 or SW 12	2 cts: Exterior Semi-Transparent Wood Preservative Stain	none	none	A14	56
	alkyd	clear gloss	SW 23 or SW 12	2 cts: Exterior Varnish (not on hardboard)	1.8	45.0	A67	60
plywood	latex	primer flat satin semi-gloss	SW 23 or SW 12	1 ct: A-100 Exterior Latex Wood Primer 2 cts: A-100 Exterior Latex Flat, or 2 cts: A-100 Exterior Latex Satin, or 2 cts: A-100 Exterior Latex Gloss	1.4 1.3 1.3 1.3	35.0 32.5 32.5 32.5	B42 A6 A82 A8	22 21 21 21
	acrylic	flat stain	SW 23 or SW 12	2 cts: WoodScapes Acrylic Solid Color Stain	1.3	32.5	A15	136
	alkyd	flat stain	SW 23 or SW 12	2 cts: Exterior Alkyd Solid Color Stain	2.0	50.0	A14	57
	polyurethane	flat stain	SW 23 or SW 12	2 cts: WoodScapes Polyurethane Semi-Transparent Stain	none	none	A15	135
	alkyd	flat stain	SW 23 or SW 12	2 cts: Exterior Semi-Transparent Wood Preservative Stain	none	none	A14	56

108F08.TIF

Figure 8. Example Of One Manufacturer's Coating System Data

4.1.0 GENERAL CLASSIFICATIONS OF PAINTS AND COATINGS

There are many technical and chemically-based ways to classify the different kinds of paints and coatings. For practical purposes, the most common architectural and industrial paints and coatings are grouped into four broad categories (*Figure 9*):

- Alkyd
- Latex
- Epoxy
- Urethane

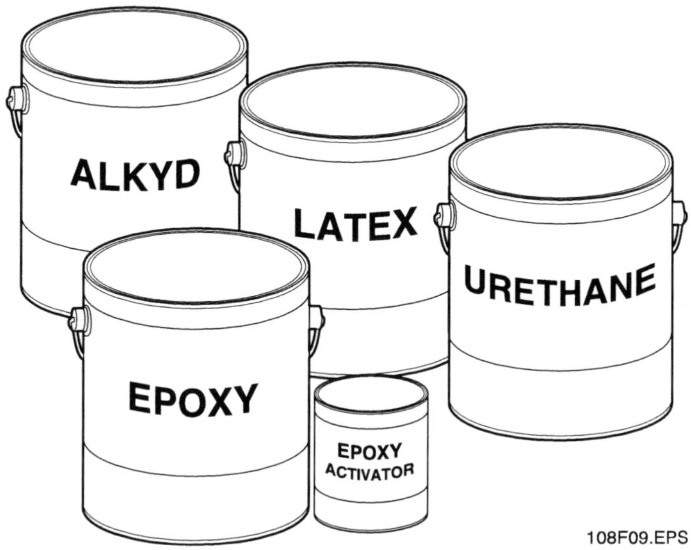

Figure 9. General Categories Of Paint

These are derived from the vehicles or binders used in the paints or coatings. They are also those typically used by paint manufacturers on paint can labels and in their product literature.

4.1.1 Alkyd Paints And Coatings

Alkyd paints and coatings are made by the chemical reaction of a drying oil acid with an alcohol. Drying of the paint surface film occurs by evaporation of the solvent. Curing occurs by oxidation. The specific properties of a particular alkyd paint or coating are determined mainly by the properties of the drying oil used to make it. The more oil used in the formula, the longer it takes to dry, the better the wetting properties, and the better the elasticity. Alkyds are typically used as interior or exterior trim paints, machinery enamels, or durable wall finishes. Alkyd coatings should not be used directly on masonry or other alkaline surfaces except over an alkali-resistant primer or sealer. They also should not be used over a galvanized metal surface except with the right primer.

4.1.2 Latex Paints And Coatings

Latex paints and coatings are also called *water-based* paints and coatings. Typically, they are made from compositions of acrylic or vinyl acrylic synthetic resins and pigments kept dispersed in water by surfactants. They may contain small amounts of **coalescing solvents**. Water-based paints dry by evaporation of the water. As the water evaporates from the film, small amounts of a coalescing solvent contained in the paint mixture allow the resin particles to fuse together (coalesce) forming a continuous coating.

4.1.3 Epoxy Paints And Coatings

There are two classes of epoxy paints and coatings: catalyzed epoxies and single-component epoxy esters. Catalyzed epoxies are coatings made by combining an epoxy resin with a curing agent. A **curing agent** is an additive that helps a coating film harden and cure by chemical reaction. Catalyzed epoxies usually come in two containers. One includes the epoxy resin, pigments, and some solvent. The other contains the curing agent. Evaporation of the solvent causes the surface to dry while a chemical crosslinking process causes it to cure. When mixing, the manufacturer's recommended sweat in or induction time must elapse before the mixture can be used. The mixture has a limited time of workability, which is called its *pot life*. Pot life may vary from a few minutes to several hours, depending on the product.

When properly cured, catalyzed epoxy coatings have excellent solvent and chemical resistance. They are excellent coatings for walls, producing a surface that is highly resistant to abrasion, chemicals, and cleaning. Water-based acrylic and solvent-based catalyzed epoxies are available.

Single-component epoxy esters are coatings produced by reacting a drying oil with an epoxy resin. This type of coating dries by solvent evaporation and cures by oxidation. Single-component epoxy ester coatings are not as hard or chemically resistant as catalyzed epoxies. However, they are easy to apply, and do not require the use of a hardener or catalyst.

4.1.4 Urethane (Polyurethane) Coatings

Urethane coatings, commonly called *polyurethane coatings*, may contain an active isocyanate compound. They form tough, hard, flexible films by one of two methods: moisture curing or catalyzation. Moisture curing urethanes dry by solvent evaporation and cure by reacting with the water vapor in the air. This normally requires that the relative humidity level of the air exceed 20%.

Like catalyzed epoxies, catalyzed urethanes are two-part urethanes. They cure by the addition of a catalyst to the isocyanate-containing component. Urethanes are light stable, gloss retentive, and non-yellowing. For best performance, they are often applied over epoxy primers or zinc-rich primers with epoxy intermediate coats to protect chemical plants, bridges, water and waste water facilities, and other industrial sites.

4.2.0 FUNCTIONAL CLASSIFICATIONS OF PAINTS AND COATINGS

Manufacturers' coating system data and product literature also describe coating products by the function or end use they serve. These classifications are:

- Primers/undercoats
- Tie coats
- Finish coats
- Sealers
- Shellacs, varnishes, and lacquers
- Stains
- Special purpose coatings

4.2.1 Primers/Undercoats

Primers are the first coat in a multicoat system. Their purpose is to provide adhesion and to protect the substrate. Primers are made to be compatible with and aid in the adhesion of the coats of paint applied on top of them. The properties of various primers differ and are determined by the type of substrate over which they are to be applied. Different types of primers are made for use on wood, metal, masonry, concrete, etc.

An undercoat, sometimes called an *intermediate coat*, is any coat applied after the primer and before the finish coat. It is often used on wood trim. It is also used on some metals where exceptional smoothness is desired.

4.2.2 Tie Coats

Tie coats are paints made to provide a transition from a primer or undercoat to a finish coat. Examples of their use is to seal the surface of a zinc-rich primer, to bond different types of coatings, or to improve the adhesion of the next coating.

4.2.3 Finish Coats

A finish coat, also called the *top coat*, is the last layer of coating applied in a multicoat system. It is made to protect a properly prepared and primed substrate from conditions such as abrasion, heat, cold, sunlight, moisture, water immersion, or damaging chemicals. Finish coats also give the finished surface the desired appearance.

Finish coats of latex and oil-based paints for both interior and exterior use are made in different sheens or gloss levels. The **sheen** is the amount of luster the finish coat has when fully cured. The degree of sheen is determined by the proportion of resin in the paint. The resin determines the degree to which the paint is absorbed into the painted surface and how much pigment is left to form a film on the surface. In short, the more resin, the less absorption, thus the more sheen or glossier the paint (*Figure 10*). Terms commonly used by manufacturers to give the sheen of their paint, from lowest to highest, are: flat or matte, eggshell or satin, semigloss, and gloss. Note that there is no industry manufacturing standard for these terms. Semigloss or gloss paints made by one manufacturer very often have a different level of sheen than those made by others. *Enamel* is another term used by manufacturers that is loosely defined and open to interpretation.

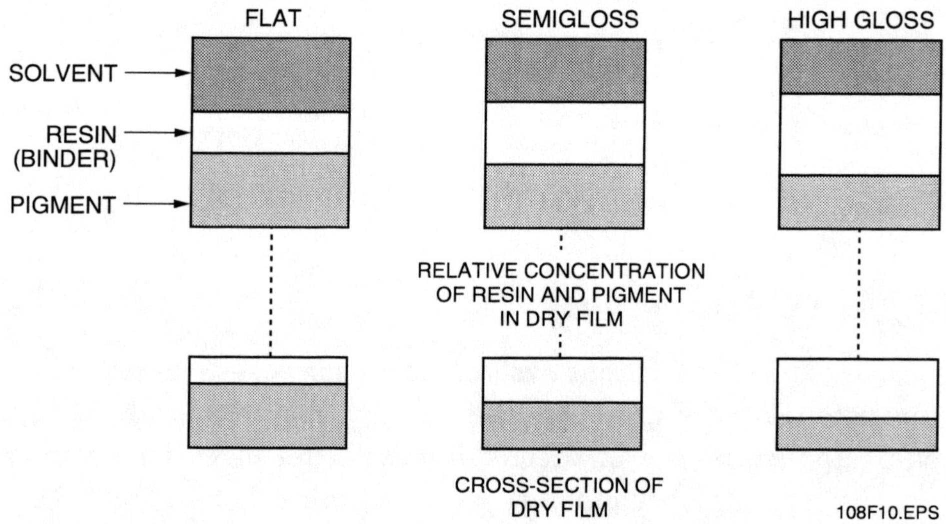

Figure 10. Example Of How The Proportion Of Resin In Paint Determines The Sheen

4.2.4 Sealers

Sealers are used to prevent excessive absorption of the finish coat into a porous surface. They are also used to prevent bleeding in pine and other wood, plaster, etc. that may stain, discolor or affect adhesion. The properties of various sealers are different and are determined by the type of substrate over which they are to be applied. Different types of sealers are made for use on wood, plaster, drywall, etc. Sealers are usually applied to a substrate before the primer and finish coats.

4.2.5 Shellacs, Varnishes, And Lacquers

Shellac is a lacquer made from lac, a natural resin. It is sometimes used as a wood finish. However, it is used mainly as a sealer and liquid filler on trim and furniture and as a sealer for knots in wood and **hot spots** in plaster. A hot spot is an area of improperly mixed plaster. If paint is applied over a hot spot, a chemical reaction causes *burning*, which damages the paint film.

Varnish is a clear or amber coating used to provide a hard, durable protective finish to a wide range of interior and exterior surfaces. When compared to interior varnish, exterior varnish is more resistant to weather and sunlight. It is also more flexible.

In its transparent form, lacquer is a fast-drying coating used to provide a smooth, clear finish to furniture, wood, and other metal surfaces. Some lacquers are modified with pigments to provide hiding power. Lacquers are generally low in solids content and require several coats to achieve the same film thickness as a single coat of varnish.

4.2.6 Stains

Stains are used to equalize or change the color of different types of wood and to enhance the beauty of the grain. They are applied before varnishes or other clear finishes. Another type of stain is varnish stain. This is a varnish into which a stain has been mixed. Use of varnish stain eliminates the need to perform two tasks, staining and varnishing. However, some painters consider the results less satisfactory than when a stain and varnish are used separately.

4.3.0 SPECIAL PURPOSE COATINGS

There are many special purpose coatings made to meet specific conditions (*Figure 11*). Some of these are:

- Fillers
- Texture coatings
- Elastomeric coatings
- Dry fall/fog coatings
- Floor coatings
- Fire-retardant/high heat-resistant coatings
- Encapsulants

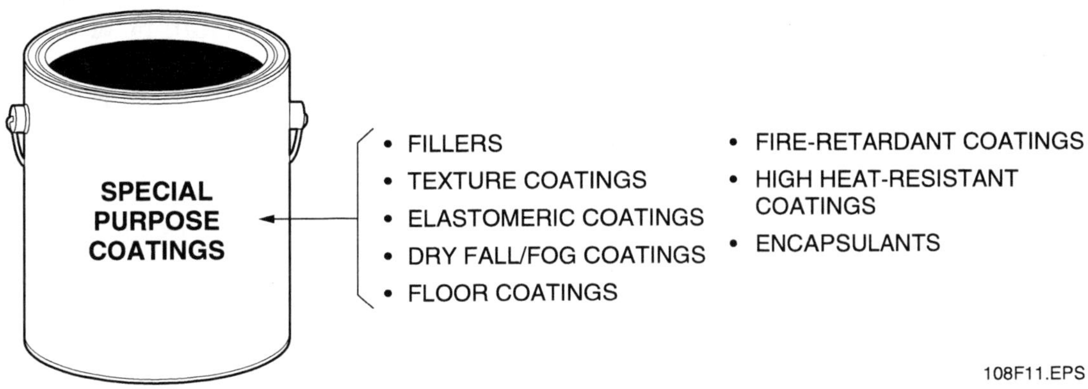

Figure 11. Special Purpose Coatings

4.3.1 Fillers

Fillers are used for filling the pores or irregularities in a surface before application of other finishes. Wood fillers are commonly used on open-grain woods like oak or walnut to avoid having a surface with a pitted appearance. Block fillers are used to fill cracks and other voids in concrete and masonry.

4.3.2 Texture Coatings

Texture coatings are thick, highly pigmented coatings applied to exterior and interior surfaces to produce various textured patterns and effects. Texture coatings are usually applied with a roller but may also be sprayed. Following their initial application, they are then worked with a brush, trowel, roller, or other tool to achieve the final pattern or effect. Texture coatings dry to a relatively thick, soft finish that acts to fill or bridge cracks, pinholes, etc. in poor surfaces. They may be used as the finish coat or they can be topcoated with suitable paint.

4.3.3 Elastomeric Coatings

Elastomeric coatings have elastic properties much like rubber. They are made to cover or hide narrow surface cracks, expand and contract with changes in temperature, and provide a protective seal against moisture. Elastomeric coatings are typically used to coat exterior concrete and masonry surfaces, roofs, and industrial steel surfaces.

4.3.4 Dry Fall/Fog Coatings

Dry fall/fog coatings are made with a solvent that will evaporate after spray application in the time required for the overspray to freely fall about 9 to 13 feet to the floor or ground. These paints are used on interior and exterior surfaces where overspray or contamination from paint fallout can be a problem.

4.3.5 Floor Coatings

Floor coatings are formulated for use on wood, concrete, or steel floors. Most serve to seal pores, resist dirt and stains, prevent marring or scratching, and allow for easy cleaning of the floor. Some floor coatings include an aggregate filler to provide a non-skid finish.

4.3.6 Fire-Retardant/High Heat-Resistant Coatings

Fire-retardant coatings help to confine a fire that might otherwise spread and cause more damage. They either act to reduce the spread of a flame over a substrate covered by the coating; resist ignition of the substrate when exposed to high temperatures; and/or insulate the substrate to increase its time to reach ignition and prevent it from melting or suffering other structural damage. High heat-resistant coatings are coatings specifically made to coat substrates subjected to high temperatures, typically 200°F (93°C) and above.

4.3.7 Encapsulants

Encapsulants are liquid coatings use to cover and seal an asbestos surface or a surface previously coated with a lead-based paint. They provide a protective barrier against the environmental and health hazards associated with exposure to lead paint dust and chips or asbestos fibers.

5.0.0 GUIDELINES FOR THE USE OF WATER-BASED AND OIL-BASED PAINTS

A professional paint job depends on the proper selection of the paint, good surface preparation, and correct application. The use of water-based versus oil-based paints depends on the type of substrate to be painted and under what environmental conditions the paint will be applied. *To make the proper selection, always follow the manufacturer's recommendations given in their coating systems data, other product literature, and container labeling.* Some other factors that can influence the selection and use of a water-based or oil-based paint follow.

5.1.0 WATER-BASED PAINTS

Water-based paints, commonly called *latex paints*, consist of pigments and additives dispersed in the vehicle, which is an emulsion of water and suitable resins. The main advantages of using latex paints are that they:

- Can be applied to a variety of exterior and interior surfaces
- Are easy to apply
- Dry rapidly (2 to 4 hours)
- Clean up with soap and water
- Have little odor
- Are VOC-compliant, making them easier on the eyes, nose, lungs, and skin
- Are flexible and non-chalking
- Resist peeling and blistering
- Retain their color and gloss

Exterior latex paints adapt well to changing weather conditions because microscopic pores allow water vapor to pass through the dried latex film. Because of this "breathing" ability, moisture is less likely to build up between the paint and surface, making it much less prone to peeling than an exterior oil-based paint. Because latex paints remain water-sensitive, avoid their use in damp, wet, or high humidity interior areas such as kitchens and baths. This is because these conditions can cause the paint to peel or become a breeding ground for mildew. As described earlier, manufacturers use additives in some of their interior water-based paints to enable their use under damp and wet conditions.

Latex paint should not be applied over an existing oil paint unless the surface has first been carefully sanded or chemically de-glossed. However, latex paints can, and often are, applied as a finish coat over an oil-based primer. This is because oil primers tend to soak into unpainted surfaces better than latex primers.

The main disadvantage of many latex paints is that they must be kept from freezing and applied at temperatures of 50°F (10°C) and above. These latex paints will form a poor film when applied in cold, damp weather because these conditions slow down water evaporation and the drying and coalescence of the paint film. The result can be a cracked or brittle paint with poor adhesion. Note that some paint manufacturers have developed a new formulation of latex paints that can be applied at temperatures below 50°F (10°C); typically, they can be applied at 35°F (1.6°C).

An excess evaporation rate, such as can occur on a hot surface, can lead to a powdery, poorly adherent paint or to a film with many voids, pinholes, or pores.

5.2.0 OIL-BASED PAINTS

Oil-based paints consist of pigments and additives dispersed in a vehicle of natural or synthetic (alkyd) oil resins dissolved or dispersed in a solvent. Most modern oil-based paints are oil-based alkyd paints. A recent development is water-reducible alkyd paints. These new paints contain emulsifiers and surfactants to make them water-soluble so that they can be cleaned up with water and better meet environmental standards. An **emulsifier** is a material added to dissimilar materials such as oil and water to produce a stable, uniform emulsion. The main advantages of using alkyd paints are that they:

- Provide good hiding
- Tolerate poor surface preparation
- Are excellent at penetrating unpainted surfaces, such as unprimed wood
- Perform well at lower temperatures
- Produce a durable, long lasting finish

Alkyd paints typically require about 24 to 48 hours to dry. Because of their longer drying time, they have a greater tendency to create lap marks, drips, and sags than latex paints. Alkyd paints repel water and dry to a hard, nonporous finish, making them easy to clean. However, should moisture manage to work its way behind an alkyd paint film, it can rot the substrate and/or cause the paint to peel. Alkyd paints should not be used on galvanized metal, concrete, or masonry because of the alkalinity of these surfaces.

The disadvantage of using alkyd and other oil-based paint is that the vehicle typically consists of about 30% to 40% solvents, so it is high in VOC emissions. It also requires the use of solvents high in VOCs for thinning and cleanup. Environmental laws and regulations to limit the amount of VOCs in paint have become increasingly restrictive. This has caused an overall reduction in the use of alkyd and other solvent-based paints for most applications in favor of latex (water-based) paints.

The solvents used in oil-based paints and for cleanup of brushes, rollers and other equipment generally smell bad and are irritating to the eyes, nose, lungs, and skin. To prevent personal injury, the MSDSs and container labeling for the specific paint and solvent(s) being used should always be read and all directions followed concerning the use of personal safety and respiratory equipment.

5.3.0 CLEANUP AND DISPOSAL OF WATER-BASED AND OIL-BASED PAINTS

Some general guidelines for the cleanup and disposal of water-based and oil-based paints are given here. The specific methods for cleaning brushes, rollers, sprayers, tools, etc. are given where appropriate in other modules.

5.3.1 Cleanup And Disposal Of Water-Based Paints

General guidelines for the cleanup and disposal of water-based paints are:

- Consolidate leftover paint into undamaged containers with good lids and seal them properly.
- Clearly label each container as to its contents in plain, understandable language.
- Dispose of all unusable latex and water-based paint and any wash water according to the local regulations and laws. Never dump in a storm sewer or in the ground.
- Empty containers should contain less than one percent paint by volume before disposal. Scrape out all used cans, buckets, potliners, etc. using a spatula or paint brush, rinse with water, then allow to dry prior to disposal.
- After drying, consolidate containers by smashing flat or filling with other disposable waste prior to placing them into trash bags for proper disposal.

5.3.2 Cleanup And Disposal Of Oil-Based Paints And Cleaning Solvents

General guidelines for the cleanup and disposal of oil-based paints and cleaning solvents are:

- Consolidate leftover paint into undamaged containers with good lids and seal them properly.
- Clearly label each container as to its contents in plain, understandable language.
- Clean empty cans, buckets, potliners, etc. using a spatula, brush, and the proper solvent to rinse leftover paint from the container. Air dry the empty containers.
- After drying, consolidate containers by smashing flat or filling with other disposable waste prior to placing them into trash bags for proper disposal.
- Use solvents sparingly when cleaning brushes and tools. Dirty solvents can often be used for the first and second rinse. Collect used solvent waste and unusable solvent-based paints in a properly-labeled metal container with a lid. Once the waste solvent is no longer usable, pour it into a suitable waste solvent barrel. Do not mix other items, such as used roller covers, rags, stir sticks, etc. or any non-liquid material in with the waste solvent. Clean, dry, and dispose of these items as dry waste.

5.3.3 Cleanup And Disposal Of Waste Solvents

Waste solvents should be kept in metal waste solvent containers or drums and placed in a holding area. These containers should be clearly marked. When full, the containers can be either shipped to a waste processing facility for disposal or the waste solvent may be recycled. When shipped to a disposal facility, the contents of the various waste containers are usually poured into a 55-gallon drum prior to pickup by the waste disposal company. Any sludge and solid residue remaining in the individual waste containers should be consolidated into a single container with a lid for pickup and disposal. All containers should be clearly labeled as to the contents.

Solvents can be recycled by a process of distillation and condensation using a solvent recovery unit (*Figure 12*). The contents of waste containers are first strained to remove any solids, then are poured into the recycling unit. The distillation process for recycling a solvent involves evaporating it, causing the impurities and tiny dirt particles to settle out into a residue. The vaporized solvent is then cooled, causing it to condense back into a clean, usable liquid solvent. The dirty residues separated from the solvent by the distillation process are removed from the recycle unit and shipped to a waste disposal facility.

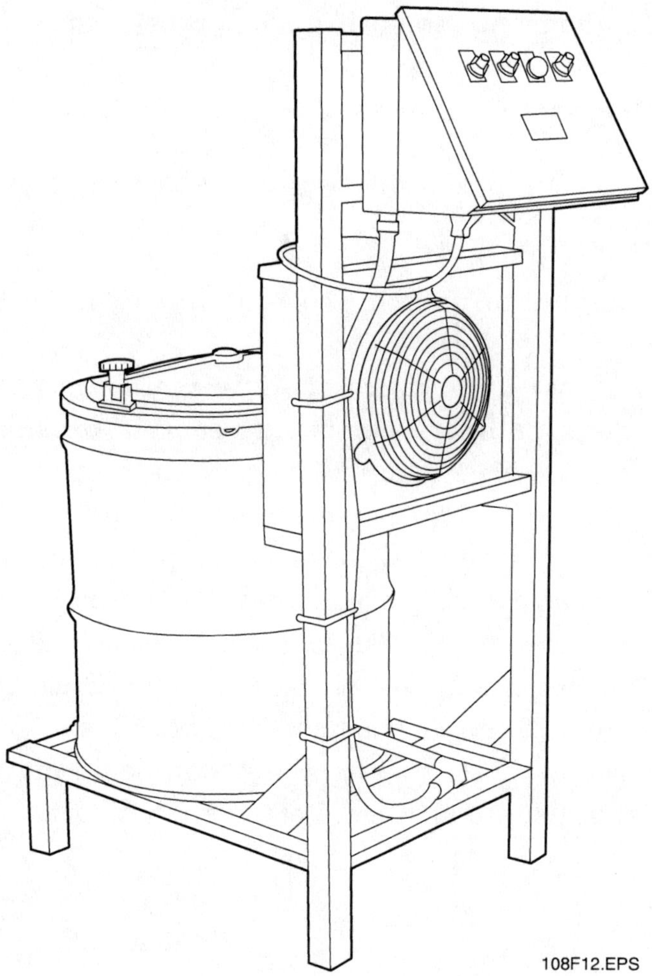

Figure 12. Solvent Recycling Unit

SUMMARY

As a result of the Clean Air Act and other new laws and regulations, water-based and oil-based paints have been re-engineered to eliminate or reduce the use of volatile organic compounds (VOCs) and lead. Today's paints are quite different in their composition and performance than the paints of just a few years ago. Even the terms *latex* and *oil-based* paint have taken on new meanings. In both cases, improved synthetic resins have yielded better performance, especially in water-based paints.

All coatings contain some or all of these four basic ingredients: pigments, resins, solvents, and additives. Depending on the specific combination of the four basic ingredients, paints and coatings can be grouped into four general categories: paints, stains, clear coatings, and special coatings.

Almost immediately after liquid paints and coatings are applied to a surface, they begin a film-forming process to become a dry, stable, and solid surface. The process by which a paint drys and cures is determined mainly by the characteristics of its resins. Paints and coatings dry and cure by one of three processes: oxidation, evaporation, or chemical crosslinking.

There are many technical and/or chemically-based ways to classify the different kinds of paints and coatings. For practical purposes, the most common architectural paints and coatings can be grouped into four broad categories: alkyd, latex, epoxy, and urethane. Paint and coating manufacturers group their products into coating systems based on the type of substrates onto which a particular group of paints or coatings can be applied and under what conditions. The paints or coatings grouped into a particular coating system can have different properties and serve different functions, but they are all designed to be compatible when used with one another. Always consult the manufacturer's coating system data to select the right system of coatings for a specific job. A professional job depends on the proper selection of the paint or coating, good surface preparation, and correct application.

Federal, state, and local laws and regulations regarding the disposal of paints and solvents are strict. These laws can vary from one locality to another. Always be aware of and obey the specific laws that apply to your area.

References

For advanced study of topics covered in this task module, the following books are suggested:

Painting & Coating Systems, Sherwin-Williams Company, Cleveland, Ohio.

Painting and Decorating Craftsman's Manual and Textbook, Painting and Decorating Contractors of America, Fairfax, Virginia.

Painting and Decorating Encyclopedia, The Goodheart-Willcox Company, Inc., Tinley Park, Illinois.

Store Training and Reference Tool - Paints, Stains, and Clear Coatings, Sherwin-Williams Company, Cleveland, Ohio.

ACKNOWLEDGMENTS
Figures 6, 7, and 8 courtesy of Sherwin-Williams Company.
Figure 12 courtesy of Painting and Decorating Contractors of America.

REVIEW/PRACTICE QUESTIONS

1. One difference between a paint and a coating is that a:
   a. coating always contains resin but a paint may not.
   b. paint always contains resin but a coating may not.
   c. coating always contains pigment but a paint may not.
   d. paint always contains pigment but a coating may not.

2. The ingredient that gives a coating its color and other properties is the:
   a. resin.
   b. solvent.
   c. pigment.
   d. additive.

3. The ingredient that gives a coating opacity is the:
   a. pigment.
   b. solvent.
   c. resin.
   d. additive.

4. The ingredient in a coating that binds the pigments and additives together and holds them to the substrate is the:
   a. resin.
   b. solvent.
   c. binder.
   d. both a and c.

5. The label on a coating can states that the coating contains 20% ketones. This is a reference to the amount of _____ in the product.
   a. pigment
   b. solvent
   c. resin
   d. additives

6. A coating is modified with a UV-absorbent material. This material is:
   a. a pigment.
   b. an additive.
   c. a solvent.
   d. a resin.

7. An additive is included in an interior, water-based paint to help retard mildew. The type of additive used is classified as a:
   a. plasticizer.
   b. thixotrope.
   c. fungicide.
   d. emulsion aid.

8. The process by which a coating dries and cures is determined mainly by the characteristics of its:
   a. resins.
   b. pigments.
   c. solvents.
   d. additives.

9. The duration of the curing process for coatings that form their films by oxidation is:
   a. days.
   b. weeks.
   c. months.
   d. years.

10. Film-forming for a water-based coating happens by a process called:
    a. transpiration.
    b. coalescence.
    c. both a and b.
    d. oxidation.

11. An alkyd paint made with long-oil alkyd resins:
    a. takes longer to dry than a paint made with short-oil alkyd resins.
    b. dries faster than a paint made with short-oil alkyd resins.
    c. takes the same amount of time to dry as a paint made with short-oil alkyd resins.
    d. dries faster than a paint made with medium-oil alkyd resins.

12. Catalyzed epoxy coatings are made by combining the epoxy resin with a:
    a. solvent.
    b. curing agent.
    c. flow agent.
    d. drier.

13. Use of a moisture curing urethane coating usually requires that the relative humidity of the surrounding air be at least:
    a. 10%.
    b. 20%.
    c. 30%.
    d. 35%.

14. The coat of paint that provides a transition between an undercoat and a finish coat is a:
    a. sealer.
    b. filler coat.
    c. tie coat.
    d. primer.

15. The term *enamel* means:
    a. different things to different paint manufacturers.
    b. the paint is a water-based paint.
    c. the paint is an oil-based paint.
    d. glossy.

16. Hot spots on a plaster surface are commonly sealed with:
    a. varnish.
    b. lacquer.
    c. varnish stain.
    d. shellac.

17. Heat-resistant coatings are typically used to:
    a. reduce the spread of a flame.
    b. resist ignition of a substrate when exposed to high temperatures.
    c. coat substrates subjected to temperatures of 200°F (92°C) and above.
    d. coat substrates subjected to temperatures of 150°F (65.55°C) and above.

18. The performance of latex paints is usually _____ alkyd paints at lower temperatures.
    a. poorer than
    b. better than
    c. the same as
    d. somewhat better than

19. Alkyd primers are better suited for use on unprimed _____ than latex paints.
    a. galvanized metal surfaces
    b. masonry surfaces
    c. wood surfaces
    d. both a and b

20. Waste turpentine should be disposed of:
    a. in a toilet.
    b. in a sink.
    c. in a storm sewer.
    d. at a disposal facility or recycled.

NCCER CRAFT TRAINING USER UPDATES

The NCCER makes every effort to keep these manuals up-to-date and free of technical errors. We appreciate your help in this process. If you have an idea for improving this manual, or if you find an error, a typographical mistake, or an inaccuracy in the NCCER's Craft Training Manuals, please write us, using this form or a photocopy. Be sure to include the exact module number, page number, a description of the problem, and the correction, if possible. Your input will be brought to the attention of the Technical Review Committee. Thank you for your assistance.

Instructors – If you found that additional materials were necessary in order to teach this module effectively, please let us know so that we may include them in the Equipment/Materials list in the Instructor's Guide.

Write: Curriculum and Revision Department
National Center for Construction Education and Research
P.O. Box 141104
Gainesville, FL 32614-1104

Fax: 352-334-0932

Craft ____________ Module Name ____________

Module Number ____________ Page Number(s) ____________

Description of Problem ____________

____

____

____

(Optional) Correction of Problem ____________

____

____

(Optional) Your Name and Address ____________

____

Brushing and Rolling Paints and Coatings

Module 07109

Painting Trainee Task Module 07109

BRUSHING AND ROLLING PAINTS AND COATINGS

PAINTING AND DECORATING CONTRACTORS OF AMERICA

NATIONAL CENTER FOR CONSTRUCTION EDUCATION AND RESEARCH

Objectives

Upon completion of this module, the trainee will be able to:

1. Recognize the various types of paint brushes and select the proper paint brush for the application.
   - Wall brushes
   - Varnish brushes
   - Sash and trim brushes
   - Stain brushes
   - Special purpose brushes
   - Decorative brushes
2. Recognize the different kinds of rollers and roller covers and select the proper roller and cover for the application.
   - Dip rollers
   - Self-feeding rollers
   - Special purpose rollers
3. Demonstrate how to properly mix paint.
4. Demonstrate how to properly apply paint to surfaces using the brush and the roller.
5. Demonstrate how to clean and store paint brushes and rollers.

Prerequisites

Successful completion of the following Task Modules is required before beginning study of this Task Module: Core Curricula, Painting Level 1, Modules 07101 through 07108.

Required Trainee Material

1. Trainee Module
2. Appropriate Personal Protective Equipment

Course Map

This course map shows all of the *Wheels of Learning* task modules in the first level of the Painting curricula. The suggested training order begins at the bottom and proceeds up. Skill levels increase as a trainee advances on the course map. The training order may be adjusted by the local Training Program Sponsor.

Course Map: Painting, Level 1

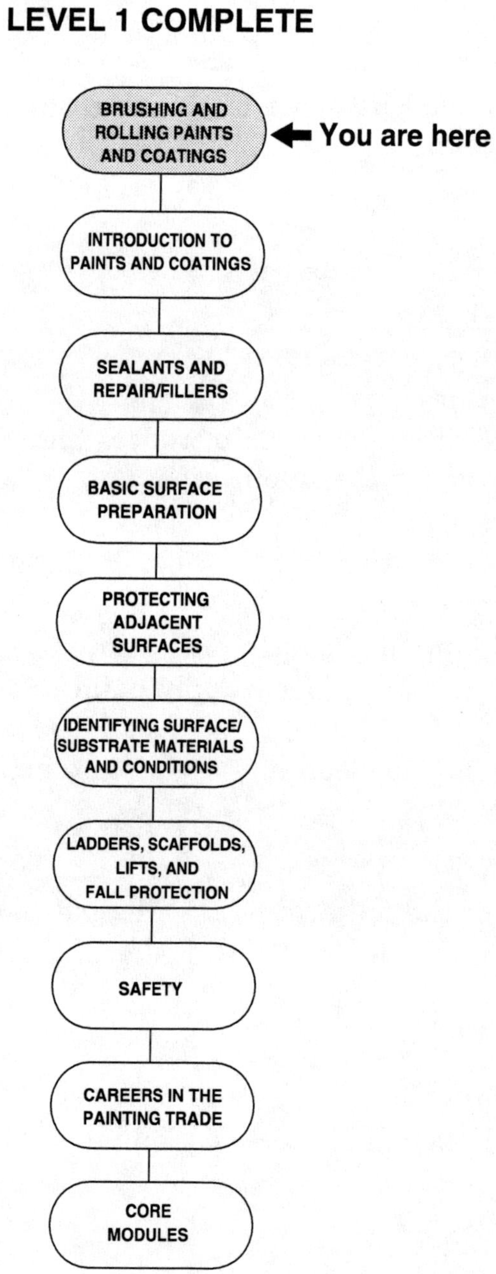

CMAP109.EPS

TABLE OF CONTENTS

Trade Terms Introduced In This Module

Boxing: Boxing is the process of pouring paint back and forth from one pail to another to achieve uniform consistency after the paint has first been thoroughly mixed by stirring.

Cutting in: Painting corners and the perimeter of windows and doors with a brush or trim roller before roller application of paint to the walls or ceiling. Also used when painting a clean edge, such as needed on a window sash.

Feathering: Tapering off the edges of a coat of paint by gradually pulling the bristles of the brush off the painted surface.

Fingering: A condition in which the bristles of a brush bunch together and do not return to their natural position. It can result from breaking in the brush improperly, allowing paint to harden in the heel, or painting with the edge rather than the flat surface of the brush.

Flag: The end of a brush bristle which divides into two or more branches. Flagging gives the brush the ability to hold paint.

Keeping a wet edge: The process of working from an unpainted area into a previously painted, still-wet area.

Laying off: Going over an entire section of a wall or other surface while the paint is still wet with light, quick strokes in one direction in order to smooth brush marks, laps, and joints, and to cover any thin or missed spots.

Stippling: A decorative finish made by using a stippling brush or roller to apply paint in a random pattern to a surface with a base coat of a different color.

1.0.0 INTRODUCTION

A professional paint job depends on the selection of the correct paint, proper preparation of the surface before painting, and applying the paint with skill in a prescribed manner. The focus of this module is on the application of paint using brushes and/or rollers. The different kinds of brushes and rollers are described, including their uses and maintenance. Also covered are the proper methods for using brushes and rollers to apply paint.

2.0.0 BRUSHES

All paint brushes consist of bristles, a heel, a ferrule, and a handle (*Figure 1*). The portion of the bristles nearest the ferrule is called the *butt end*; the lower part of the brush is its *heel*; and the tip of the brush is called the *flag end*. The bristles of the brush are held in position at the lower end of the ferrule by an epoxy adhesive. Normally, the bristles are separated by wooden or aluminum spacer plugs. These spacer plugs provide support for the bristles and help keep them from fanning out when paint is applied.

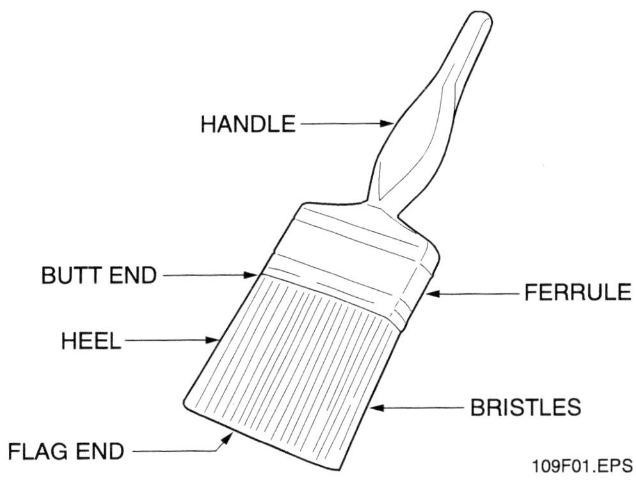

Figure 1. Parts Of A Paint Brush

2.1.0 NATURAL AND SYNTHETIC BRISTLE BRUSHES

Either natural or synthetic bristles can be used in brushes. Hogs' hair bristles are used in good brushes made with natural bristles. There are two benefits of using hogs' hair bristles over other animal hairs: they are tapered and their tips are ragged or **flagged** (*Figure 2*), a natural feature that helps the bristles to hold more paint and to spread it evenly onto a surface. Flagging also helps the brush to keep its shape. Synthetic bristles, also called *filaments*, can be made from nylon, polyester, or a blend of these materials. Synthetic fibers can be made in three basic shapes: solid, cross-sectional, and hollow. The solid shape is the easiest to clean and lasts the longest. The cross-sectional filament is an economical choice; the hollow shape is the least desirable. Synthetic filaments have no natural flags, but are usually tapered and flagged in the manufacturing process to closely match the features of hogs' hair bristles.

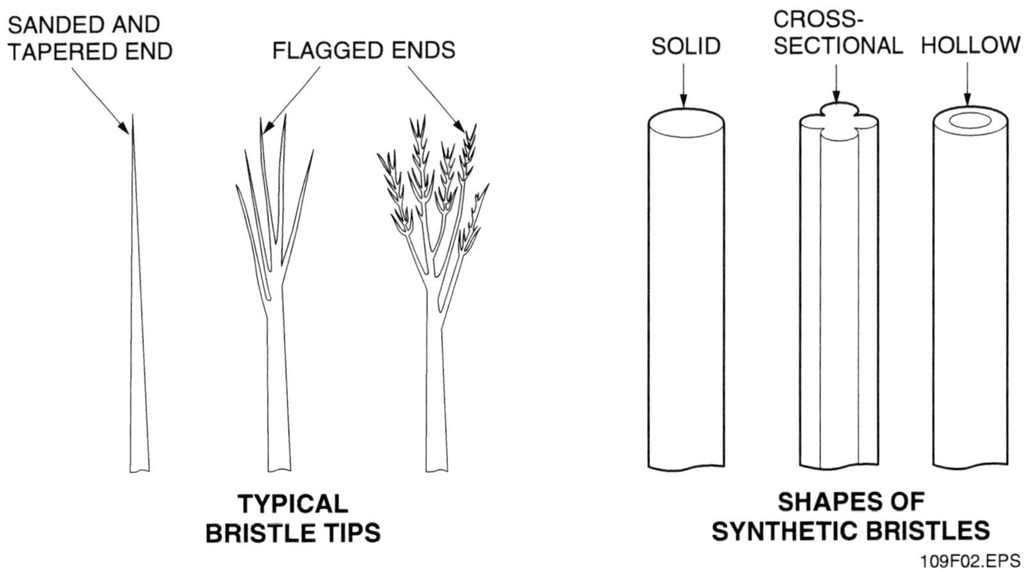

Figure 2. Characteristics Of Bristles

Note Many new reduced-solvent paints are heavier and resist flowing with flag-tipped brushes. For this reason, certain synthetic brush tips are made with sanded and tapered bristles to make the paint flow more easily.

A general rule of thumb is that brushes made with natural bristles are used only with oil or alkyd paints, while those made with synthetic bristles are used with water-based paints. This is because natural bristles can be softened by the water in water-based paints, causing the natural bristles to puff up and become unruly. Another reason is that natural bristles can tolerate the harsh solvents used to clean them, while some synthetic brushes will deteriorate in the solvents. In recent years, brush manufacturers have developed synthetic bristle brushes designed for use with both oil-based and water-based paints.

2.2.0 TYPES OF BRUSHES

Selection of a brush is determined by the type of finish being applied, the size and shape of the area being covered, and the quality and speed of application desired. Many types of brushes are made, each with a specific purpose and performance benefit. Some common brushes are described in the following paragraphs.

2.2.1 Wall Brushes

The typical wall brush (*Figure 3*) is 4 inches wide, but the range is from about 3 to 6 inches, or sometimes even larger. These brushes are used to paint large, flat, regular surfaces, such as walls, exterior siding, ceilings, and floors.

Figure 3. Wall And Enamel/Varnish Brushes

2.2.2 Varnish Brushes

Varnish brushes are available in sizes ranging from 2 to 4 inches wide. They may have a chiseled or tapered tip for cutting in (discussed later), and they are made with high-quality bristles to assure even spreading of finish coatings. They are used to paint trim, doors, baseboards, and other woodwork.

2.2.3 Sash And Trim Brushes

Sash and trim brushes (*Figure 4*) may be flat, angular, or oval. They have the following features in common: they are small, usually from 1 to 3 inches wide; and they have a fluted handle for precise control. Angular brushes allow the painter to work at different angles more comfortably. Flat sash and trim brushes can be used for flat, narrow areas, such as baseboards. They can also be used for finish work on doors, door frames, and other trim. Oval brushes can be used to coat curved or irregular surfaces, such as carved or detailed molding.

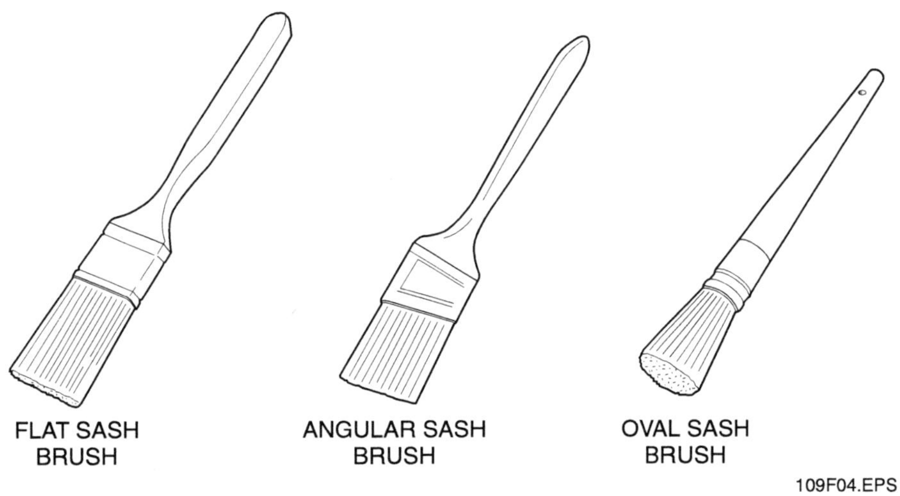

Figure 4. Sash And Trim Brushes

2.2.4 Stain Brushes

Stain brushes generally range from 4 to 6 inches in size. These brushes are designed for the application of stain coatings.

2.2.5 Special Purpose Brushes

Some areas are very difficult to paint using conventional brush shapes and handles. Some brushes, such as the bent ferrule or *hockey-style brush* (*Figure 5*) are designed to reach otherwise inaccessible places.

Figure 5. Special Purpose Brushes

Masonry and roofing brushes are two other types of special purpose brushes. Masonry brushes are designed for tough, abrasive wear on irregular surfaces. They have long-wearing bristles and come in sizes from 5 to 6 inches wide with bristle lengths of 4 to 6 inches.

Roofing brushes normally have a long handle set at an angle like a push broom for ease of use on large surfaces. They have short lengths of fibrous bristles that are often bound to a block in three bunches.

2.2.6 Decorative Brushes

Decorative brushes include the wide range of artist's brushes made from the hair of various animals, stippling brushes, stenciling brushes, bronzing brushes, etc. The sizes of these brushes range from very small up to one inch wide or more.

2.3.0 GUIDELINES FOR SELECTING QUALITY BRUSHES

The quality of brushes can range from very poor to excellent. Brush quality is determined both by its materials and its construction. Only quality brushes should be used because they hold more paint, leave a better finish, and do not shed their bristles. They also last longer and generally make painting go easier and faster. Price and feel are both factors in selecting a quality brush. Normally, the higher the price, the better the brush. Other guidelines for testing the quality of a brush are:

- Pull gently on the bristles. An inferior brush will leave several bristles in your palm, whereas a good brush will lose only one or two.

- Bounce and wiggle the bristles in your palm. A good brush has bristles that feel soft and springy, and bounce back into shape quickly when let go.
- Make sure the bristles are thick and plentiful. Good brushes will have bristles that are at least 2 inches long. The wider the brush, the longer the bristles. For example, a brush that is 1 or 2 inches wide should have bristles about 3 inches long; a brush that is 3 or 4 inches wide should have bristles about 4 inches long. The longer the bristles and the thicker their mass, the more paint the brush can hold.
- Good brushes have a substantial handle that is usually made of lacquered beech, birch, or maple wood. It should be balanced, comfortable, and shaped to fit the hand.
- The metal ferrule that holds the handle should be made of a rust-resistant metal and fastened to the handle with brads, screws, or rivets. Make sure it does not rock when the bristles are wiggled back and forth. Poor brushes usually have larger spacer plugs, and therefore have fewer bristles than better brushes.

3.0.0 ROLLERS

Rollers are used to apply paints and coatings to large, regular surfaces such as floors, ceilings, and walls. They transfer and spread paint onto a surface by rolling it with a fabric covered roller cover. Rollers can be classified by the method of supplying paint to the roller cover or the function they serve. These classifications are:

- Dip rollers
- Self-feeding rollers
- Special purpose rollers

3.1.0 DIP ROLLERS

Dip rollers (*Figure 6*) are the most common type of roller. They are so named because the paint is supplied to the roller cover by dipping it into a roller tray or a five-gallon can. Dip rollers consist of a handle, a frame, and a cage or cylinder which holds the roller cover. Normally, the cage has nylon end caps or bearings that help it to rotate smoothly. The diameter and width of the roller cover is selected to match the diameter and width of the cage on which it is to be mounted.

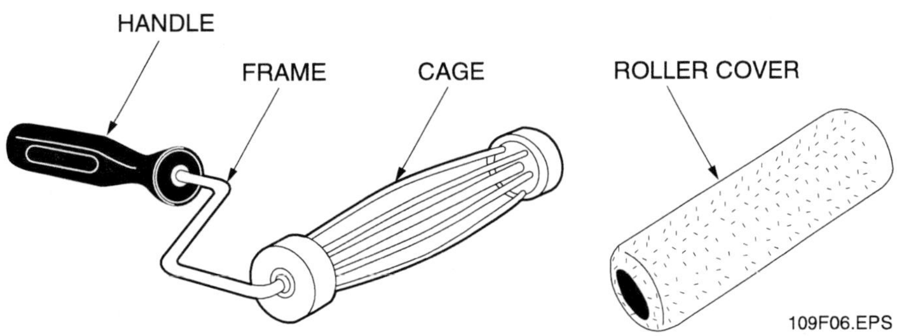

Figure 6. Parts Of A Dip Roller

Roller covers consist an of inner cylinder, called the *core*, which is commonly made of baked phenolic fiberboard to which a fabric is bonded. The fabric holds the paint much as a sponge holds water. Standard size roller covers have a core with an inside diameter of 1½ inches. Common widths range from 3 to 18 inches. However, 9 inches is the most popular width because it is easy to control, is readily available, and fits easily into a paint tray or five-gallon pail for dipping and cleaning. Narrower roller covers are used for trimming and similar applications. Wider widths are used to paint large, flat areas such as floors.

Note The use of wider width rollers for overhead work is impractical because they are heavy when filled with paint and clumsy to use in close or busy areas.

The roller cover fabric can be made of knitted or woven natural or synthetic fibers. Natural fibers include lamb's wool, sheepskin, and mohair (goat hair). Synthetic fibers include polyester, acrylic, and nylon. Special purpose covers are also made of foam for use on smooth or textured surfaces. Others are made with plastic or carpet loop to create special finish effects such as stippling. **Stippling** is a decorative finish made by using a stippling brush or roller to apply paint in a random pattern to a surface that has been painted with a base coat of a different color.

The length of the roller cover fibers, called the *nap*, ranges from ⅛ inch to 1½ inches long. The nap is important because it determines the kinds of surfaces on which the cover can be used and its overall performance on those surfaces. For practical purposes, naps can be classified as short, medium, and long. Short naps down to ⅛ inch provide a smooth finish on very flat and smooth surfaces such as wood and metal. Medium naps ranging between ¼ and ¾ inch are typically used on walls and ceilings made of drywall, rough wood, cement block, etc. Long naps between 1 inch and 1½ inches are suitable for use on rough surfaces such as textured masonry and concrete blocks where the paint must be worked into crevices, pits, and other irregularities.

The goal in rolling is to put as much paint on the surface as quickly as possible without leaving roller marks, craters, bubbles, and other surface imperfections. A roller cover with a long nap will leave waves and craters on a smooth surface, while a nap that is too short holds relatively little paint so more effort goes into frequent reloading of the roller and keeping a wet edge. For this reason, manufacturers label their roller cover packages to show the types of coatings and surfaces for which a roller cover is designed. When choosing a roller cover, always follow the manufacturer's recommendations.

3.2.0 DIP ROLLER ACCESSORIES

Common accessories used with dip rollers include paint trays, paint grids, and extension poles (*Figure* 7). A paint tray is used to hold the paint supply for the roller. It has a well for the paint and a ramp that leads into the well. As the roller is moved back and forth over the roughened surface of the ramp, it works the paint into the roller cover and removes excess

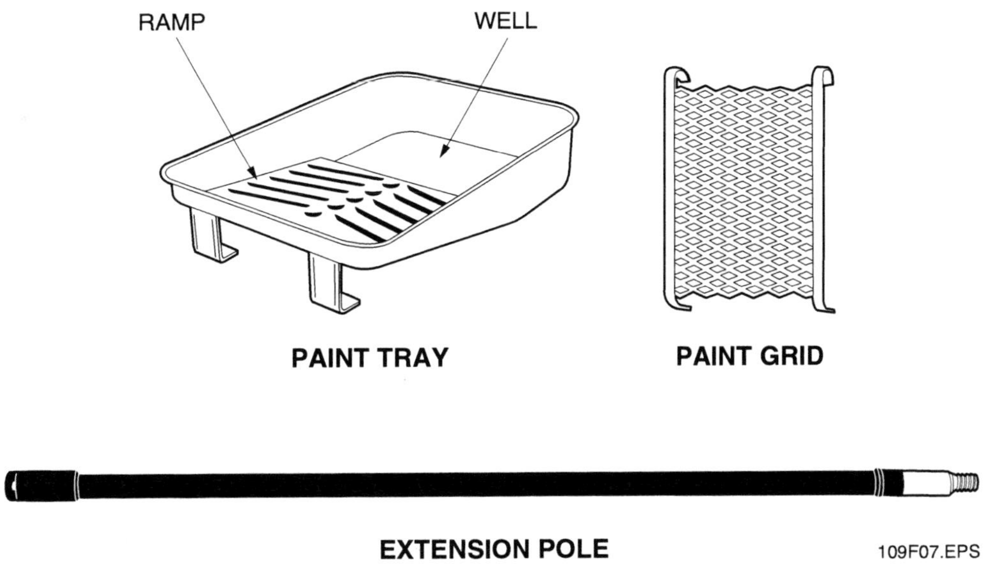

Figure 7. Roller Accessories

paint from the cover before it is applied to the surface. Quality paint trays are made from sturdy metal or plastic. In general, the use of a deeper pan is best because it allows more work to be done between refills.

A paint grid or mesh screen is used when rolling paint from a five-gallon pail. The grid is used to work the paint into the roller cover and remove excess paint from the cover before it is applied to the surface. Paint grids or screens hook over the rim of a five-gallon pail that is about half-full of paint, allowing the grid to hang from the top without being immersed in paint.

An extension pole can be screwed into the threaded end of most roller handles, allowing ceilings and upper wall areas to be painted without using a ladder. A pole can also be used to eliminate bending or kneeling when painting floors, sealing driveways, etc. Extension poles are made of lightweight wood, fiberglass, or metal in telescoping or fixed lengths. Fiberglass or metal poles are often telescoped for easy storage and range in length from 4 feet up to 16 feet or more.

3.3.0 SELF-FEEDING ROLLERS

Self-feeding rollers have a continuous supply of paint applied to the roller cover, eliminating the need for dipping the roller into a pail or pan. They are the most efficient types of rollers for use on very large rolling jobs, such as painting several rooms or the exterior of a building.

One popular type of pressure-fed roller uses a hose and pump to draw paint directly from a five-gallon pail or other container (*Figure 8*). With the hose connected to a fitting on the roller handle, paint flows through the inside of the tubular handle and into the core of the roller cover. The core is perforated to allow the paint to pass into the fabric of the cover.

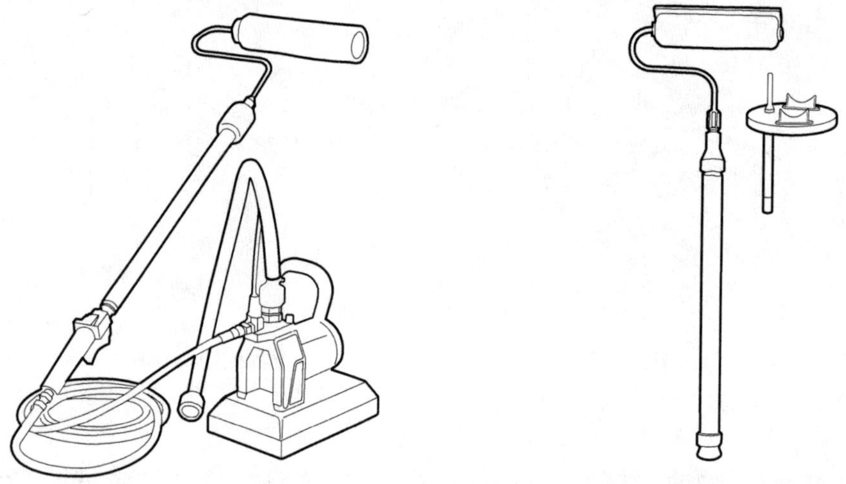

Figure 8. Self-Feeding Rollers

This type of pressure-fed roller is normally equipped with a flow control switch located on the roller handle. It is used to start and stop the flow of paint.

Another type of self-feeding roller is manually operated. It has a special lid and fill tube that mounts on a paint can. The roller's extension pole serves as a paint reservoir. This extension pole with attached roller cover is filled with paint by placing the end of the pole on the lid adapter. Then, the sections of the pole are operated to siphon paint from the can into the extension pole. Once filled, the pole is detached from the lid, and the pole sections are pushed to pump paint to the roller cover core. Pores in the roller cover core allow the paint to saturate the roller cover's outer fabric.

A third type of self-feeding roller is called a *fountain roller*. It is similar in construction to a dip roller except that the roller cover's hollow core serves as a reservoir for holding paint. The paint is poured into the core, then the core is sealed with a cover. As with the other self-feeding rollers, pores in the roller cover core allow the paint to saturate the roller cover's outer fabric.

3.4.0 SPECIAL PURPOSE ROLLERS

There are many kinds of rollers that have been developed for special purposes (*Figure 9*). A few common ones are:

- *Trim rollers* – Small rollers of various types, shapes, and sizes designed for painting trim, corners, edges, and other narrow surfaces.
- *Pipe rollers* – Roller unit that has a spring spindle with a number of narrow rollers that move under pressure to conform to contoured surfaces such as pipes.

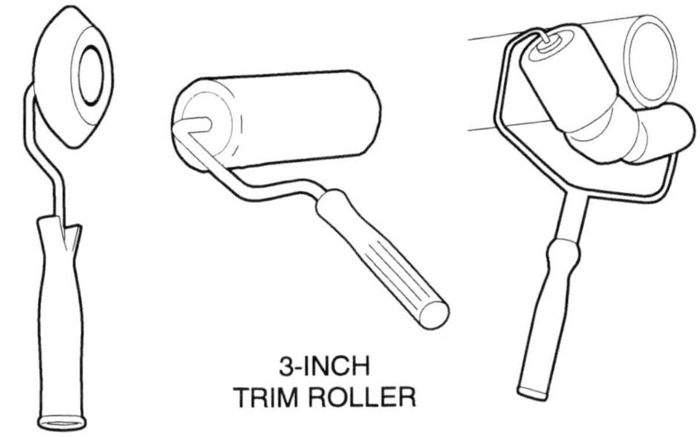

Figure 9. Special Purpose Rollers

- *Paint stripers* – Roller unit consisting of a series of rollers, one behind the other, mounted in a steel frame. Stripers are used to apply directional boundaries and lines on roads and parking lots.
- *Spike rollers* – Rollers made with sharp metal spikes that are used on thick floor film coatings after application, but before drying, to make sure that the coating has no entrapped solvent or air bubbles.

3.5.0 GUIDELINES FOR SELECTING QUALITY ROLLERS AND ROLLER COVERS

Like brushes, the quality of dip rollers and roller covers can range from poor to excellent. Only quality rollers and roller covers should be used because they hold more paint, leave a better finish, and do not shed fabric. They also last longer and generally make painting go easier and faster. As with brushes, price and feel are both factors in selecting a good quality roller and cover. Normally, the higher the price, the better the roller or cover. A cheap roller cover will tend to split and/or drip paint. It will also have a tendency to skid along the wall without turning because the nap is unable to grip the surface. Other guidelines for testing the quality of a roller and roller cover are:

- The roller handle, frame, and cage should be substantial. Cheap handles and frames will bend with pressure. The roller should have a balanced feel and the handle should feel comfortable in your hand.
- The cage should have nylon sleeves or bearings on the ends that allow it to rotate freely.
- A good roller cover will have a regular consistency through the nap as you roll it in your palm. Poor covers often have small lumps and bumps.
- The roller cover nap should be soft and pliable, but when rolled in the palm of your hand, it should not feel loose. No fibers should pull loose when the nap is softly pinched.
- The core of a roller cover should be sturdy. Its strength is an indication of durability and quality.

4.0.0 OTHER PAINT APPLICATION TOOLS

Paint mitts and paint pads are two common application tools. Paint mitts (*Figure 10*) are useful when painting wrought iron railings, pipe, and similar surfaces that are too small for rolling, too irregular for brushing, or impractical for spraying. Paint mitts are fabric mitts worn on the hand. They can be made of natural and synthetic fibers and with long or short naps for painting rough or smooth surfaces. Mitts are usually shaped in either a mitten (with thumb) or square-end style. When worn, the paint mitt is dipped into the paint, then used to spread the paint by rubbing the mitt over the surface. Protective gloves should always be worn under a paint mitt to prevent the paint absorbed by the mitt from getting on your skin.

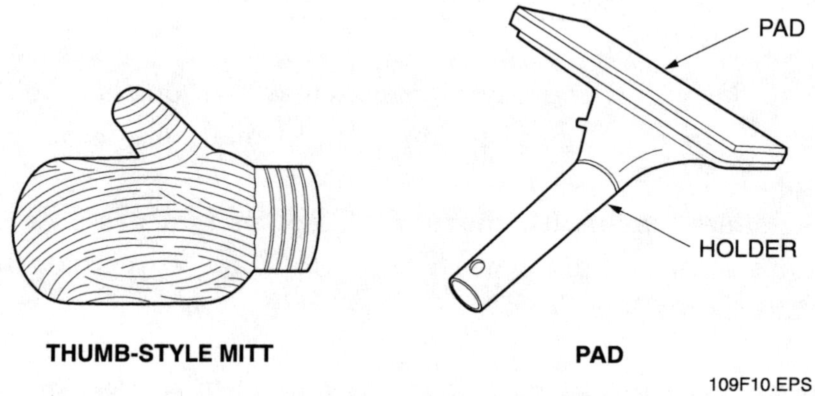

Figure 10. Paint Mitt And Pad

Paint pads can be used for general purpose interior or exterior painting. The use of paint pads instead of brushes and rollers is largely a matter of personal preference. The basic paint pad (*Figure 10*) usually has a square or rectangular painting surface covered with napped synthetic fibers. These pads fit into a holder which has a spring-loaded handle so that the pad is always parallel to the surface being painted. The handle is normally threaded to accept an extension pole for greater reach.

A paint pad is loaded with paint by dipping it into a tray designed for use with paint pads. One common pad tray is designed to hold ½ gallon of paint in a pail-shaped reservoir beneath a framed grid. The grid is filled by tipping the tray and allowing paint to flow from the reservoir onto a portion of the grid. A large amount of paint can be allowed to flow onto the grid as any excess will flow back into the sump. The pad is dipped into the paint in the wet grid area and the excess is scraped off by using a light wiping action across a dry portion of the grid.

There are many special purpose paint pads, each with different naps and shapes. Some are made with long naps for painting rough exterior surfaces such as shakes, shingles, and concrete block. Other pads are made with two surfaces at right angles for painting inside corners of walls and between walls and ceilings. Mini-sized pads are made for painting trim, sashes, and so on.

Generally, paint can be applied faster using a pad than with a brush, but a pad is usually slower than a roller. Paint applied with a pad generally does not penetrate as well as paint applied with a brush. One advantage of using pads over rollers is that pads do not splatter the way that rollers sometimes do.

5.0.0 PREPARING PAINT BEFORE APPLICATION

Before painting can begin, the surface must be clean, dry, and otherwise properly prepared. For guidelines about the preparation of various surfaces, refer to the paint manufacturer's instructions and the basic surface preparation module studied earlier. Equally important is the preparation of the paint before use. It must be thoroughly mixed so that all the ingredients blend together to provide uniform color and consistency. Mixing also affects the durability of the paint coat. Poor adhesion, uneven color, inadequate film thickness, poor hiding ability, cracking, and other paint failures can result from incomplete mixing. The process for preparing paint can involve:

- Mechanical shaking
- Manual or mechanical stirring
- Boxing
- Straining
- Thinning
- Catalyzing

5.1.0 MECHANICAL SHAKING

Shaking or agitating the paint using a mechanical shaker is usually done at the paint distributor at the time of purchase. It can also be done at the job site if the equipment is available. Mechanical shaking of paint at the job site most likely would be done if considerable time has elapsed between the time the paint was purchased and when it is used. Shaking machines vigorously mix large or small containers of paint without having to open them. When some latex paints are mixed with a shaking machine, they should be allowed to settle for at least an hour or more to eliminate the foaming that occurs in the emulsifier during the shaking process. Shaking partially-filled cans of previously used paint may also cause foaming.

To avoid foaming and bubbles, varnishes, shellac, and other clear finishes should never be shaken on a mechanical shaking machine.

WARNING! Never shake an aluminum-filled coating on a mechanical shaking machine because doing so can cause the paint container to explode.

5.2.0 STIRRING PAINT

Paint stirring can be done either manually or mechanically. Mixing by hand with stir sticks works well for smaller containers (under two gallons). When mixing larger containers, a mechanical mixer should be used because it is faster and achieves more uniform color and consistency. Mechanical mixers are made in various sizes (*Figure 11*) for stirring paint in containers ranging from one gallon up to 55 gallons. Small units attach to a hand-held power drill or drill press; large units operate with air-driven or electric motors.

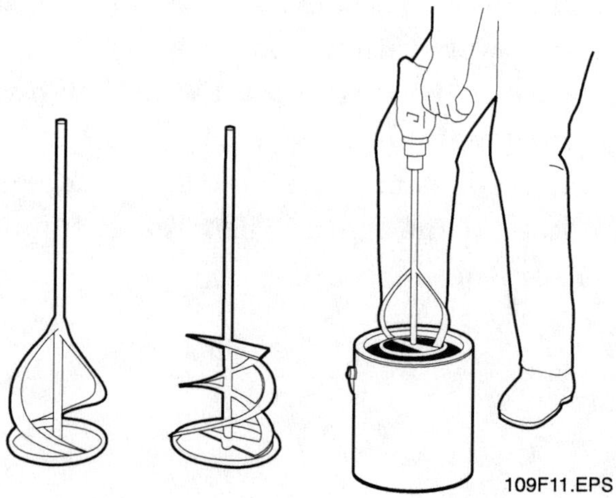

Figure 11. Mixing Paint With A Portable Drill-Powered Mixer

Mixing paint too fast can cause foaming or air to enter the paint, which is difficult to remove and leaves bubbles or blisters in the film. The mixer speed should be set so the mixer creates only a small vortex or whirlpool in the paint as it stirs. It is better to mix at a slow speed for a long time than at a high speed for a short time. Good practice is to let the paint settle after mixing so that any trapped air can rise to the surface.

If there is heavy separation and settlement of pigment in a container of paint, the top liquid should be poured off into another clean container. Following this, the remaining liquid and the settled pigment should be stirred until it is smooth. Then, slowly blend in the liquid that was removed, and stir the mixture thoroughly.

5.3.0 BOXING AND STRAINING

After the paint has been stirred, it should be boxed. **Boxing** is the process of pouring paint back and forth from one pail to another to achieve uniform color and consistency. This is always good practice but is especially important when mixing containers of the same color paint that were manufactured during different production runs.

During the boxing process, any lumps, skins, and other particles in the paint should be strained out through a wire screen, sieve, double-thick cheesecloth, or commercial paint strainer. Straining is especially important if using paint that has been used or stored previously. Straining freshly mixed paint improves its brushing ability and other qualities.

5.4.0 THINNING

Most of today's paints do not require thinning. However, some paints need to be thinned for jobs such as priming new, unpainted, or severely weathered wood with oil-based paint. Thinning is sometimes required when partially-used paint has been stored, causing some of the paint vehicle to evaporate from the container during storage. If paint thinning is required, always make sure to comply with the prevailing EPA and local area laws and regulations concerning VOC restrictions.

When using paints that must be thinned, remember that the type of thinner used varies according to the type of paint. Latex paints are thinned with water; oil-based paints are thinned with mineral spirits, turpentine, or other thinners. Use only the type and amount of thinner recommended on the manufacturer's product data sheet or container label. Also, use just enough to achieve the desired consistency, not the maximum amount allowed. Before thinning, make sure the paint is mixed well. Add the thinner slowly and continue mixing until the paint has a uniform consistency.

5.5.0 CATALYZING

Some catalyzed paints and coatings, such as epoxies and urethanes, are made by combining a resin with a curing agent. Catalyzed coatings usually come in two containers. One includes the resin, pigments, and some solvent. The other contains the curing agent or catalyst. The contents of the two containers must be mixed together before application. When mixing multiple-component coatings, make sure to carefully follow the instructions on the manufacturer's label and application data sheets to achieve the proper mixing ratio and formula. Usually, each component is mixed separately, and then they are blended together until the color and consistency are uniform. Some multi-component paints have an induction time or sweat-in time for the components to start to react before the paint can be applied. Sweat-in time may be affected by temperature and humidity. If the product is to be thinned, it should be done after sweat-in.

The manufacturer's instructions will also specify pot life, which is how long the paint can be applied before the continuing chemical reaction of the components makes it unusable. Pot life may vary from a few minutes to several hours, depending on the product. Once the paint reaches the end of its pot life, it will not cure properly and must be discarded. Therefore, mix only as much paint as can be used during the pot life.

6.0.0 APPLYING PAINT WITH A BRUSH

Applying paint with a brush is typically done when painting small areas, cutting in, and/or when painting windows and doors. **Cutting in** is the term used for painting inside corners and around the trim edges of windows, doors, etc. with a brush or trim roller before applying paint to a wall or ceiling with a roller. It also means painting a clean edge such as needed at wall/ceiling joints or on a window sash.

Painting begins by choosing the correct brush for the paint being used. Use natural bristles for oil-based paints and synthetic bristles for water-based paints. As a rule, choose the largest brush that is suitable for the job, typically a 4-inch brush for cutting in and painting large surfaces and an oval or angled 2½-inch to 3-inch sash brush for details such as windows and doors. Make sure the brush is free of dust, dirt, loose bristles, or other particles that may show up on the surface.

6.1.0 BASIC BRUSHING TECHNIQUES

When painting with a brush, work from a partially-filled container (bucket or pot). When dipping the brush into the container, the bristles should be dipped no more than halfway (*Figure 12*). Dipping deeper than halfway can cause paint to build up in the heel of the brush and also increases the chances of dripping and splattering. To set the paint in the brush and remove the excess paint, gently tap or slap one side of the brush against the inside of the container. Do not remove excess paint by wiping the bristles against the rim of the container because this removes too much paint from the brush and may damage some types of bristles.

Figure 12. Dip Brush Halfway Into The Paint

The method for holding the brush while painting is one of personal preference. Obviously, painting is easier if the brush is held comfortably. *Figure 13* shows three common methods. Apply the paint with two or three light strokes, then spread it smoothly and evenly using long, even, parallel strokes. Whenever possible, paint with the tip and flat sides of the brush and not the side edges. Use just enough pressure to bend the bristles slightly.

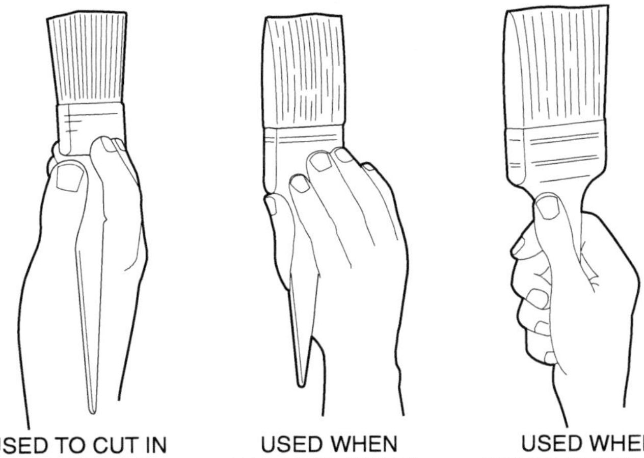

Figure 13. Methods Of Holding A Paint Brush

Finish off the leading edge of the paint by gradually pulling the bristles of the brush off the painted surface (*Figure 14*). This is called **feathering**. Whenever possible, paint vertically because up and down strokes are less tiring. On a vertical surface, paint from top to bottom, and don't stop until you reach a natural stopping point such as a door, window, trim, siding joint, etc. As you paint, **keep a wet edge**. This means working from an unpainted area into a previously painted, still-wet area. To avoid lap marks, feather the edges. When the surface is covered and while the paint is still wet, the final step is to **lay it off**, which means to go over the entire section you have just painted using light, quick strokes all in one direction in order to smooth brush marks, laps, and joints, and to cover any thin or missed spots.

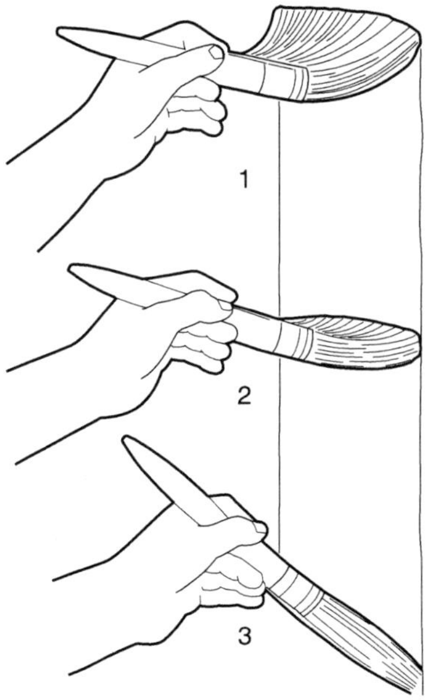

Figure 14. Feathering With A Brush

Note Glossy paint shows brush marks more than flat paint. When using glossy paint, work in smaller areas near the wet edge and lay it off frequently.

6.2.0 CUTTING IN WITH A BRUSH

Cutting in involves using a brush to apply about two inches of paint to the edges surrounding windows and doorways, at inside wall corners, and around floors before applying paint on the adjacent surfaces with a roller. When painting inside corners or wall-ceiling joints that will be painted the same color, cutting in can be done by painting four or five overlapping brush strokes at right angles to the edge of the wall or ceiling, then smoothing this strip with one long stroke. Leave a slight space near the baseboard, then smooth the paint all the way down when you begin the long stroke (*Figure 15*). Use the flat, wider side of the brush rather than the narrow part (edge). Painting with the narrow side of the brush may cause **fingering**. Fingering occurs when bristles bunch together and do not return to their natural position. Some painters prefer to use the edge of the brush to cut in 1/8 inch to 1/4 inch from the corner, then finish by making an additional pass with the wide side of the brush.

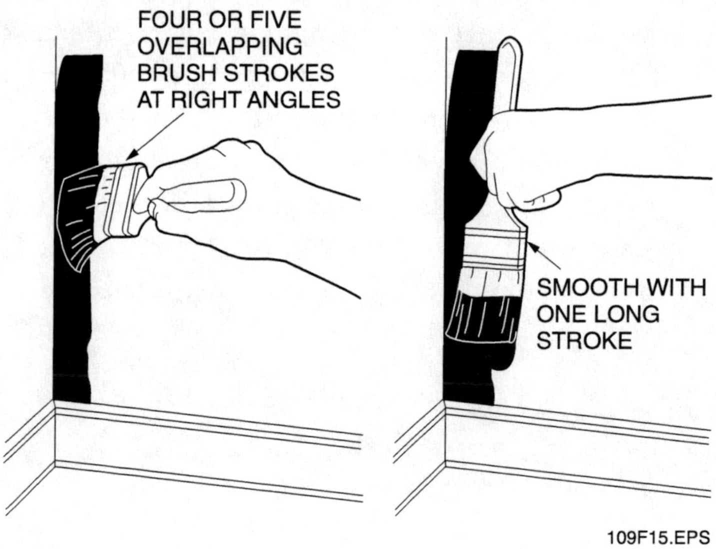

Figure 15. Cutting In Before Rolling

Cutting in inside corners or wall-ceiling joints where two different colors will meet, walls at baseboards, walls next to any other trim, and windowpane dividers (muntins) requires brushing using a beading technique. This is done by first pressing the brush flat against the surface, forcing a thin line of paint (the bead) to float along the bristle tips. Then, with a smooth, steady motion, a bead of paint is drawn along the edge of the surface you are painting (*Figure 16*).

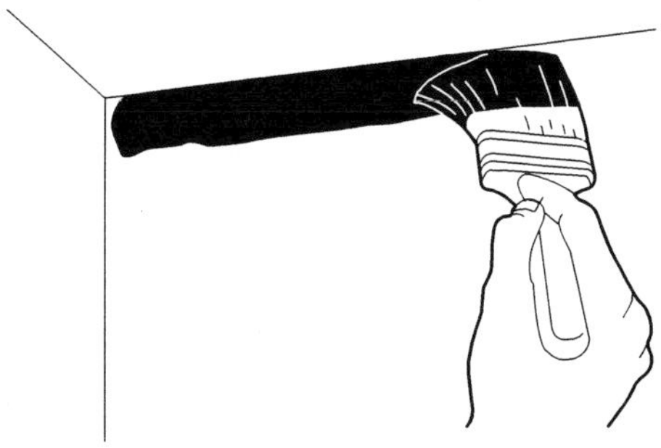

109F16.EPS

Figure 16. Drawing A Paint Bead

7.0.0 APPLYING PAINT WITH A ROLLER

Rolling is typically done when painting large, flat surfaces such as walls, ceilings, and floors. It is also done when painting large, rough surfaces such as brick, stucco, or concrete block. These surfaces can be painted much faster with a roller than with a brush. As previously mentioned, cutting in corners or trim must be done with a brush or special purpose roller before beginning to roll the surface. Cutting in should be done as the rolling progresses in order to maintain a wet edge to prevent lapping. Because rolling can be done faster than painting with a brush or trim roller, it is common for two or more painters to do the cutting in at the same time that a third painter is doing the rolling.

The rolling procedure begins by following the manufacturer's recommendations to choose the correct roller cover for the type of paint being used and the correct nap for the type of surface being painted. Make sure the roller cover is free of loose nap, dust, dirt, and any other particles that may show up on the surface. Before use, a new roller cover should always be "rubbed down" by wringing the cover in your bare hands vigorously to remove all loose nap. The roller and paint tray or pail and paint grid should also be checked and wiped clean to prevent any dirt, sand, dirty solvent, or previously used paint from mixing with the new paint.

When using a dip-type roller, fill the paint tray with paint until the well is full and about one-third of the ramp is covered. Do not overfill the tray. Dip but do not submerge the roller into the paint, then roll it back and forth over the roughened surface of the ramp several times until the cover is well coated. If using a five-gallon pail with a paint grid or mesh screen, fill the pail about halfway. Dip the roller into the paint, and roll it back and forth over the paint grid or mesh screen until the cover is well coated. When done, the roller cover should be full of paint, but not dripping. If using a pressure-fed or other type of self-feeding roller, fill it with paint and prepare it for use according to the manufacturer's instructions. A useful way to break in and wet the roller cover fibers is to roll out the first load of paint onto kraft paper, scrap drywall, or another disposable surface. This also helps to remove any air bubbles that may be trapped in the roller cover.

Apply the paint from a fully-loaded roller to the surface being painted in a zigzag pattern like the letter N or M (*Figure 17*). The first stroke should be in the upward or forward direction. This helps to prevent dripping and splattering. Be careful not to roll too fast because this also helps to avoid splattering. Use an even pressure on the roller to achieve an even coating. Without lifting the roller from the surface, cross roll to spread the paint. Then finish with light roller strokes all in the same direction but at a right angle to the cross roll. As the roller nears the end of the area being painted, feather the paint by slowly pulling the spinning roller off the surface.

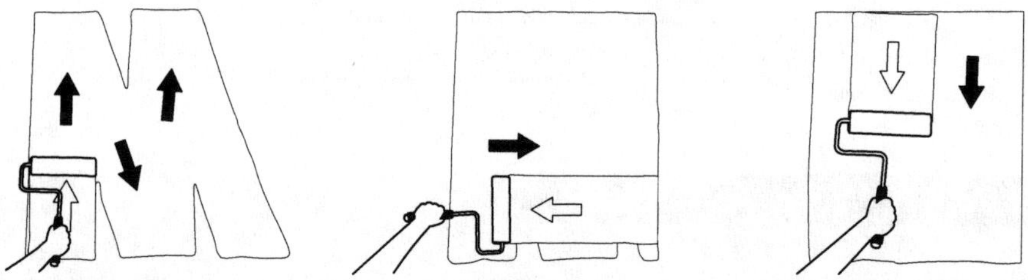

Figure 17. Rolling Technique

When painting a surface, work from the wet edge to prevent lapping. For example, when painting an 8-foot high wall, a right-handed painter should work from left to right, keeping the leading edge of paint wet by working from top to bottom across the surface. Don't stop until reaching a natural break such as a corner. If you can't go from top to bottom because of scaffolding, make sure that the edges are not allowed to dry before the scaffolding is moved because this will most likely cause lap marks.

8.0.0 GUIDELINES FOR PAINTING USING BRUSHES AND ROLLERS

Before starting to paint an interior or exterior area or surface, always make sure that all adjacent surfaces have been properly protected with drop cloths and tarps and masked against unwanted painting or paint spills as covered in an earlier module. The paints being used should be those recommended by the manufacturer for use on the surfaces to be painted. If applying multiple coats such as a primer, intermediate, and/or final coat to a surface, all coatings must be compatible. This section describes painting interior and exterior surfaces using paint brushes and rollers. Methods for painting interior and exterior surfaces using a paint sprayer are covered in later modules.

8.1.0 PAINTING INTERIOR SURFACES

The basic methods for painting interior surfaces such as ceilings, walls, and floors using paint applied with brushes and rollers are described here. Methods used to paint doors, windows, and other surfaces are covered later in this module.

When painting interior surfaces, cutting in corners and trim must be done with a brush or special purpose roller before the surface can be rolled with a roller. Cutting in should be done as the rolling progresses in order to maintain a wet edge to prevent lapping. For small jobs, cutting in and rolling are usually done by the same person. On larger jobs, it is common for two or more painters to do the cutting in while at the same time a third painter does the rolling. Regardless of the method, the procedures for cutting in and rolling rooms are basically the same.

The general rule of thumb is to start at the top and work toward the bottom. In a room, this means starting with the ceiling and ending with the baseboards. As shown in *Figure 18*, begin painting in a corner of the ceiling. Use a brush or trim roller to cut in a 2-inch or wider strip of paint at the edges of the ceiling where it meets the walls. If working with a helper, rolling can begin on the adjacent section of the ceiling (*Figure 19*). If working alone, cut in a few feet at a time to avoid lap marks, then roll the adjacent section while the cut-in strip is still wet. Because a roller leaves a different appearance and texture than a brush, roll as close as possible to the edges of the ceiling. When rolling, work across the narrow dimension of the room, rolling a strip about 3 feet wide, then feather the surface to even out coverage. Keep a wet edge by painting from a dry section into a wet one, overlapping edges to prevent streaks. Roll slowly to minimize splattering. Continue this process of cutting in and rolling until the entire ceiling is painted. Be sure that the first coat is dry before applying a second coat.

The walls of a room are painted in basically the same way as described for the ceiling. Cut in around the trim edges of doors and windows, in the corners of the walls, at the baseboards, and at the wall-ceiling joints. Start painting in the upper corner of a wall. Roll paint in about 2 x 4 foot sections, beginning at the ceiling and working down to the floor, usually moving

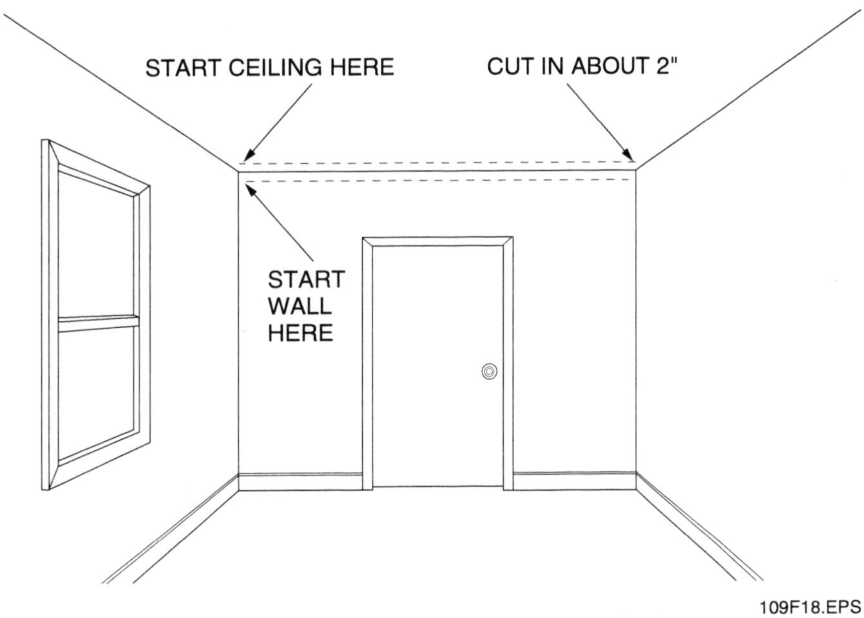

Figure 18. Painting And Cutting In Takes Organization

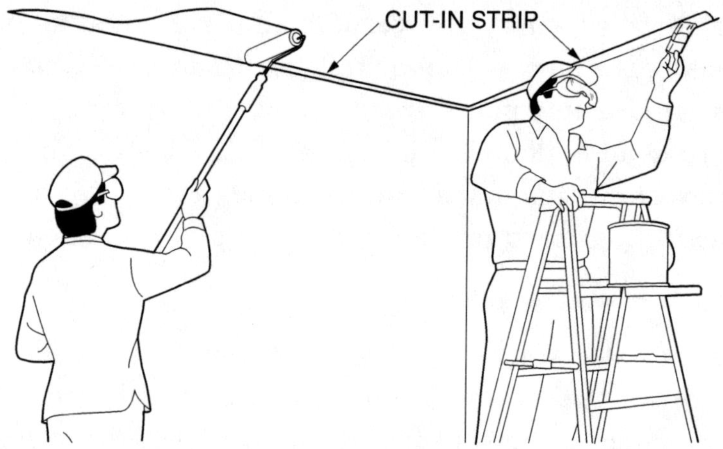

Figure 19. Team Painting

from the left side of the wall to the right. Complete one wall before starting another and be sure that the first coat is dry before applying a second coat. If a second coat of paint is applied on either walls or ceilings, cut in and roll the second coat in the same way as previously described for the first coat.

8.2.0 PAINTING EXTERIOR SURFACES

The general sequence for painting exterior surfaces is to paint the main wall surfaces first. Next, paint trim, railings, porches, steps, and the foundation. Last, paint any removable items such as shutters or similar items. The basic methods for painting exterior surfaces such as walls using paint applied with brushes and rollers are described here. Methods used to paint doors, windows, etc. are covered later in this module.

Always begin by painting at the top of walls and work down. Paint flat walls made of concrete, masonry, or cinder block in the same way as for interior walls, working from top to bottom and usually horizontally from left to right. Paint horizontal siding in horizontal strips; vertical siding in vertical strips. You should try to work from one natural break point to the next such as from a window to a door. Where there are no natural breaks, paint across the width of the wall, coating only as many courses as will allow you to keep blending the paint before it dries. Some other general considerations are:

- Follow the paint manufacturer's directions concerning temperatures. Be cautious of sudden drops in temperature. Paint only when the weather moderates to a fairly even temperature.
- Don't start painting too early in the morning. Give the sun a chance to dry the surface. Don't paint in the direct rays of hot summer sun because paint does not spread well on a hot surface and it can dry too fast. Ideally, you should paint in shaded areas recently bathed in the sun. This allows the surface to be free of moisture and also allows the paint to dry slower.

- If painting with oil-based paints, don't paint after a rain until the wood has dried thoroughly. After an extended wet period, allow a few days of good drying weather before painting. If you must paint on a damp surface, make sure to use an appropriate water-based (latex) paint.
- Don't paint in strong winds; fresh paint can collect leaves, dirt and dust particles, and insects.

Depending on the nature of the exterior surface, painting may be done with a brush (or pad) or a roller. Use of a roller speeds painting on larger flat surfaces that are typically made of brick, stucco, concrete, or masonry. Before rolling, cut in around windows, doorways, and corners using a brush following the same basic procedure as that described for painting interior rooms with a roller. Cover the surface in narrow bands or sections to avoid drying of the paint between sections that overlap. Remember to roll paint up to the edged strip, working in as close as possible for a uniform appearance. Paint should be rolled in even strokes with crisscrossing to achieve uniform coverage.

Much exterior painting is done on overlapping siding such as clapboards or shingles. Begin by using a brush to paint the bottom edges of the siding first. Then paint the adjacent flat surface of the siding, always leading from an unpainted area back into the last wet area. Failure to keep a wet edge can result in lap marks that even a second coat may not be able to hide. Spread the paint with smooth, even horizontal strokes. On shingles and similar vertical surfaces, the final brush strokes should be done vertically along the grain of the wood.

If painting siding with a paint pad, the siding edge can be cut in first using the pad face rather than the edge of the pad. Following this, wipe the pad across the adjacent flat surface of the siding to deposit an even, smooth coat. Overlap the previous wet edge by about 25%.

Regardless of whether using a brush or paint pad, always work to a logical stopping point, such as a door or window. *Never stop painting in the middle of a large area.* Drips often form on the bottom edges of siding so be sure to look back over your work from time to time to pick up any drips or runs.

If coating the siding with stain, the basic procedure is the same as when brushing and rolling with paint. However, stain brushes and rollers must be used instead of the standard paint brushes and rollers. A stain brush is generally wider and stubbier and has shorter bristles that keep the brush from holding too much stain. This helps to control dripping. Its wider bristle mass helps compensate for its shorter bristles by increasing the area each stroke will cover. Apply stain systematically, the same as paint. Since stain is less viscous than paint, it has a greater tendency to drip and is messier to work with. Typically, two coats of stain are required to cover a new surface because a single coat may not produce the desired color tone and there is an increased chance of lap marks.

9.0.0 PAINTING DOORS, WINDOWS, AND CABINETS

Before painting doors, windows, or cabinets, they must be carefully prepared to remove old paint drips, fat edges, etc. All easily removed hardware should be removed and other hardware masked to prevent unwanted painting. The exact sequence for painting doors, windows, and cabinets can vary depending on the painter's personal preferences and the surrounding conditions. Some typical methods are described here.

9.1.0 PAINTING DOORS

Depending on the situation, doors can be painted in place or removed from the door frame. Door removal is often done on large jobs where all doors are being painted at the same time. The procedure given here describes how to paint a door while it is hung in place.

Step 1 If it is a new wooden door, begin by painting the top and bottom edges of the door (*Figure 20*). Painting these edges is necessary to prevent the wood from absorbing moisture and warping. The bottom of the door can be painted either before the door is hung or it can be painted in place using a brush specially made for that purpose. Note that the top and bottom edges of a wooden door need only be painted once in a lifetime to seal the wood. Following this, prop the door open with wedges and paint the hinge and latch edges. If the door opens into the room you are painting, use the same color on the latch edge as on the door. If it opens into the next room, do not paint the latch edge because it should be the same color as the other room.

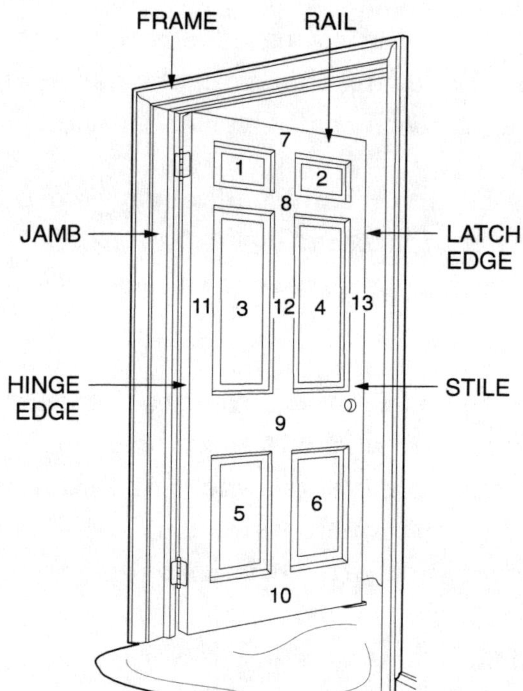

Figure 20. Painting A Panel Door

Unless otherwise required, it is preferable to leave the hinges unpainted. Be sure to thoroughly wipe off any excess paint that may get onto the adjacent surfaces of the door while painting the edges.

Step 2 When painting the front or back surfaces of the door, always work from top to bottom. If painting a flat slab-type door, the remaining surfaces can be painted using a roller, if desired. If painting a panel door, use a brush and begin by first painting the panels in the sequence (1 through 6) shown in *Figure 20*. At each panel, paint the moldings or borders first, then paint the inside of the panel. Brush in several directions to spread the paint, then smooth by brushing and laying off. If it is a wooden door, finish brushing in the direction of the wood grain. After all the panels are painted, paint the horizontal sections, which are called the *rails* (7 through 10). Following this, paint all the vertical sections, which are called the *stiles* (11, 12, and 13).

Step 3 Next, paint the door frames and jambs. Paint the top of the door frame, then the two sides. Then paint the part of the jamb between the frame and the doorstop. If the door opens into the room you are painting, paint only the side of the doorstop that directly faces you. If the door opens into the next room, paint both the side of the doorstop that directly faces you and the side next to it.

9.2.0 PAINTING WINDOWS

There are three common types of windows: double-hung, casement, and fixed-sash. The procedures given here describe how to paint double-hung and casement windows as they are usually more difficult to paint. The fixed-sash window does not open and is painted in a similar way to double-hung and casement windows.

9.2.1 Painting Double-Hung Windows

Double-hung windows consist of two sashes that move up and down (*Figure 21*). To paint a double-hung window, begin by raising the inside sash and lowering the outside sash. Paint the outside sash first. Paint the muntins, then paint the sash frame. *Do not paint the bottom edge, since you will use it to move the sash.* Next paint the inside sash; muntins first, then the frame. *Do not paint the top edge.* Move the sashes back about one inch from their closed positions. Complete the sashes by painting the parts of the inside sash that were obstructed and the bottom edge of the outside sash. Finally, paint the jamb, apron, sill, and trim.

9.2.2 Painting Casement Windows

Casement windows open like doors; they move in or out rather than up or down (*Figure 22*). Some are controlled by cranks, others by push rods or handles. Begin by opening the window wide. Paint the top, side, and bottom edges first. Finish with the muntins, frames, casings, and sills. Work from the inside out, doing the horizontal parts of the window first, followed by the vertical parts.

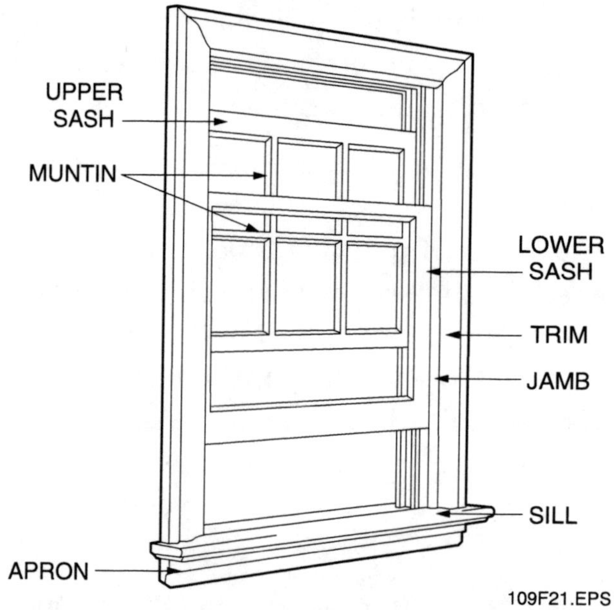

Figure 21. Typical Double-Hung Window

Figure 22. Casement Window

9.3.0 PAINTING CABINETS

Painting cabinets is somewhat more difficult than painting doors and windows because of their interior surfaces. Before painting cabinets, remove the doors, drawers, and hardware. Cabinets are painted by working from the inside to the outside. Paint the interior wall surfaces followed by the shelves and any partitions. When painting the shelves, do the bottoms first, then the tops and edges. Unless otherwise directed, it's best to leave the inside of the drawer openings unpainted to allow the drawers to move freely.

Paint the exterior next, working from the top down. This is often done using a smooth nap roller, followed by laying it off with a brush. Paint the cabinet doors inside first, then outside. If the doors are paneled, paint the panels first. Finish by painting the fronts of the drawers. It is not necessary to paint the bottoms or sides as the paint would prevent smooth sliding.

10.0.0 CLEANING AND STORING PAINT BRUSHES AND ROLLER COVERS

10.1.0 BRUSHES

If a brush is to be used again, it must be well cleaned immediately after use, before the paint dries. Brushes used with oil-based paints must be cleaned with solvents such as mineral spirits or turpentine. Those used with shellac should be cleaned with alcohol or household ammonia. Most brushes used with water-based paints can be cleaned with warm, soapy water after being flushed well to remove any remaining wet or dry, hardened paint. This applies to water-based paints that have a polyvinyl acetate (PVA) vehicle or an acrylic resin base with no additional resins.

Always refer to the paint manufacturer's cleanup directions because some water-based paints contain emulsified oil components that water alone will not clean. These paints have PVA or an acrylic resin base and an additional resin base of linseed alkyd or polyester or other emulsified resins. As the paint dries, the additional emulsified resins undergo chemical changes that will leave a crust on the bristles if the brush is not cleaned properly. Normally, a brush used with these water-based paints must first be cleaned with a solvent such as lacquer thinner, paint thinner, or turpentine before it can be cleaned with soap and water.

Clean brushes used in oil-based paints according to the guidelines given below.

WARNING! To prevent injury, always refer to and follow the instructions given on the paint and solvent manufacturer's MSDSs and product labels concerning the use of proper personal protection equipment and the handling of oil-based paints and related solvents.

CAUTION Collect used solvent waste in a container with a lid. Once waste solvent is no longer usable, pour it into a suitable waste solvent barrel. Do not mix other items such as used roller covers, rags, stir sticks, or any non-liquid or paint solid material in with the waste solvent. Clean, dry, and dispose of these items as dry waste. Dispose of dirty solvents and dry waste in accordance with all federal, state, and local laws.

Step 1 Use a rag to wipe off as much paint as you can from the brush and ferrule.

Step 2 Let the brush soak in a bucket of solvent for a few minutes. Then use a brush comb or wire brush (*Figure 23*) to comb the flat sides and edges of the brush. Comb from the ferrule toward the end of the bristles, forcing the paint out. Repeat this process while dipping the brush into the solvent to loosen any paint trapped inside the bristles near the ferrule.

Figure 23. Using A Wire Brush To Clean Paint From A Brush

Step 3 Hold the brush upside down, and bend the bristles back and forth, forcing solvent into the upper heel and ferrule of the brush. Repeat this action several times while pouring thinner into the bristles. Note that dirty solvent can often be used for the early rinses but clean solvent must be used for the later rinses. Continue the process until clear solvent runs out of the brush.

Step 4 Use a brush spinner (*Figure 24*) to remove the solvent and dry the brush. Make sure to hold the spinner inside a paint bucket or other suitable container while pumping it to prevent getting solvent on yourself, others, and the surrounding area.

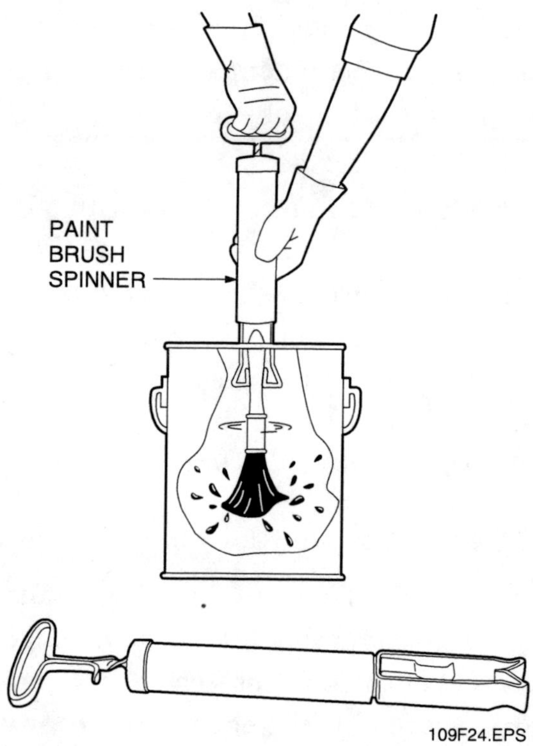

Figure 24. Using A Brush Spinner To Clean Paint From A Brush

Step 5 Use a clean brush comb or wire brush to return the bristles to their original shape, with the bristles straight and parallel.

The procedure for cleaning brushes used with water-based paints is basically the same as those used with oil-based paints. The difference is that most water-based paints are cleaned using soap and water rather than solvents.

CAUTION Make sure to dispose of all unusable latex and water-based paint and any wash water in accordance with all federal, state, and local laws that apply to your area. Do not dispose of this material in a storm sewer or in the ground. The cleanup and handling of water and solvent wastes was covered in detail in *Introduction To Paints And Coatings.*

10.2.0 CLEANING ROLLERS

If a roller cover is to be used again, it must be cleaned well immediately after use, before the paint dries. Roller covers used with oil-based paints are cleaned with mineral spirits or turpentine. Those used with water-based paints are cleaned with soap and water. Clean roller covers according to the following guidelines:

Step 1 Without removing the roller cover from the roller, use the semi-circular groove of a painter's tool (*Figure 25*) or the edge of a putty knife to scrape the wet paint from the roller cover into the pan or five-gallon pail.

Step 2 Remove the roller cover from the roller and soak it in solvent or water, as applicable. Use the painter's tool to scrape the cover again after it has been soaked.

Figure 25. Using A Painter's Tool To Clean Paint From A Roller

Step 3 Use a roller spinner (*Figure 26*) to clean out as much solvent or water from the cover as possible. Make sure to hold the spinner inside a paint bucket or other suitable container while pumping it to prevent getting solvent on yourself, others, and the surrounding area.

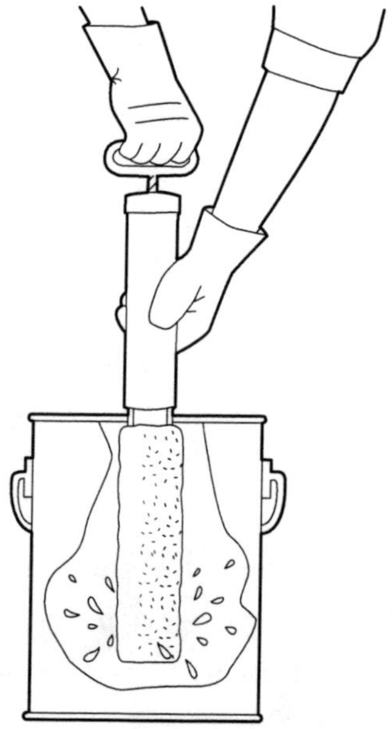

109F26.EPS

Figure 26. Using A Roller Spinner To Clean And Dry A Roller

Step 4 Repeat steps 1 through 3 as necessary until the cover is clean.

Step 5 Stand the cover on end and allow it to dry thoroughly. Do not allow anything to lean against the cover as it dries.

Step 6 Clean, rinse, and dry the roller, paint tray, and other accessories.

Step 7 Make sure to dispose of waste paint, solvent, and waste water in accordance with all federal, state, and local laws and regulations that apply to your local area.

10.3.0 STORAGE OF BRUSHES AND ROLLERS

Brushes and rollers can be stored wet or dry, usually depending on how soon they are going to be reused. After cleaning and drying, brushes and rollers can be wrapped in their original cover, brown kraft paper, aluminum foil, or wax paper. This helps to keep the brush or roller in shape. After wrapping, brushes can be stored either flat or by hanging. Brushes that are going to be reused soon are often suspended in a solvent base (not water because water rots and warps the handles). While suspended in the solvent, they must not sit on their bristles as this can cause the bristles to lose their shape. Brushes can be suspended wet or dry in a can by placing a rod through a hole drilled near the upper part of the ferrule and laying the rod across the top of the can (*Figure 27*). Special brush containers, called *brush keepers*, are also available. They allow several brushes to be suspended for storage in the container at the same time.

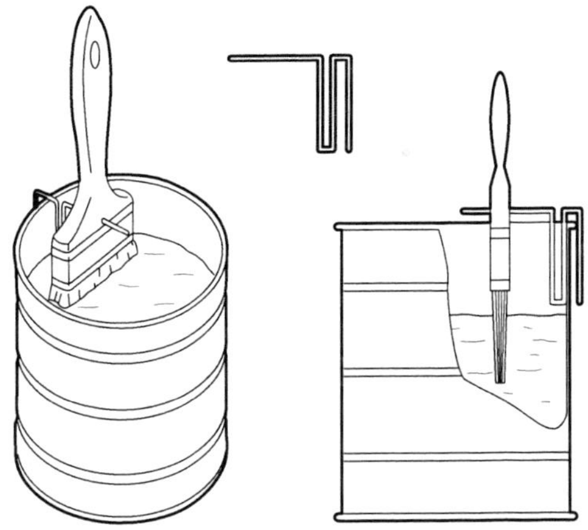

Figure 27. Suspending A Wet Brush In A Paint Can Or Pail

Overnight storage of roller covers can be done by first stripping the paint out of the roller cover, then removing it from the frame. Following this, the cover is wrapped in plastic or aluminum foil and left to float in a container of solvent or water.

SUMMARY

All paint brushes consist of bristles, a heel, a ferrule, and a handle. Either natural or synthetic bristles can be used. Hogs' hair bristles are used in good brushes made with natural bristles. Synthetic bristles, also called *filaments,* can be made from nylon, polyester, or a blend of these materials. A general rule of thumb is that brushes made with natural bristles are used only with oil or alkyd paints, while those made with synthetic bristles are used with water-based paints.

Rollers are used to apply paints and coatings to large, regular surfaces such as floors, ceilings, and walls. They transfer and spread paint onto a surface by rolling it with a fabric covered roller cover. The most common type of roller is the dip roller. It consists of a handle, a frame, and a cage or cylinder upon which the roller cover can be put on and taken off. The roller cover fabric can be made of knitted or woven natural or synthetic fibers. Natural fibers include lamb's wool, sheepskin, and mohair (goat hair). Synthetic fibers include materials such as polyester, acrylic, and nylon. The roller cover nap length determines the kinds of surfaces on which the cover can be used and its overall performance on those surfaces.

The selection of a brush or roller is determined by the type of finish being applied, the size and shape of the area being covered, and the quality and speed of application desired.

References

For advanced study of topics covered in this task module, the following books are suggested:

Painting and Decorating Craftsman's Manual and Textbook, Painting and Decorating Contractors of America, Fairfax, Virginia.

Painting and Decorating Encyclopedia, The Goodheart-Willcox Company, Inc., Tinley Park, Illinois.

Store Training and Reference Tool - Paints, Stains, and Clear Coatings, Sherwin-Williams Company, Cleveland, Ohio.

REVIEW/PRACTICE QUESTIONS

1. The elements of a professional paint job include:
   a. use of the correct paint.
   b. proper preparation of the surface.
   c. applying the paint in the prescribed manner.
   d. all of the above.

2. Brushes made for use with latex paint normally have bristles made of:
   a. nylon.
   b. polyester.
   c. both a and b.
   d. hogs' hair.

3. The most common wall brush is _____ inches wide.
   a. 4
   b. 7
   c. $2^1/_2$
   d. 8

4. A hockey-style brush is a type of:
   a. wall brush.
   b. special purpose brush.
   c. sash brush.
   d. decorative brush.

5. A stenciling brush is a type of:
   a. wall brush.
   b. special purpose brush.
   c. sash brush.
   d. decorative brush.

6. A 2-inch sash brush should have bristles that are at least _____ inches long.
   a. $2^1/_2$
   b. 2
   c. 3
   d. $3^1/_2$

7. The size of a roller cover used with a particular roller is determined mainly by the:
   a. size of the area to be painted.
   b. type of paint being used.
   c. size of the roller cage or cylinder on which it is to be mounted.
   d. nap length of the roller cover.

8. When painting a smooth wood cabinet, a roller cover with a _____ nap should be used.
   a. short
   b. medium
   c. long
   d. extra long

9. The most efficient roller to use when painting several rooms is the:
   a. fountain roller.
   b. self-feeding roller.
   c. dip roller.
   d. special purpose roller.

10. A roller used to paint contoured surfaces is called a:
   a. trim roller.
   b. paint striper roller.
   c. spike roller.
   d. pipe roller.

11. A _____ can be used for painting surfaces that are too irregular for brushing or rolling.
   a. paint mitt
   b. paint pad
   c. paint striper
   d. paint grid

12. Generally, paint can be applied faster using a:
   a. pad instead of a brush.
   b. pad instead of a roller.
   c. roller instead of a pad.
   d. both a and c.

13. Mixing latex paint with a mechanical shaker causes:
   a. stippling.
   b. foaming.
   c. poor adhesion.
   d. uneven color.

14. Mixing paint too fast with a mixing machine can cause:
   a. blisters in the film.
   b. separation of the pigments.
   c. sludge to occur.
   d. both a and b.

15. When dipping a paint brush into the paint in a container, the bristles should be dipped no more than _____ their length.
    a. 1/4
    b. 1/3
    c. 1/2
    d. 3/4

16. Whenever possible, you should paint with the _____ of a brush.
    a. tip
    b. flat side
    c. side edge
    d. both a and b

17. When cutting in and rolling paint, the cutting in should be done as the rolling progresses in order to:
    a. reduce splattering.
    b. prevent lapping.
    c. reduce dripping.
    d. match the color and texture.

18. When painting an interior room, the _____ should be painted first, then the _____, then the _____.
    a. baseboard; ceiling; walls
    b. walls; baseboard; ceiling
    c. ceiling; walls; baseboard
    d. baseboard; walls; ceiling

19. When painting exterior clapboards, paint the _____ first, then the _____.
    a. shutters; main surfaces
    b. main surfaces; trim
    c. shutters; trim
    d. trim; main surfaces

20. Waste mineral spirits should be:
    a. disposed of in a toilet.
    b. disposed of in a sink.
    c. disposed of in a storm sewer.
    d. collected and/or stored in a waste solvent container.

NCCER CRAFT TRAINING USER UPDATES

The NCCER makes every effort to keep these manuals up-to-date and free of technical errors. We appreciate your help in this process. If you have an idea for improving this manual, or if you find an error, a typographical mistake, or an inaccuracy in the NCCER's Craft Training Manuals, please write us, using this form or a photocopy. Be sure to include the exact module number, page number, a description of the problem, and the correction, if possible. Your input will be brought to the attention of the Technical Review Committee. Thank you for your assistance.

Instructors – If you found that additional materials were necessary in order to teach this module effectively, please let us know so that we may include them in the Equipment/Materials list in the Instructor's Guide.

Write: Curriculum and Revision Department
National Center for Construction Education and Research
P.O. Box 141104
Gainesville, FL 32614-1104

Fax: 352-334-0932

Craft ______________ Module Name ______________

Module Number ______________ Page Number(s) ______________

Description of Problem ______________

____

____

____

(Optional) Correction of Problem ______________

____

____

(Optional) Your Name and Address ______________

____